Microsystems

Series Editors

Roger T. Howe
Department of Electrical Engineering
Stanford University
Stanford, California

Antonio J. Ricco
Small Satellite Division
NASA Ames Research Center
Moffett Field, California

For further volumes:
http://www.springer.com/series/6289

Dan Zhang
Editor

Advanced Mechatronics and MEMS Devices

Editor
Dan Zhang
Faculty of Engineering and Applied Science
University of Ontario Institute of Technology
Oshawa, ON, Canada

ISSN 1389-2134
ISBN 978-1-4899-9745-6 ISBN 978-1-4419-9985-6 (eBook)
DOI 10.1007/978-1-4419-9985-6
Springer New York Heidelberg Dordrecht London

Printed on acid-free paper

Springer is part of Springer Science+Business Media (www.springer.com)

Preface

As the emerging technologies, advanced mechatronics and MEMS devices will be the revolutionary measures for the extensive applications in modern industry, medicine, health care, social service, and military. Research and development of various mechatronic systems and MEMS devices is now being performed more and more actively in every applicable field. This book will introduce state-of-the-art research in these technologies from theory to practice in a systematic and comprehensive way.

The book entitled "Advanced Mechatronics and MEMS Devices" describes the up-to-date MEMS devices and introduces the latest technology in electrical and mechanical microsystems. The evolution of design in microfabrication, as well as emerging issues in nanomaterials, micromachining, micromanufacturing, and microassembly are all discussed at length within this book.

Advanced Mechatronics also provides readers with knowledge of MEMS sensor arrays, MEMS multidimensional accelerometers, digital microrobotics, MEMS optical switches, micro-nano adhesive arrays, as well as other topics in MEMS sensors and transducers. This book will not only include the main aspects and important issues of advanced mechatronics and MEMS devices but also comprises novel conceptions, approaches, and applications in order to attract a broad audience and to promote the technological progress. This book will aim to integrate the basic concepts and current advances of micro devices and mechatronics with interdisciplinary approaches. Different kinds of mini mechatronics and MEMS devices are designed, analyzed, and implemented. The novel theories, modeling methods, advanced control algorithms, and unique applications are investigated. This book is suitable as a reference for engineers, researchers, and graduate students who are interested in mechatronics and MEMS technology.

I would like to express my deep appreciation to all the authors for their significant contributions to the book. Their commitment, enthusiasm, and technical expertise are what made this book possible. I am also grateful to the publisher for supporting this project, and would especially like to thank Mr. Steven Elliot, Senior Editor for Engineering of Springer US, Mr. Andrew Leigh, Editorial Assistant of Springer US, and Ms. Merry Stuber, Editorial Assistant of Springer US for their constructive assistance and earnest cooperation, both with the publishing venture in general and the editorial details. We hope the readers find this book informative and useful.

This book consists of 12 chapters. Chapter 1 introduces a new concept for a six-degrees-of-freedom silicon-based force/torque sensor that consists of a MEMS structure including measuring piezo resistors. Chapter 2 presents a piezoelectrically actuated robotic end-effector based on a hierarchical nested rhombus multilayer mechanism for effective strain amplification. Chapter 3 analyzes the critical aspects of various calibration methods and autocalibration procedures for MEMS accelerometers with sensor models and a principled noise model. Chapter 4 reviews the current methods in micro-nanomanipulation and the difficulties associated with this miniaturization process. Chapter 5 discusses the new bottom-up approach called "digital microrobotics" for the design of microrobot architectures that is based on elementary mechanical bistable modules. Chapter 6 describes the flexure-based parallel-kinematics stages for assembly of MEMS optical switches based on a low-cost passive method. Chapter 7 introduces the sensing approach for the measurement of both contact force and elasticity of micro-tactile sensors. This is conducted with the spring-pair model and the more precise contact model. Chapter 8 investigates the gripping techniques with physical contact, including friction microgrippers, pneumatic grippers, adhesive gripper, phase changing, and electric grippers. The development of a variable curvature microgripper is presented as a case study. Chapter 9 provides a prototype of a wall-climbing robot with gecko-mimicking adhesive pedrails and micro-nano adhesive arrays. Chapter 10 develops the biomimetic flow sensor inspired from natural lateral line and draws the artificial cilia from a polymer solution. Chapter 11 introduces an interesting jumping mini robot with a bio-inspired design including the dynamically optimized saltatorial leg that can imitate the motion characteristics of a real leafhopper. Chapter 12 studies the modeling and design of the H-infinity PID plus feedforward controller for a high-precision electrohydraulic actuator system.

Finally, the editor would like to sincerely acknowledge all the friends and colleagues who have contributed to this book.

Oshawa, ON, Canada Dan Zhang

Contents

Contributors

Gabriella Bonsignori CRIM Lab, Polo Sant'Anna Valdera, Pontedera, Italy

N. Alberto Borghese Computer Science Department, University of Milan, Milano, Italy

Torgny Brogårdh ABB Automation Technologies, Västerås, Sweden

Richard Burton Department of Mechanical Engineering, University of Saskatchewan, SK, Canada

Vincent Chalvet Automatic Control and Micro-Mechatronic Systems Department (AS2M), CNRS-UFC-UTBM-ENSMM, FEMTO-ST Institute, Besançon, France

Brandon K. Chen University of Toronto, Toronto, ON, Canada

Qiao Chen Automatic Control and Micro-Mechatronic Systems Department (AS2M), CNRS-UFC-UTBM-ENSMM, FEMTO-ST Institute, Besançon, France

Wenjie Chen Mechatronics Group, Singapore Institute of Manufacturing Technology, Singapore, Singapore

Paolo Dario CRIM Lab, Polo Sant'Anna Valdera, Pontedera, Italy

Jörg Eichholz Fraunhofer Institute for Silicon Technology (FhG—ISIT), Itzhehoe, Germany

Irene Fassi National Research Council of Italy, Roma, Italy

Iuri Frosio Computer Science Department, University of Milan, Milano, Italy

Xin Fu The State Key Laboratory of Fluid Power Transmission and Control, Zhejiang University, Hangzhou, China

Yassine Haddab Automatic Control and Micro-Mechatronic Systems Department (AS2M), CNRS-UFC-UTBM-ENSMM, FEMTO-ST Institute, Besançon, France

Da Li State Key Laboratories of Transducer Technology, Institute of Intelligent Machines, Chinese Academy of Sciences, Hefei, Anhui, China

Fei Li The State Key Laboratory of Fluid Power Transmission and Control, Zhejiang University, Hangzhou, China

Wei Lin Mechatronics Group, Singapore Institute of Manufacturing Technology, Singapore, Singapore

Yang Lin Department of Mechanical Engineering, University of Saskatchewan, SK, Canada

Weiting Liu The State Key Laboratory of Fluid Power Transmission and Control, Zhejiang University, Hangzhou, China

Philippe Lutz Automatic Control and Micro-Mechatronic Systems Department (AS2M), CNRS-UFC-UTBM-ENSMM, FEMTO-ST Institute, Besançon, France

Tao Mei State Key Laboratories of Transducer Technology, Institute of Intelligent Machines, Chinese Academy of Sciences, Hefei, Anhui, China

Claudia Pagano National Research Council of Italy, Roma, Italy

Federico Pedersini Computer Science Department, University of Milan, Milano, Italy

Peng Peng Department of Mechanical Engineering, University of Minnesota, Minnesota, USA

Rajesh Rajamani Department of Mechanical Engineering, University of Minnesota, Minnesota, USA

Umberto Scarfogliero CRIM Lab, Polo Sant′Anna Valdera, Pontedera, Italy

Yang Shi Department of Mechanical Engineering, University of Victoria, Victoria, BC, Canada

Cesare Stefanini CRIM Lab, Polo Sant′Anna Valdera, Pontedera, Italy

Yu Sun University of Toronto, Toronto, ON, Canada

Jun Ueda Mechanical Engineering, Georgia Institute of Technology, Atlanta, GA, USA

Dapeng Wang State Key Laboratories of Transducer Technology, Institute of Intelligent Machines, Chinese Academy of Sciences, Hefei, Anhui, China

Xuan Wu State Key Laboratories of Transducer Technology, Institute of Intelligent Machines, Chinese Academy of Sciences, Hefei, Anhui, China

Department of Precision Machinery and Precision Instrumentation, University of Science and Technology of China, Hefei, Anhui, China

Guilin Yang Mechatronics Group, Singapore Institute of Manufacturing Technology, Singapore, Singapore

Aiwu Zhao State Key Laboratories of Transducer Technology, Institute of Intelligent Machines, Chinese Academy of Sciences, Hefei, Anhui, China

Chapter 1
Experience from the Development of a Silicon-Based MEMS Six-DOF Force–Torque Sensor

Jörg Eichholz and Torgny Brogårdh

Abstract A six-DOF (Degrees Of Freedom) force–torque sensor was developed to be used for interactive robot programming by so-called lead through. The main goal of the development was to find a sensor concept that could drastically reduce the cost of force sensors for robot applications. Therefore, a sensor based on MEMS (Micro Electro Mechanical System) technology was developed, using a transducer to adapt the measuring range needed in the applications to the limited measuring range of the silicon MEMS sensor structure. The MEMS chip was glued with selected epoxy adhesive on a planar transducer, which was cut by water jet guided laser technology. The transducer structure consists of one rigid cross and one cross with four arms connected to the rigid cross by springs, all in the same plane. For this transducer a German utility patent [Weiß M, Eichholz J Sensoranordnung. Pending German utility patent] is pending. The MEMS structure consists of one outer part and one inner part, connected to each other with beams obtained by DRIE (Deep Reactive Ion Etching) etching. On each beam four piezoresistors are integrated to measure the stress changes used to calculate the forces and torques applied between the outer and inner part of the MEMS structure. The inner part was glued to the mentioned rigid cross of the transducer and the outer part was glued to the four arms including the transducer springs. FEM (Finite Element Modeling) was used to design both the MEMS- and transducer part of the sensor and experimental tests were made of sensitivity, temperature compensation, and glue performance. Prototypes were manufactured, calibrated, and tested, and the concept looks very promising, even if more work is still needed in order to get optimal selectivity of the sensor.

J. Eichholz (✉)
Fraunhofer Institute for Silicon Technology (FhG—ISIT), Fraunhoferstr. 1, Itzhehoe 25524, Germany
e-mail: joerg.eichholz@isit.fraunhofer.de

T. Brogårdh
ABB Automation Technologies, Västerås 72168, Sweden
e-mail: torgny.brogardh@se.abb.com

D. Zhang (ed.), *Advanced Mechatronics and MEMS Devices*, Microsystems, DOI 10.1007/978-1-4419-9985-6_1,

1.1 Introduction

In the EU-project SMErobot™ (www.smerobot.org) one task was to develop easy-to-use robot programming methods. One such method is lead through programming, whereby the robot operator directly interacts with the tool carried by the robot and moves the tool to the positions that the robot is expected to go to. In order to make this concept useful a six Degrees Of Freedom (DOF) force-and-torque sensor is needed. Unfortunately the sensors available for this are very expensive, which hinders a broader use of these sensors for robot programming, especially for SMEs (Small- and Medium-sized Enterprise) where the cost of robot automation is often too high. In order to find a concept for a six-DOF force-and-torque sensor with the potential to have a lower manufacturing cost, development was started in collaboration between Fraunhofer ISIT and ABB Robotics. The idea was to develop a MEMS sensor element, which could be mounted on a low cost steel transducer, which transformed the force and torque levels of the application to the force and torque levels that a MEMS structure can handle. There are some suitable MEMS structures for six-DOF force and torque measurements in the literature [1–6, 7] but no solution could be found on how to combine a low cost easy to scale steel transducer with a MEMS structure for six-DOF force and torque measurements. Main problems for this combination were to find a suitable transducer structure, to find structures for mating the MEMS element with the transducer, to compensate for the difference in temperature expansion between Silicon and Steel, and to make sure that the bonding between Silicon and Steel has the performance needed. In the following section the measurement concept is first described, then the results of the design and simulations of both the MEMS- and transducer structure are presented. The fabrication and mounting of the MEMS chip is then described followed by some information about the measurement electronics, sensor tests, and sensor calibration.

1.2 Measurement Concept

In order to be able to measure three force and three torque components a MEMS structure based on crystalline silicon with integrated silicon piezoresistors according to Fig. 1.1 was developed. This type of approach is widely used, see for example [1–6, 7]. Here forces and torques are measured between the inner part and the outer part of the sensor element by means of 16 piezoresistors integrated according to the left part of Fig. 1.1 in groups of 4 on 4 beams connecting the outer and inner parts of the sensor element. A silicon wafer with the surface oriented in [1 1 0] crystal direction was used, and the sensor piezoresistors were applied in the same orientation, which resulted in a high piezoresistive effect.

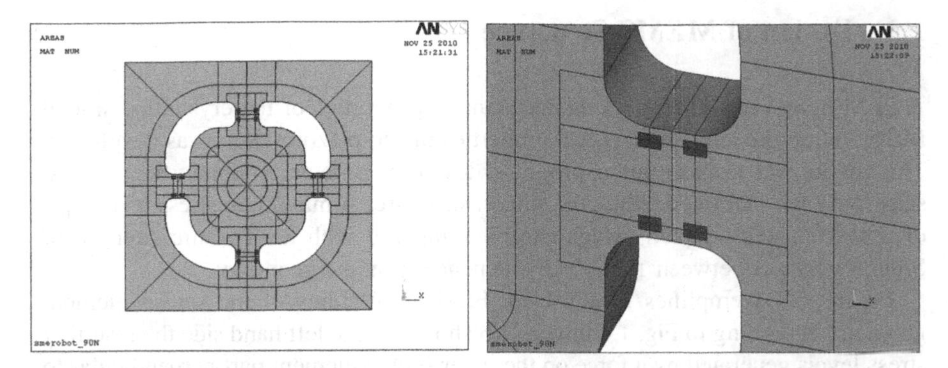

Fig. 1.1 The layout of the MEMS sensor element. On the *left* it shows the MEMS structure with four beams connecting the outer and inner parts of the sensor element and on the *right hand side* one of the beams with four integrated piezoresistors can be seen

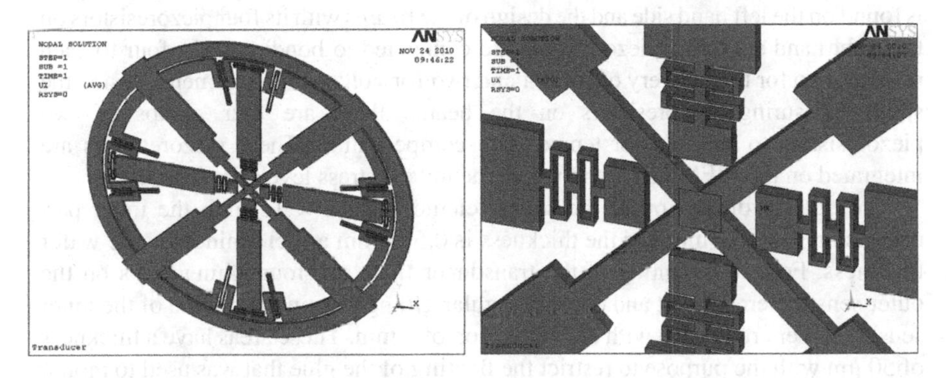

Fig. 1.2 The transducer design with one rigid cross and one cross including springs for the connection of the MEMS structure to the transducer. The *left figure* shows the whole transducer and the *right figure* the central part where the MEMS structure is mounted

For the low cost transducer, a sheet of steel was used and the spring system, which adapts the MEMS sensor element to the applications, was manufactured by laser cutting. A critical problem was then how to design a laser cut transducer in order to make the mounting of the MEMS sensor on the transducer as simple as possible. Figure 1.2 illustrates the solution that was found to this problem. According to Fig. 1.2 on the left the transducer is divided into two cross like shapes, where one cross is rigid and connects to the inner part of the sensor element while the other cross is provided with springs that connect to four spots on the outer part of the sensor element. Figure 1.2 shows on the right a detail of the transducer where the MEMS structure is mounted. Notice that the transducer is designed in such a way that it can be cut from a single sheet of metal.

1.3 Design of MEMS Structure

A FEM model including both the mechanical properties of the crystalline silicon MEMS structure and the electrical properties of the piezoresistors was developed. This model was used to find appropriate design parameters to obtain the targeted sensitivity without overloading the silicon structure. A maximum stress level σ_{max} of 300 MPa was adopted, which can be compared with the fracture strength of Silicon which is between 1 and 5 GPa dependent on the geometry.

Figure 1.3 exemplifies results from FEM simulations of the sensor element geometry according to Fig. 1.1. Figure 1.3 shows on the left hand side the resulting stress levels generated by a force on the inner sensor element part perpendicular to the sensor plane (in the z-direction), in the middle it shows the result from an in plane force (in the x-direction) and on the right hand side the result of a torque around the y-axis in the sensor plane. The MEMS structure in Fig. 1.3 is designed for 100 N max force. In the project also a MEMS structure for 10 N was developed.

The final design of a sensor element is shown in Fig. 1.4, where the complete layout is found on the left hand side and the design of the beams with its four piezoresistors on the right hand side. The piezoresistors are connected to bonding pads, four to each resistor, two for the delivery of current and two for voltage measurement. Beside the stress measuring piezoresistors on the beams there are four groups of two piezoresistors to be used for temperature compensation. These piezoresistors are integrated on the MEMS structure where the lowest stress levels are found.

The outside dimension of the sensor element is 12 × 12 mm, the inner part dimension is 6 × 6 mm and the thickness is 0.508 mm as determined by the wafer thickness. For the mounting of the transducer there are four gluing areas on the outer sensor element part and one rectangular gluing area in the middle of the inner sensor element part, each with the dimension of 9 mm. These areas have a thickness of 50 μm with the purpose to restrict the floating of the glue that was used to mount the sensor element on the transducer. The piezoresistive areas seen in Fig. 1.4 to the right are 25 × 40 μm and are implanted to a depth of 0.5 μm. The beams that form

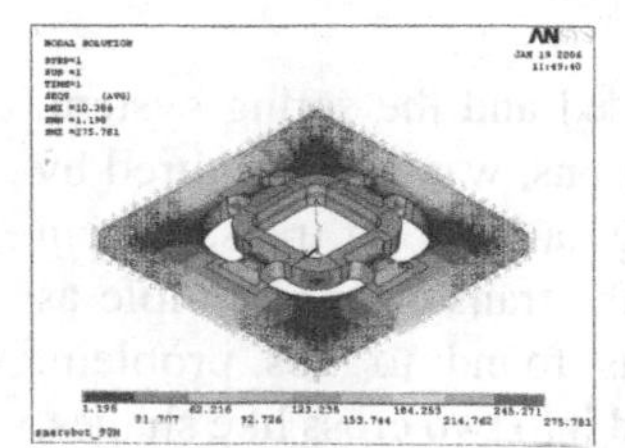
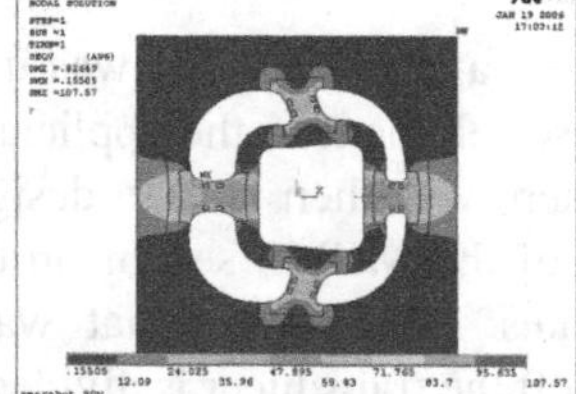
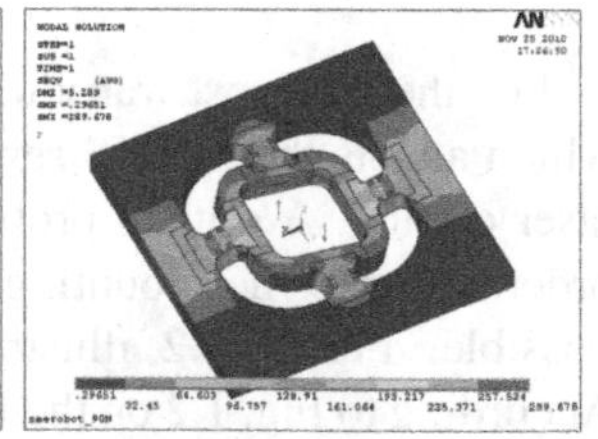

Fig. 1.3 Examples of FEM results for the sensor element in Fig. 1.1. On the *left* for a force of 10 N in the z-direction (perpendicular to the sensor plane), in the *middle* for 10 N in the x-direction and to the *right* for a torque of 0.15 N m around the y-axis. The scales below the models indicate the stress in MPa, for which the limit is set to 300 MPa. Since the point forces are known to produce stress singularities, the regions where the forces are applied are omitted in Fig. 1.3

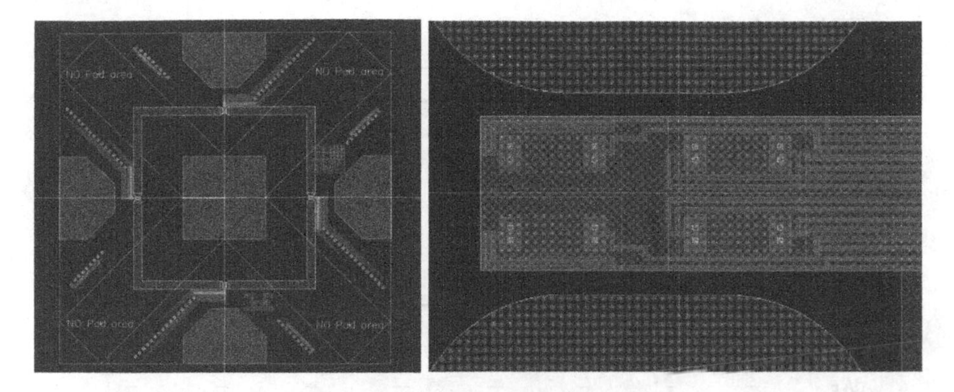

Fig. 1.4 Final design of the sensor element with a detail in the *right* figure showing one of the beams with four piezoresistors and their electrical connections

the bridges between the outer and inner sensor element parts have for the 100 N sensor element the dimension of 400 × 400 μm with the thickness the same as for the silicon chip, i.e., 508 μm. The 10 N sensor element had a beam width of 100 μm. Using DRIE (deep reactive ion etching) it would have been possible to reduce the thickness of the beams and thereby obtaining a more isotropic sensor. But this was not done because the sensor will then become more fragile. So the sensor anisotropy had to be compensated for by the transducer. In order to reduce the risk of stress concentrations the beam ends were designed with an outer curvature radius of 100 μm.

The purpose of the MEMS structure is to give well defined relations between applied forces/torques (F_x, F_y, F_z/M_x, M_y, M_z) and changes of the resistance values of the 16 piezoresistors mounted on the four beams. Because of the symmetry of the sensor element the expressions for F_x and F_y will have the same coefficients (but for different resistors) and the same situation is found for M_x and M_y. Therefore, only the expressions for F_x, F_z, M_x and M_z are shown. Table 1.1 lists the resistor combinations, which were used.

As an example the first row of the table denotes the resistor combination $\mathrm{d}R_\mathrm{a}$ as:

$$\mathrm{d}R_\mathrm{a} = \mathrm{d}R_1 + \mathrm{d}R_2 + \mathrm{d}R_9 + \mathrm{d}R_{10} - \mathrm{d}R_5 - \mathrm{d}R_6 - \mathrm{d}R_{13} - \mathrm{d}R_{14}.$$

The dR parameters give relative changes in resistance from the state of no force or torque on the sensor element. The resistors are named clockwise starting at the positive x-axis and with at first the eight outer resistors (R_1–R_8) and then the eight inner resistors (R_9–R_{16}). Thus the resistors shown in the right Fig. 1.4 are to the right R_1 and R_2 and to the left R_9 and R_{10}. The values of $\mathrm{d}R_\mathrm{a}$, $\mathrm{d}R_\mathrm{b}$, $\mathrm{d}R_{F_z}$, $\mathrm{d}R_{M_z}$, and $\mathrm{d}R_\mathrm{T}$ are used to calculate forces, torques and temperature in the following way:

$$\frac{F_x}{F_{x_0}} = k_1 \times \mathrm{d}R_\mathrm{a} + k_2 \times \mathrm{d}R_\mathrm{b},$$

Table 1.1 Resistor combinations with sign used for the calculation of the force and torque components

	R_1	R_2	R_3	R_4	R_5	R_6	R_7	R_8
dR_a	+	+			−	−		
dR_b			+	−			−	+
dR_{F_z}	+	+	+	+	+	+	+	+
dR_{M_z}	+	−	+	−	+	−	+	−
dR_T	+	+	+	+	+	+	+	+
	R_9	R_{10}	R_{11}	R_{12}	R_{13}	R_{14}	R_{15}	R_{16}
dR_a	+	+			−	−		
dR_b			+	−			−	+
dR_{F_z}	−	−	−	−	−	−	−	−
dR_{M_z}	−	+	−	+	−	+	−	+
dR_T	+	+	+	+	+	+	+	+

$$\frac{M_y}{M_{y_0}} = k_3 \times dR_a + k_4 \times dR_b,$$

$$\frac{F_z}{F_{z_0}} = k_5 \times dR_{F_z} + k_6 \times dR_T,$$

$$\frac{dT}{T_0} = k_7 \times dR_{F_z} + k_8 \times dR_T,$$

$$\frac{M_z}{M_{z_0}} = k_9 \times dR_{M_z}.$$

Here F_{x_0}, M_{y_0} etc are scale factors corresponding to max values and k_1–k_9 are parameters that are identified at sensor calibration. These parameters depend on the piezoresistive effect, which in the [1 1 0] crystal direction can be calculated according to:

$$dR = \frac{\Delta R}{R} = \frac{1}{2}(\pi_{11} + \pi_{12} + \pi_{44})S_{11} + \frac{1}{2}(\pi_{11} + \pi_{12} - \pi_{44})S_{22} + \pi_{12}S_{13},$$

where dR is the relative change of the piezoresistance, where $\pi_{11} = 0.066$, $\pi_{12} = 0.011$ $\pi_{44} = 1.38$ [9] (dimension 1/GPa) are the piezoresistive coefficients of silicon, and S_{ij} are stress components at the positions of the piezoresistors.

Table 1.1 shows results obtained for the 10 N sensor when forces of 10 N, torques of 0.5 N m and temperature increase of 100° were simulated on the FEM model of the sensor element in Fig. 1.4. Using the formulas for calculating the forces and torques and applying identified calibration constants seem to give a good selectivity (low scattering value) for the sensor. Later it was however shown that this was not the case but since there was no time in the project to fabricate a sensor structure without this problem, the decision was to make use of the existing MEMS structure to validate the technology involved. To calibrate a sensor in a reasonable time another approach was chosen that is described in the chapter test- and calibration results.

1.4 Design of Transducer Structure

The role of the transducer is to adapt the force and torque ranges of the applications to the measurement intervals of the MEMS structure. Moreover, it is used to reduce the temperature sensitivity of the force sensor, to make the sensor more isotropic and to protect the MEMS structure from overload and mechanical shock [8]. Figure 1.5 illustrates the mounting of the MEMS structure on the transducer. The central grey region of the sensor element (see Fig. 1.5 to the right) is glued on the rigid central cross of the transducer, which acts as a base structure. The four outer grey regions (stud bumps) are glued on four arms of the transducer, which contain the spring system. Together with the sensor element these arms form a second cross.

Beside the spring close to the sensor element, each of the four arms also has springs in its other end as illustrated in Fig. 1.6. According to the right of Fig. 1.6 there are three springs for each arm. The two springs at the sides of the arm end connect to the rigid cross, making it possible to fabricate the transducer in a single sheet of metal. The constellation of the three springs increases the isotropy of the sensor by to some extent compensate for the earlier mentioned lack of isotropy of the sensor elements. The ring above the transducer in Fig. 1.6 is used to simulate the rigid mounting of the spring arms on the sensor flange. The rigid cross is mounted in the sensor housing.

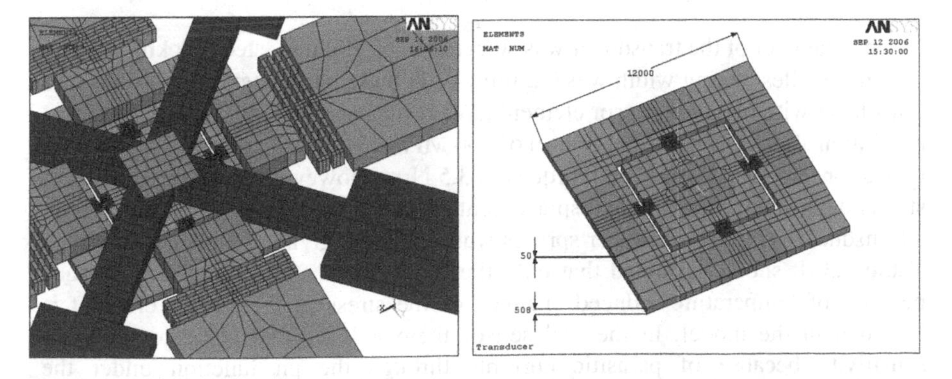

Fig. 1.5 Mounting of the sensor element (to the *right*) on the central part of the transducer (to the *left*)

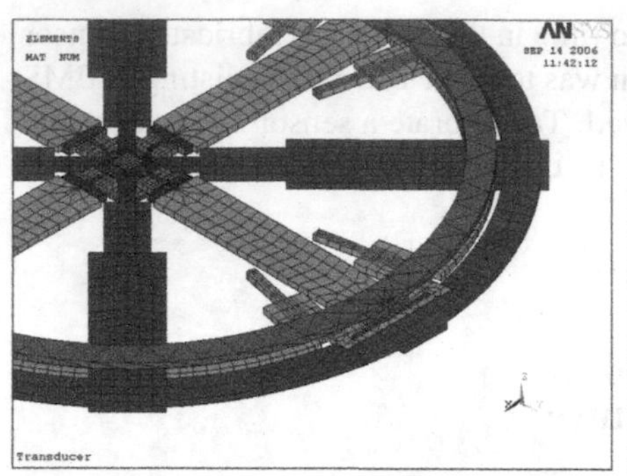

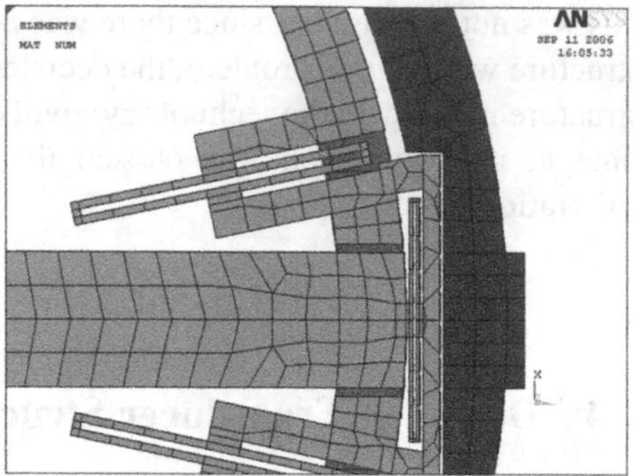

Fig. 1.6 Illustration of the spring system of the transducer. The ring above the transducer (to the *left*) is mounted on the spring supported arms. These arms are connected to the rigid cross via three springs as shown in the *right* figure

Table 1.2 Simulated max stress in the piezoresistors, max stress in the silicon structure and the calculated resistance changes according to the previous formula when forces are applied to the sensor including transducer in the *x*- and *z*-directions, torques around the *y*- and *z*-axes and when temperature is increased by 100 °C

FEM-Simulation of maximum stress applied to the 10 N MEMS-structure						
	Unit	$F_x = 10$ N	$F_z = 10$ N	$M_y = 0.5$ Nm	$M_z = 0.5$ Nm	$dT = 100$ °C
Stress in piezo	MPa	80	85	67	186	42
Stress in Silicon	MPa	133	157	188	295	138
dR_a	%	6	0	6,8	0	0
dR_b	%	1.8	0	1.2	0	0
dR_{F_z}	%	0	0.9	0	0	0
dR_{M_z}	%	0	0	0	5.9	0
dR_T	%	0	6.4	0	0	2.8

Notice the lack of isotropy dependent on the low width/height ratio of the beams in the 10 N sensor, which results in a higher silicon stress for M_z than for M_y

The diameter of the transducer was selected to be 100 mm, steel thickness 1 mm and the smallest spring width was 0.2 mm. With these parameters simulations of a transducer with mounted sensor element gave results as shown in Table 1.2. With a maximum allowed Silicon stress level of 300 MPa, this transducer could be possible to use for forces up to 40 N and torques to 3.5 N m. However, the springs were not stiff enough and the maximum displacement got too large at these levels. Therefore, a transducer version with stiffer springs was simulated giving the results shown in Table 1.3. It should be noted that only the temperature effect on piezoresistance because of temperature induced changes of the stress in the sensor element is included in the model. In the real sensor there will also be some temperature sensitivity because of parasitic currents through the pn junction under the piezoresistor and the piezo coefficients are slightly temperature dependent.

Table 1.3 Maximum changes in resistivity, maximum stress in piezoresistors and maximum displacement in the transducer structure with stiffer springs than used in the simulations according to Table 1.4

FEM-simulation with stiff transducer			
Force/torque	Resistivity change (%)	Stress at piezo (MPa)	Maximum displacement (μm)
$F_x = 10$ N	6.0/1.8	80	35
$F_z = 10$ N	6.4	85	141
$M_y = 0.5$ N m	6.8/1.2	67	412
$M_z = 0.5$ N m	5.9	166	88
$\Delta T = 100$ K	2.8	42	134

In column 2 for the force of 10 N in x-direction and for the torque of 0.5 N m in y-direction two values are mentioned. This means that these two cases cause a similar stress to the same bridges and therefore change of resistivity of the resistors on the bridges of the F/T-sensor

Table 1.4 Maximum values of piezoresistivity, piezostress, silicon stress, steel stress, stress in the adhesive for the sensor element mounting and maximum displacement in the transducer structure as response to force, torque and temperature changes

FEM-simulation of change of resistivity including transducer						
Force/torque	Resistivity change (‰)	Stress at piezo (MPa)	Stress in silicon (MPa)	Stress in steel (MPa)	Stress at adhesive (MPa)	Max. displacement (μm)
$F_x = 10$ N	3.6/5.1	12.4	79	106	26	120
$F_z = 10$ N	8.2	12.5	41	353	24	309
$M_y = 0.5$ N m	9.8/1.2	15.3	41	335	24	633
$M_z = 0.5$ N m	7.1	21	38	107	38	121
$\Delta T = 100$ °C	18.3	29	143	229	667	150

The transducer was manufactured from sheets of steel by means of laser cutting. In order to keep the melting process under control water jet laser cutting had to be used, see Fig. 1.7.

1.5 Fabrication of MEMS Chip

The MEMS structure was fabricated using well established processes as described in Figs. 1.8 and 1.9. The substrate was n-doped 6″ Silicon wafers of thickness 508 μm. The most critical process steps are the formation of the piezoresistors and the deep reactive silicon etching (DRIE) to separate the inner sensor element part from the outer.

In order to obtain optimal piezoresistive coefficient (up to 70 Ω/MPa for p-type Si) the implantation dose and the annealing conditions were tuned and Fig. 1.10 shows the simulated Boron profiles after doping and after annealing. Also the DRIE etching process was tuned, in this case with respects to edge angles.

Fig. 1.7 Result of laser cutting of the transducer. To the *left* using standard laser cutting and to the *right* with water jet guided laser cutting

Thus, Fig. 1.11 shows the results of DRIE etching with process parameters optimized for two different etching depths.

After the DRIE etching the surfaces of the side walls get very rough, see Fig. 1.12 to the left. This roughness will increase the risk that the sensors will break, and therefore, a passivation was made of the etched walls, giving the wall structure that can be seen in figure to the right.

1.6 Mounting of MEMS chip

Critical for the sensor concept is to have a simple low cost process for the mounting of the sensor element on the transducer. Therefore, gluing was used and different adhesives were tested to make sure that the strength and the stiffness of the mountings will not degrade with time or with the number of stress cycles that it will be subjected to. Figure 1.13 on the left shows the test equipment used for stiffness measurement of the adhesive and an example of the test results can be seen in Fig. 1.13 on the right. Table 1.5 shows a list of the adhesives tested. Although he best performance was found for ABLESTIK Ablebond 84-3T, finally the glue DELO-DUOPOX_AD895 was taken because the curing could be done at room temperature so that no temperature stress was frozen into the F/T-sensor structure.

The selected adhesive was found to handle more than 50 N for an area corresponding to a gluing pad on the sensor element and FEM calculations were made to make sure that the stress on the adhesive surfaces was acceptable. Thus Fig. 1.14 shows the stress distribution on the gluing pads, if a point force of 10 N is applied to the outer transducer ring. The stress levels obtained are typically far

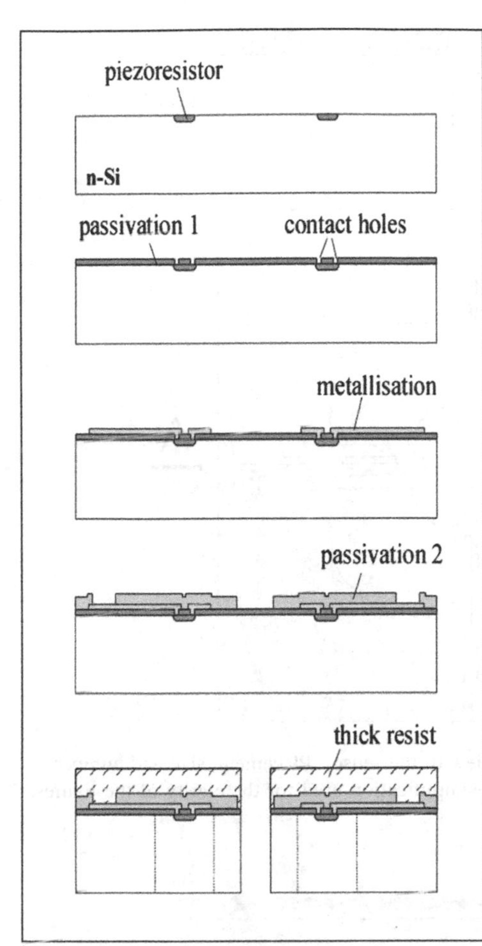

1. Lithography and implantation (Boron) of the p-type piezoresistors (through a thin thermal oxide)
2. Annealing thermal oxide and activation of Boron (drive-in),
3. Passivation of the piezoresistors with LPCVD[1] silicon oxide,
4. Lithography and dry etching of contact holes,
5. Metallization with Al alloy
6. Lithography and wet etching of narrow structures with width 3 μm and separation 6 μm.
7. Passivisation of the metallization with PECVD[2] silicon nitride. Generation of glue areas, dicing marks and light shield.
8. Lithography and opening of the pads and the DRIE areas by dry etching,
9. Lithography using a thick photo resist (25 μm), dry etching of the oxide passivation and DRIE of the Si substrate
10. Resist strip and cleaning

[1] LPCVD (Low Pressure Chemical Vapor Deposition)

[2] PECVD (Plasma Enhanced Chemical Vapor Deposition)

Fig. 1.8 Schematical description of the process flow for the fabrication of the force sensors. The cross-sections correspond to the top view shown in Fig. 1.9

below 10 MPa, the value the adhesive layer can handle. The simulated higher values in some corners will not be obtained in real life since the pads and the adhesives will have round corners.

In order to mount the sensor element on the transducer a special pick- and place tool was designed, see Fig. 1.15. The procedure when mounting the sensor element on the transducer was the following:

1. Before the wafers are diced the wafers are glued onto an elastic foil from which every single sensor element can easily be picked with the pick- and place tool.

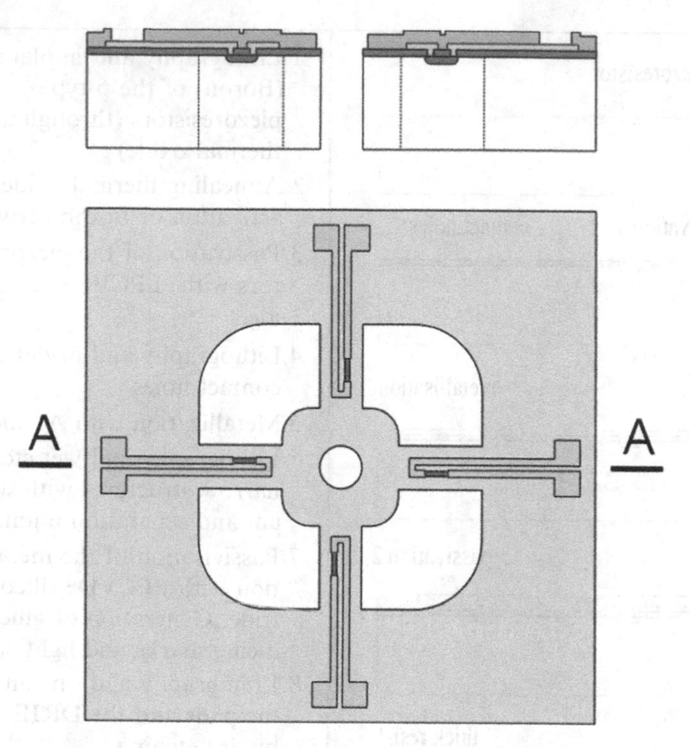

Fig. 1.9 *Lower figure* shows a schematic top view of the sensor. Placement, size and number of the piezoresistors not as in the real sensor, the simplification used for the clarity of the figures. *Upper figure* shows the cross section at A-A

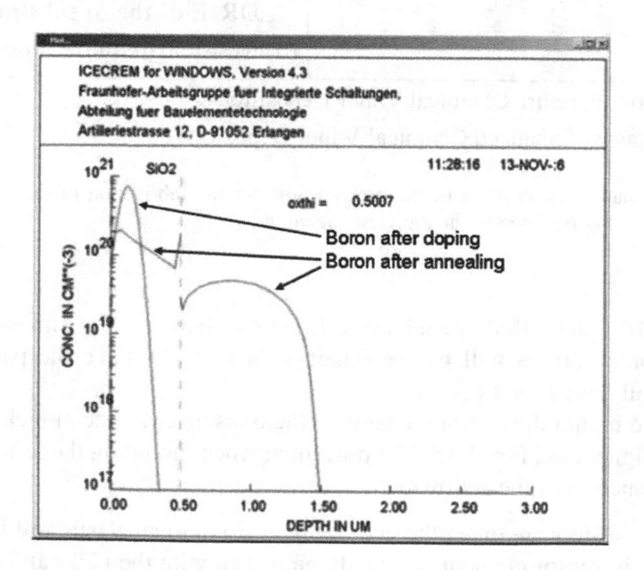

Fig. 1.10 The Boron profile after doping and after annealing of the piezoresistors

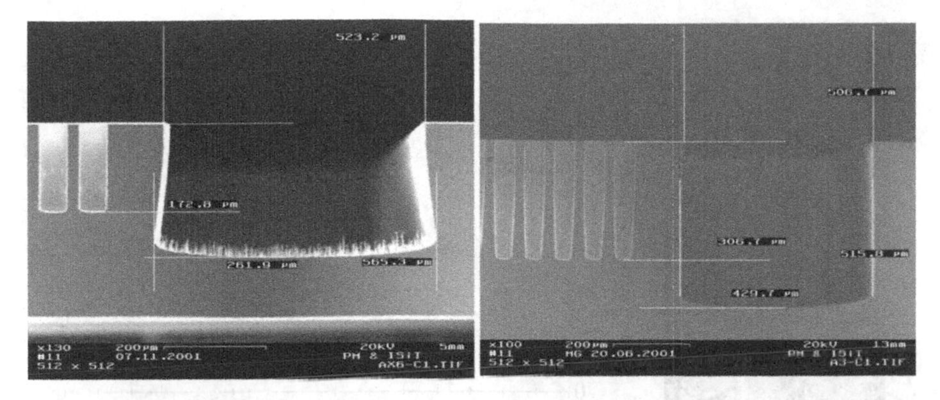

Fig. 1.11 SEM pictures of DRIE etching optimization examples. On the *left hand side* the process is optimized for an etching depth of 50 µm and on the *right hand side* for 500 µm etching depth. The etching in these tests was stopped before it had completely penetrated the silicon chip

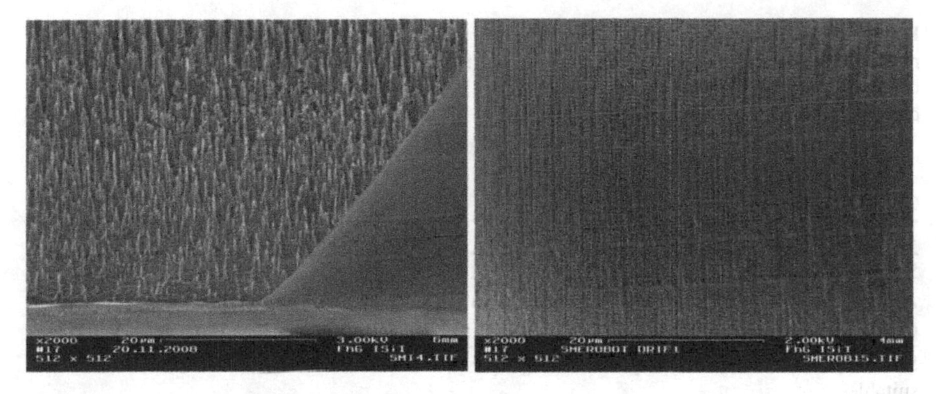

Fig. 1.12 SEM of the Silicon walls after DRIE etching (*left figure*) and after passivation (*right figure*)

2. When time for mounting the sensor element is lifted from the elastic foil with the pick- and place tool and placed precisely into a fabricated mechanical pick-up.
3. Stud bumps are placed on each gluing area of the sensor element to define the distance between the sensor element and the transducer.
4. Glue is dispensed onto all five gluing areas on the sensor element.
5. The transducer, which has previously been cleaned, is placed onto the sensor and stays there until the glue is fixed and the mounting is ready.

   After these steps the electrical connections are made:

6. Glue flexible PCB (Printed Circuit Board) to transducer (only moderate precision necessary).

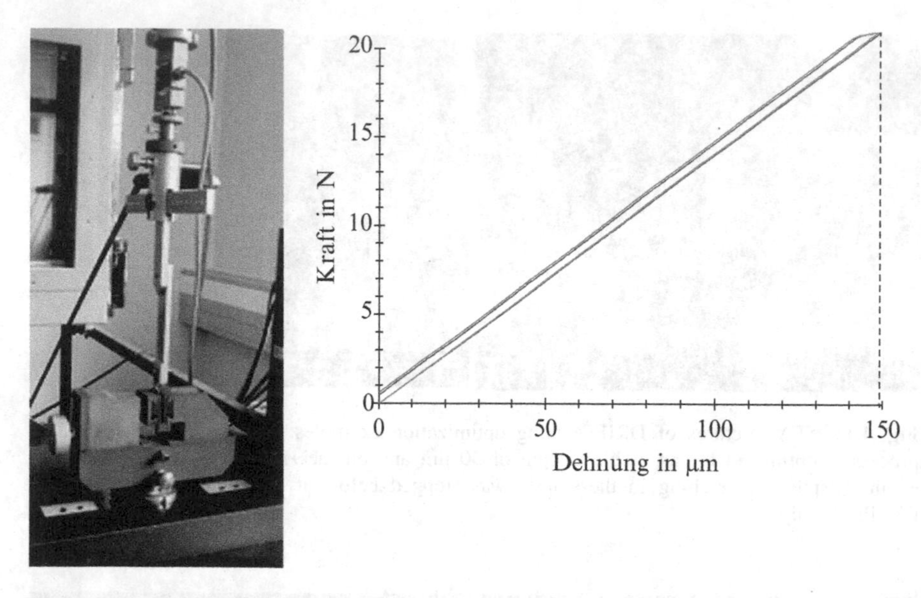

Fig. 1.13 Several types of adhesives were tested in the equipment (shown in the *left figure*) with respect to the linearity of the stiffness curve. An example of the linearity measurement results can be seen in the *right figure*. Long term test were made to make sure that the linearity was not changed after a large number of load cycles

Table 1.5 The tested nine adhesive types

Tested adhesives		
Sekundenkleber	EPO-TEK H77	UHU endfest 300
ABLESTIK Ablebond 84-3T	NAMICS 8437-2	NAMICS U8443
COOKSON Staychip 3082	COOKSON Staychip 3100	DELO-DUOPOX AD895

Sekundenkleber is an ordinary fast curing epoxy. DELO-DUOPOX_AD895 was found most suitable

7. Perform the bonding between the PCB and the sensor (sensor pads can be seen in Fig. 1.4).
8. Put glop top as a protection on all bond wires.

1.7 Measurement Electronics

The resistances of the piezoresistors are measured by separate wiring pairs for delivering the current and for measuring the voltage drop respectively. Within total 24 piezoresistors on the sensor element, 96 connections must be made between the sensor elements and the electronics. For the measurement electronics used in

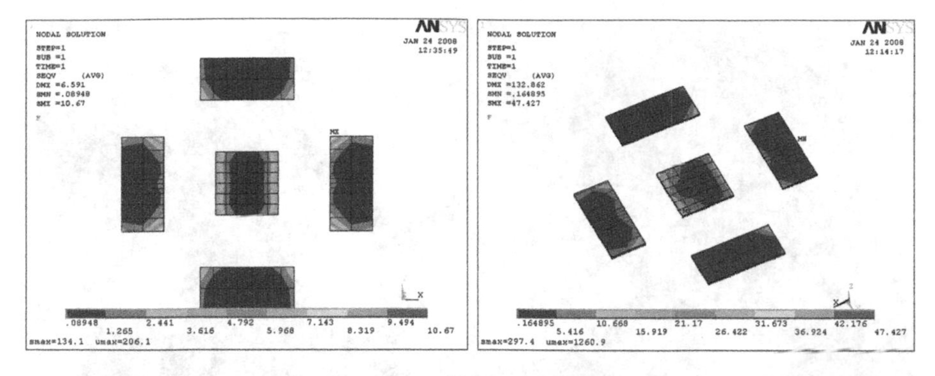

Fig. 1.14 Calculated stress on the adhesive layer at the gluing pads on the sensor element. On the *left hand side* with 10 N transducer force in the *x*-direction and on the *right hand side* with 10 N in the *z*-direction (perpendicular to the transducer plane)

Fig. 1.15 Pick- and place tool for mounting the sensor element on the transducer. Vacuum is used and the inner- and outer parts of the sensor elements are lifted by separate vacuum inlets

the test of the force-and-torque sensor, four connection PCBs were mounted symmetrically on the rigid cross of the transducer according to Fig. 1.16 to the left. At the inner end of these PCBs the bonding was made to the sensor element and at the outer end the connector for a flat cable was mounted. The flat cables were on the other side of the sensor housing connected to the measurement electronics as shown in Fig. 1.16 to the right. In Fig. 1.17 the bonding between the connection PCB and the corresponding section on the sensor element is illustrated. The right bondings are for the temperature reference piezoresistors and the left bondings are for the piezoresistors measuring the stress on the beams between the outer and inner part of the sensor element.

Fig. 1.16 The connection of the sensor element to the measurement electronics was made by means of four connection PCBs mounted on the rigid transducer cross (*left*) and flat cables connecting these PCBs to the measurement electronics (*right*) on the other side of the sensor housing

Fig. 1.17 The bondings between the connection PCB and the bonding pads for six of the piezoresistors on the sensor element

The measurement electronics consists of two multiplexers, which can multiplex four signals at a time, a counter to control the multiplexing and an SMU (Source Measurement Unit) to make the analogue measurements, see overview in Fig. 1.18.

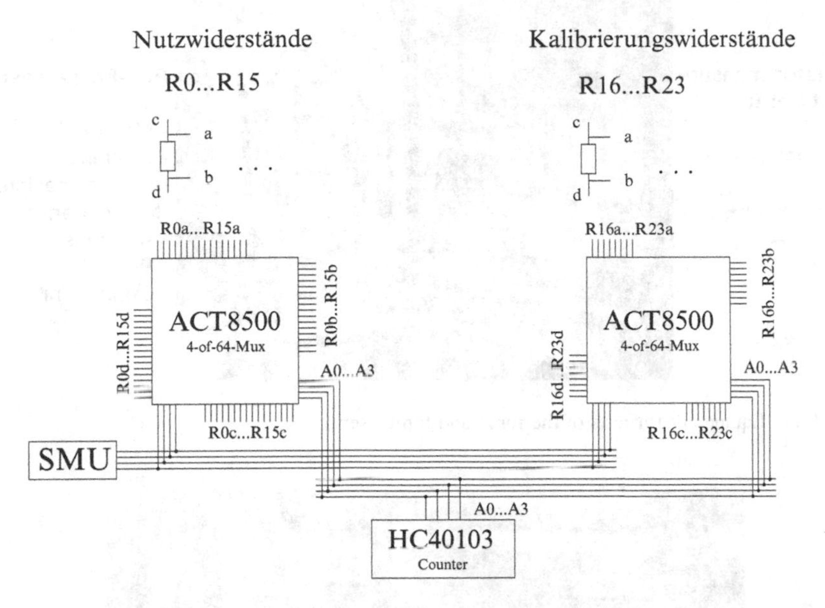

Fig. 1.18 The measurement electronics used for the testing of the force-and-torque sensor. The stress measuring piezoresistors are multiplexed by the Integrated Circuit to the left and the temperature measuring piezoresistors (to the *right*) uses the same kind of multiplexer. The multiplexers are connected to the Sensor Measurement Unit and the measurements are stored in a standard PC using an IEEE488 interface

1.8 Sensor Tests and Calibration

The assembled force-and-torque sensor was analyzed using a Pull-Force-Tester as seen in Fig. 1.19. The Pull-Force-Tester is meant for material testing but can also be used to generate accurate forces for the testing of a force sensor. By mounting the sensor in different directions it was possible to get the force and torque vectors needed for the analysis. Included in the setup of Fig. 1.19 is a commercial six-DOF Force-and-Torque sensor, which was used to check that the test equipment was working properly and to have as a reference for the sensor analysis.

For the calibration of the force sensor the setup in Fig. 1.20 was used. Underneath the PC that includes the IEEE488 interface the SMU (Keithley SMU236) is located. The SMU allows precise measurements of a voltage while a constant current is applied. A sensor holder (to the left in the figure) was used to mount the sensor in different directions and by means of weights accurate forces and torques could be applied to the sensor.

In order to facilitate the analysis of the sensor tests easy to use software interfaces were developed as shown in Fig. 1.21. At the bottom on the left hand side the schematic placement of all 16 measurement resistors and all 8 temperature resistors is plotted. The green colors indicates that they are all measured as can be seen at the 20 small windows showing the resistance change over the time.

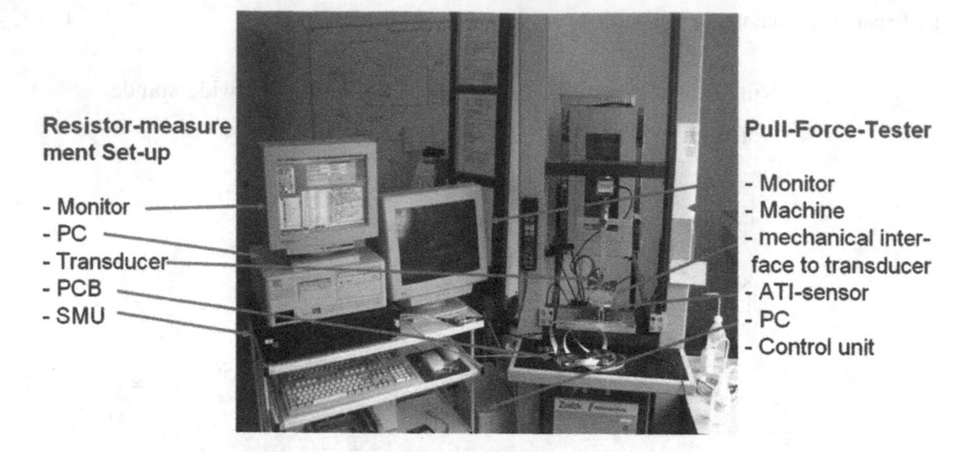

Fig. 1.19 Equipment for tests of the force-and-torque sensor

Fig. 1.20 Equipment for the calibration of the force-and-torque sensor

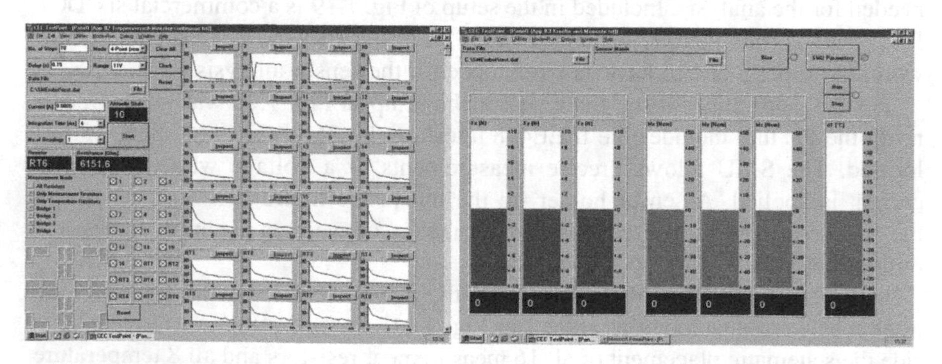

Fig. 1.21 Interfaces to the measurement system. The interface on the *left hand side* was used to analyze the behavior of the individual piezoresistors and the interface on the *right hand side* to test the calibration of the sensor

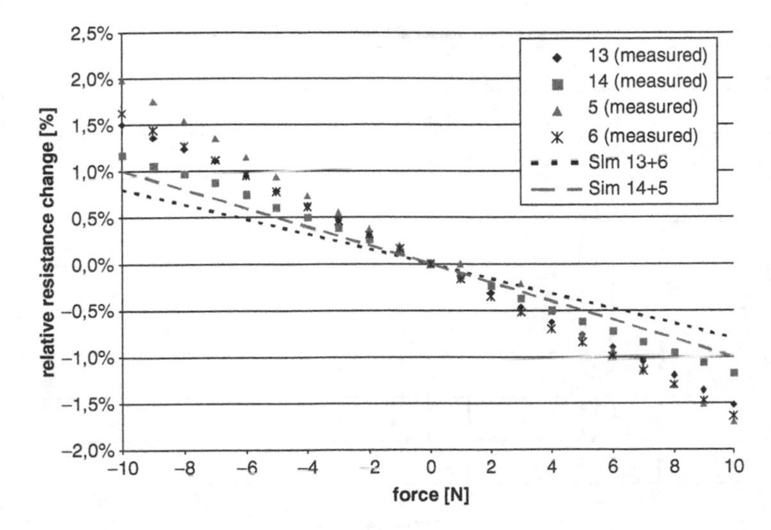

Fig. 1.22 Plot of resistance change as function of force for individual piezoresistors, simulated values *solid lines* and measured *dotted lines*

Using the equipment in Fig. 1.18 the linearity and sensitivity of the different piezoresistors were obtained. Figure 1.22 shows the result for some of the resistors and as can be seen there is quite a good agreement between with the FEM model results and of the results from the measurements on the sensor.

Beside tests of linearity and sensitivity also the temperature compensation was tested. These tests were made with sensors in a climate chamber. During these tests the chamber was at first cooled down to 0°C and then heated up stepwise by 10–80°C. Each temperature step was held for 20 min and the resistance values of all the piezoresistors of the sensor were measured (Fig. 1.23).

The calibration of the sensors was not made using the expressions described in connection with Table 1.1 because it was found out that only the Pull-Force-Tester it was possible to properly apply the calibration forces and torques and this was quite time consuming. In addition the calibration needed to be renewed when the sensor is screwed into the housing shown in Fig. 1.16. So another approach was realized using the equipment in Fig. 1.20 taking a commercial sensor and applying calibration weights made of steel with only relative precision. Taking then the orthonormalizing process by using the Gram–Schmidt Method it was possible to achieve a 16 $\times$ 6 matrix to calculate the forces and torques.

1.9 Final Demonstrator

To show the performance of the F/T-sensor a demonstrator was constructed, that

– Could to be connected to a PCB or a notebook
– Had USB as the interface

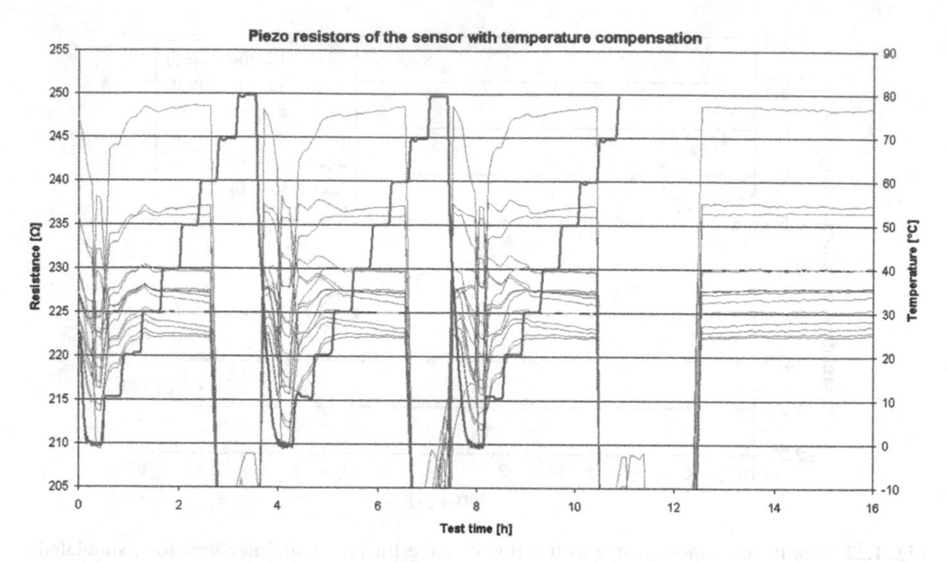

Fig. 1.23 Results from temperature compensation of individual piezoresistors. The stepwise increasing *solid curve* shows the temperature (with readings to the right) and the other curves show the temperature compensated resistance values of individual piezoresistors. After transients when cooling the temperature chamber to zero degrees the compensated resistance readings are more or less constant between 10 and 80 °C

- Was based on the PCB according to Fig. 1.24 including a Micro-Controller, an A/D-converter and a memory to store the calibration matrix and to calculate the forces and torques
- Included a manageable housing of aluminum that can be flanged to a robot and that includes an overload protection, see Fig. 1.16
- Allowed calibration and measuring via a revised graphical user interface (GUI) (see Fig. 1.25)

With this hardware and GUI it is possible to switch between measuring the single resistors as shown in Fig. 1.26, to present the calculated forces and torques.

Figure 1.25 shows the user interface for measurement and force and torque calculations. The following functionality was implemented:

Connection settings: Settings for the USB device of the measurement hardware and the PC.

"Measurement settings": The number of measurement samples, the amount of values to calculate the arithmetic mean and the delay between the measurements can be decided here. Moreover, the name of and path to the file to store the data in are defined.

"Resistors": Here one can choose which of the 24 resistors that will be measured including the resistors for temperature compensation. With "Singleshot" only one measurement will be done in contrast to "continuous." "Bias" allows to subtract the first measured values so that the offset of the resistors can be eliminated, while

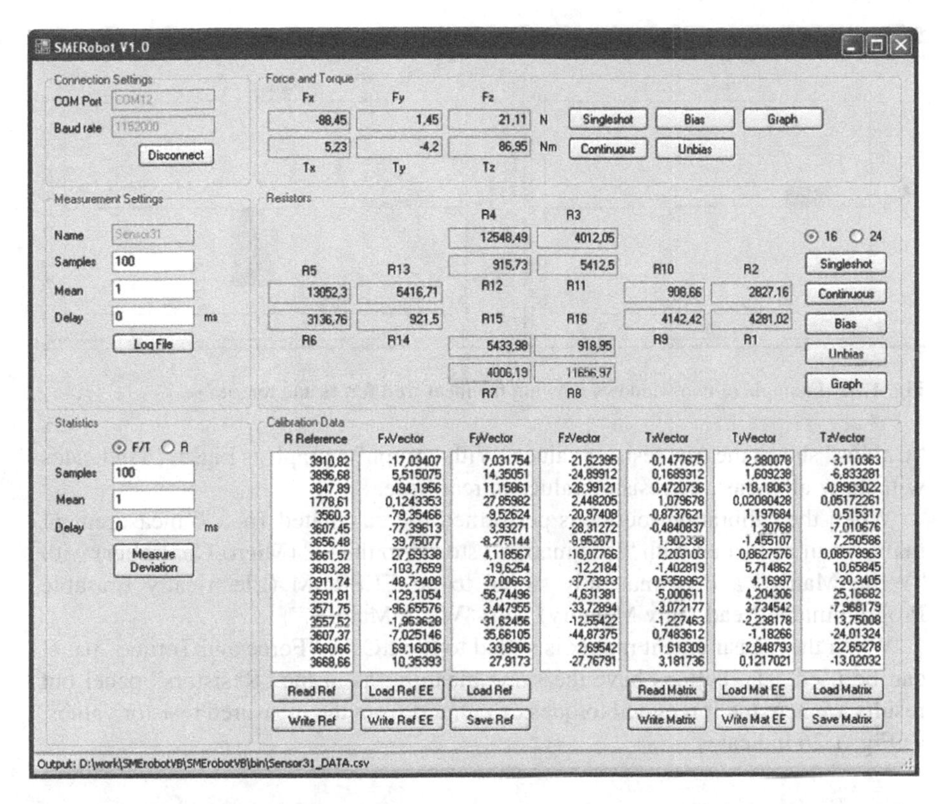

Fig. 1.24 Graphical user interface to be used for measurements on the individual resistors as well as for calculating the applied forces and torques

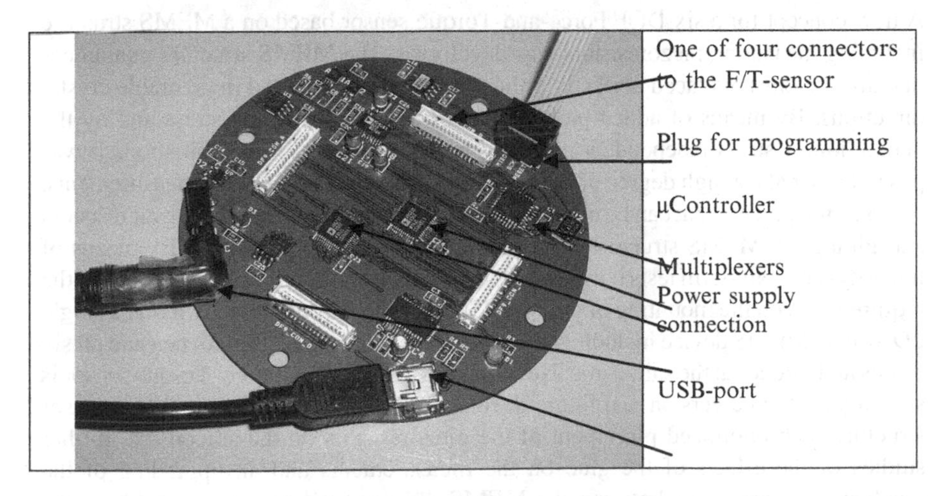

Fig. 1.25 PCB of the final demonstrator

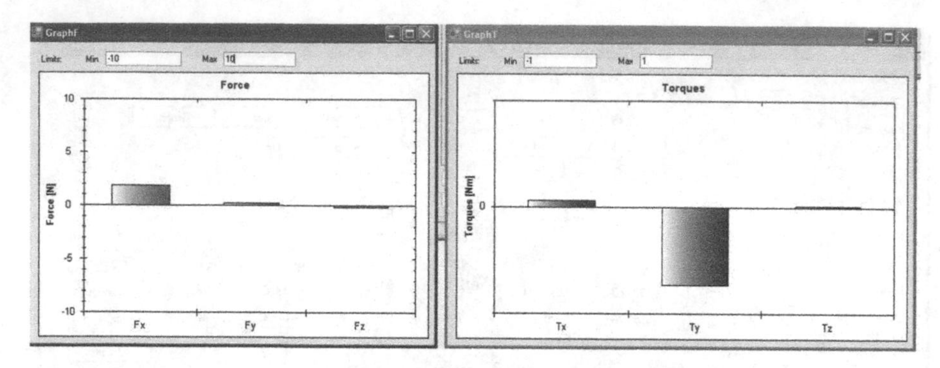

Fig. 1.26 Example of two windows showing the measured forces and torques

"Unbias" shows the true resistor values. With "graph" a graph as Fig. 1.26 indicates will show all measured resistor values in real time.

When the calibration routine is performed the calculated 16 × 6 measurement matrix, can be loaded with "load matrix," stored to the μC (Micro-Controller) with "Write Matrix" and permanently stored to an EEPROM (Electrically Erasable Programmable Read-Only Memory) with "Write Mat EE."

When the measurement matrix is stored to the μC the "Force and Torque" panel can be used. The buttons have the same meaning as in the "Resistors" panel but results are here the forces and torques calculated from the measured resistor values, as Fig. 1.26 indicates.

1.10 Conclusions

A new concept for a six-DOF Force-and-Torque sensor based on a MEMS structure including measuring piezoresistors was developed. The MEMS structure guarantees that all resistors are placed perfectly orthogonal to each other and in favorable crystal directions. By means of added piezoresistors on places with low stress and by the generation of stress independent piezoresistance measurement combinations it was possible to make a high degree of temperature compensation of stress measurements. A mounting process that can be made automatic for electrically contacting all resistors and gluing the MEMS structure on a steel transducer was developed. By means of linearity- and long-term tests it was possible to find an epoxy adhesive that fulfilled the requirements on the mounting of the MEMS structure. It could be shown that a single 2D-silicon MEMS device including a steel transducer could be used to measure all six forces and torques at the same time. Nevertheless, the six-DoF Force–Torque sensor is still in a prototype version and future development will deal with a modified sensor structure with optimized placement of the piezoresistors on the silicon die, further studies of the effects of the glue on the measurements and an upgrading of the mechanical connections between the MEMS chip and the transducer and between the transducer and the sensor housing.

Acknowledgments The sensor was developed during the European Project SMErobot (contract number 011838). The authors wish to thank Dr. Manfred Weiß for his contribution.

References

1. Okada K (1990) Flat –type six-axial force sensor. In: Tech. Digest of the 9th Sensor Symposium, pp 245–248
2. Okada K (1988) Force and moment detector using resistor. European patent application EP0311695B1
3. Dao DV et al (2003) A MEMS based microsensor to measure all six components of force and moment on a near-wall particle in turbulent flow. In: Transducers '03, Boston, June 8–12, pp 504–507
4. Hirabayashi Y, Sakurai N, Ohsato T (2007) Multiaxial force sensor chip. European patent application EP1852688A2
5. Ohsato T, Sakurai N, Hirabayashi Y, Yokobayashi H (2005) Multi-axis force sensor chip and multi-axis force sensor using same. European patent application EP1653208A2
6. Ohsato T, Hirabayashi Y (2003) Six-axis force sensor. European patent application EP1327870A2
7. Ruther P et al (2005) Novel 3D piezoresistive silicon force sensor for dimensional metrology of micro components. In: Sensors, 2005 IEEE, Irvine, CA, Oct 31–Nov 3, pp 1006–1009
8. Weiß M, Eichholz J Sensoranordnung. Pending German utility patent
9. Völklein F, Zetterer T (2006) Praxiswissen Mikrosystemtechnik. Table 3.3-2, equation 3.3-30, ISBN 3-528-13891-2, Vieweg Verlag

Chapter 2
Piezoelectrically Actuated Robotic End-Effector with Strain Amplification Mechanisms

Jun Ueda

Abstract This chapter describes a nested rhombus multilayer mechanism for large effective-strain piezoelectric actuators. This hierarchical nested architecture encloses smaller flextensional actuators with larger amplifying structures so that a large amplification gain on the order of several hundreds can be obtained. A prototype nested PZT cellular actuator that weighs only 15 g produces 21% effective strain (2.53 mm displacement from 12 mm actuator length and 30 mm width) and 1.69 N blocking force. A lumped parameter model is proposed to represent the mechanical compliance of the nested strain amplifier. This chapter also describes the minimum switching discrete switching vibration suppression (MSDSVS) approach for flexible robotic systems with redundancy in actuation. The MSDSVS method reduces the amplitude of oscillation when applied to the redundant, flexible actuator units. A tweezer-style end-effector is developed based on the rhombus multilayer mechanism. The dimensions of the end-effector are determined by taking the structural compliance into account. The assembled robotic end-effector produces 1.0 N of force and 8.8 mm of displacement at the tip.

2.1 Introduction

Recent advances in actuation technology have produced exciting new ideas in the growing field of biomechatronic and bio-robotic devices. Today there is a wide variety of choices of actuators in terms of size, material, structure, and control. For example, standard AC/DC rotary motors are widely available. Ultrasonic actuators are small in size and widely used in digital cameras. Fluid actuators are also widely used in industry for high-power applications. The aforementioned commercialized actuators

J. Ueda (✉)
Mechanical Engineering, Georgia Institute of Technology, 771 Ferst Drive, Atlanta, GA, USA
e-mail: jun.ueda@me.gatech.edu

D. Zhang (ed.), *Advanced Mechatronics and MEMS Devices*, Microsystems,
DOI 10.1007/978-1-4419-9985-6_2, © Springer Science+Business Media New York 2013

are in general reliable and low cost. However, these conventional actuators may not deliver sufficient performance for certain novel applications, in particular, for biomedical systems. Novel robotic and mechatronic devices used in such systems require novel actuators that have the following features:

- Energy efficiency
- Compactness
- Low weight
- High-speed operation
- Natural compliance (so as to be unable to harm humans or environments)
- Silent operation

When a new technology area emerges for which the corresponding consumer market is not mature, there will be a significant lack of actuation technologies that meet the requirements of the new technology. This has long been one of the driving motivations in actuator research. For example, assistive technology requires compact, lightweight, and safe (i.e., compliant) actuators. These requirements unfortunately exclude most AC/DC rotary motors. Pneumatic actuators are too large and their speed of response is too slow. For distributed camera network systems, such as for security applications, fast but silent actuation is paramount.

Piezoelectric ceramics, such as Lead Zirconate Titanate (PZT), have a high power density, high bandwidth, and high efficiency. PZT outperforms other actuator materials, including shape memory alloy (SMA), conducting polymers, and electrostrictive elastomers, with respect to speed of response and bandwidth. Its maximum stress is as large as SMA, and the efficiency is comparable to electrostrictive elastomers. Furthermore, PZT is a stable and reliable material that is usable in diverse, harsh environments. Figure 2.1 shows a qualitative comparison of materials. Note that we do not claim that conventional actuators should be replaced with piezoelectric actuators. The main focus of this research is to develop new actuator devices that cover the applications for which conventional actuators are not suitable; piezoelectric actuators could provide fast, zero backlash, silent (gearless), and energy-efficient actuation in a compact body.

Piezoelectric materials are known to have two major piezoelectric effects that make them useful for both actuation and sensing [18, 19, 21, 25, 28]. A piezoelectric material generates electric charges on its surfaces when stress is applied. This effect, called "direct piezoelectric effect," enables one to use the material as a sensor that measures its strain or displacement associated with the strain. The "converse piezoelectric effect," where the application of an electrical field creates mechanical deformation in the material, enables one to use the material as an actuator. The energy efficiency of piezoelectric actuation is in general very high [29]; the converse piezoelectric effect directly converts electrical energy to mechanical strain without generating much heat. Use of piezoelectric actuators for vibration control of structures is an active area of research in aerospace engineering. A rapid change in consumer markets also necessitates new actuators; for example, the recent increase in functionality of cellular phones has accelerated the development of compact piezoelectric actuators for auto-focus camera modules even though the market is mature.

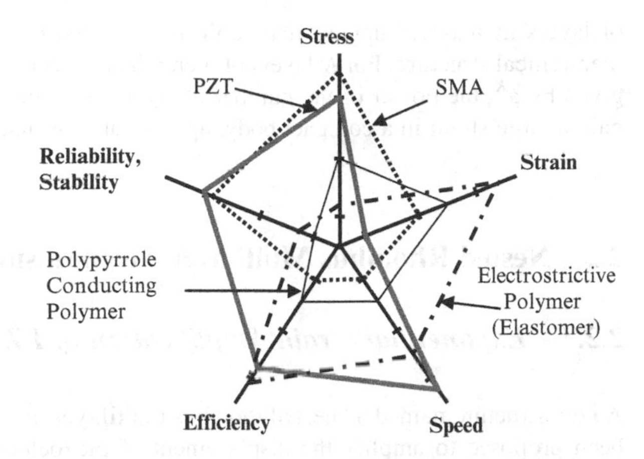

Fig. 2.1 Qualitative comparison of actuator materials

A unique "cellular actuator" concept has been presented, which in turn has the potential to be a novel approach to synthesize biologically inspired robot actuators [30–34]. The concept is to connect many PZT actuator units in series or in parallel and compose in totality a macro-size linear actuator array similar to skeletal muscles.

The most critical drawback of PZT is its extremely small strain, i.e., only 0.1%. Over the last several decades efforts have been taken to generate displacements out of PZT that are large enough to drive robotic and mechatronic systems [2, 4, 6, 7, 10, 11, 16–18, 24, 29]. These can be classified into (a) inching motion or periodic wave generation, (b) bimetal-type bending, and (c) flextensional mechanisms. Inching motion provides infinite stroke and bimetal-type mechanism [8, 24] can produce large displacement and strain, applicable to various industrial applications when used as a single actuator unit; however, unfortunately, the reconfigurability by using these types may be limited due to the difficulty in arbitrarily connecting a large number of actuator units in series and/or in parallel to increase the total stroke and force, respectively. In contrast, flextensional mechanisms such as "Moonie" [7, 17], "Cymbal" [6], "Rainbow" [10], and others [11] are considered suitable for the reconfigurable cellular actuator design.

An individual actuator can be stacked in series to increase the total displacement. Note that this simple stacking also increases the length of the overall mechanism and does not improve the strain in actuation direction, which is known to be up to 2–3%. Therefore, a more compact actuator with larger strain is considered necessary for driving a wide variety of mechatronic systems.

In this chapter, an approach to amplifying PZT displacement that achieves over 20% effective strain will be presented [33, 34]. The key idea is hierarchical nested architecture that encloses smaller flextensional actuators with larger amplifying structures. A large amplification gain on the order of several hundreds can be obtained with this method. Unlike traditional stacking mechanism [3, 18], where the gain α is proportional to the dimension of the lever or number of stacks, the amplification gain of the new mechanism increases *exponentially* as the number of layers increases. Suppose that strain is amplified α times at each layer of the hierarchical structure. For K layers of hierarchical mechanism, the resultant gain is given by α^K, the power of the number of layers. This nesting method allows us to gain a large strain in a compact body, appropriate for many robotic applications.

2.2 Nested Rhombus Multilayer Mechanism

2.2.1 *Exponential Strain Amplification of PZT Actuators*

A new structure named a "nested rhombus multilayer mechanism" [23, 33–35] has been proposed to amplify the displacement of piezoelectric ceramic actuators in order to create a novel actuator unit with 20–30% effective strain, which is comparable to natural skeletal muscles [12, 27]. This mechanism drastically mitigates the small strain of PZT itself. This large strain amplification is to hierarchically nest strain amplification structures, achieving exponential strain amplification. This approach uses rhomboidal shaped layered strain amplification mechanisms. This mechanism exhibits zero backlash and silent operation since no gears, bearings, or sliding mechanisms are used in the amplification structures.

Traditional "Moonie" flextensional mechanisms [17] will be used for the basis of the proposed nested rhombus multilayer mechanism. As shown in Fig. 2.2a, the main part of the mechanism is a rhombus-like hexagon that contracts vertically as the internal unit shown in gray expands. The vertical displacement, that is, the output of the mechanism, is amplified if the angle of the oblique beams to the horizontal line is less than 45°. Figure 2.2b illustrates how the strain is amplified with this mechanism.

A schematic assembly process of the proposed structure is shown in Fig. 2.3. A series of piezoelectric actuators with the "Moonie"-type strain amplification mechanisms are connected (in Fig. 2.3, five actuator units are serially connected) and nested in a larger rhombus amplification mechanism.

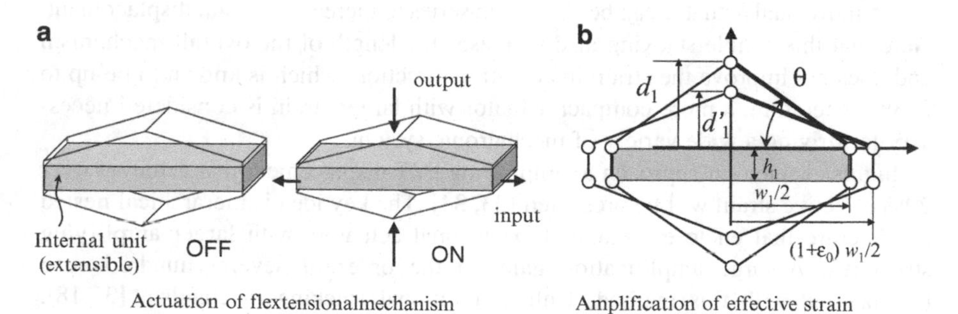

Fig. 2.2 Amplification principle of flextensional mechanisms [17] (© IEEE 2010), reprinted with permission

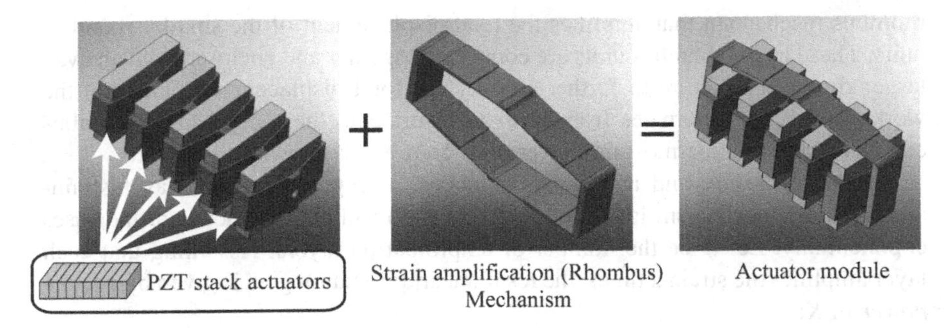

Fig. 2.3 Schematic assembly of nested rhombus multilayer mechanism

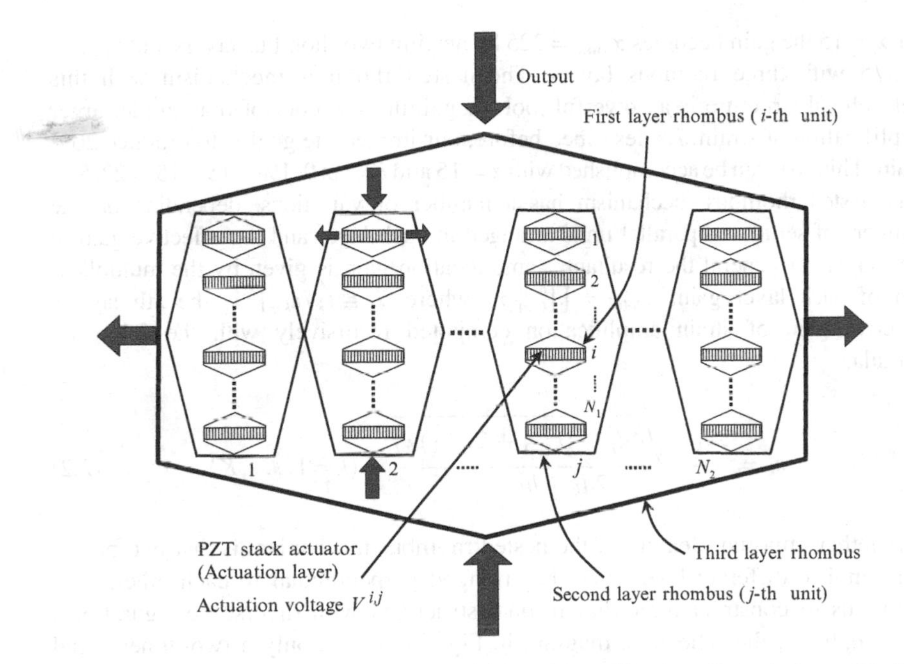

Fig. 2.4 Proposed nested structure for exponential strain amplification: the strain is amplified by three layers of rhombus strain amplification mechanisms, with the first layer, called an actuator layer, consisting of the smallest rhombi directly attached to the individual PZT stack actuators (© IEEE 2010), reprinted with permission

Figure 2.4 illustrates the "exponential" strain amplifier, which consists of the multitude of rhombus mechanisms arranged in a hierarchical structure. The innermost unit, i.e., the building block of the hierarchical system, is the standard rhombus mechanism, or conventional flextensional mechanism, described above. These units are connected in series to increase the output displacement. Also these units can be arranged in parallel to increase the output force. The salient feature of this hierarchical mechanism is that these rhombus units are enclosed with a larger rhombus mechanism that amplifies the total displacement of the smaller rhombus units. These larger rhombus units are connected together and enclosed with an even larger rhombus structure to further amplify the total displacement. Note that the working direction alternates from layer to layer; e.g., the second layer rhombus extends when the innermost first layer units contract as shown in Fig. 2.2b.

As this enclosure and amplification process is repeated, a multilayer strain-amplification mechanism is constructed, and the resultant displacement increases exponentially. Let K be the number of amplification layers. Assuming that each layer amplifies the strain α times, the resultant amplification gain is given by α to the *power* of K:

$$\alpha_{total} = \alpha^K. \tag{2.1}$$

For $\alpha = 15$ the gain becomes $\alpha_{\text{total}} = 225$ by nesting two rhombus layers and $\alpha_{\text{total}} = 3,375$ with three rhombus layers. The nested rhombus mechanism with this hierarchical structure is a powerful tool for gaining an order-of-magnitude larger amplification of strain. As described before, our immediate goal is to produce 20% strain. This goal can be accomplished with $\alpha = 15$ and $K = 2$: $0.1\% \times 15 \times 15 = 22.5\%$. This nested rhombus mechanism has a number of variations, depending on the numbers of serial and parallel units arranged in each layer and the effective gain in each layer. In general the resultant $\triangleq$amplification gain is given by the multiplication of each layer gain: $\alpha_{total} = \prod_{k=1}^{K} \alpha_k$ where $\alpha_k \triangleq \varepsilon_k / \varepsilon_{k-1}$ is the kth layer's effective gain of strain amplification computed recursively with the following formula,

$$\varepsilon_k = \frac{2d_k - \sqrt{4d_k^2 - (\varepsilon_{k-1}^2 + 2\varepsilon_{k-1})w_k^2}}{2d_k + h_k} \quad (k = 1, \ldots, K). \tag{2.2}$$

Another important feature of the nested rhombus mechanism is that two planes of rhombi in different layers may be arranged perpendicular to each other. This allows us to construct three-dimensional structures with diverse configurations. For simplicity, the schematic diagram in Fig. 2.4 shows only a two-dimensional configuration, but the actual mechanism is three dimensional, with output axes being perpendicular to the plane. Three-dimensional arrangement of nested rhombus mechanisms allows us to densely enclose many rhombus units in a limited space. Figure 2.5 illustrates a three-dimensional structure. The serially connected first-layer rhombus units are rotated 90° about their output axis x_1. This makes the rhombus mechanism at the second layer more compact as shown in Fig. 2.3; the length in the x_2 direction is reduced. Namely, the height h_1 in Fig. 2.2b, which is a nonfunctional dimension for strain amplification, can be reduced. These size reductions allow us not only to pack many PZT units densely but also to increase the effective strain along the output axis.

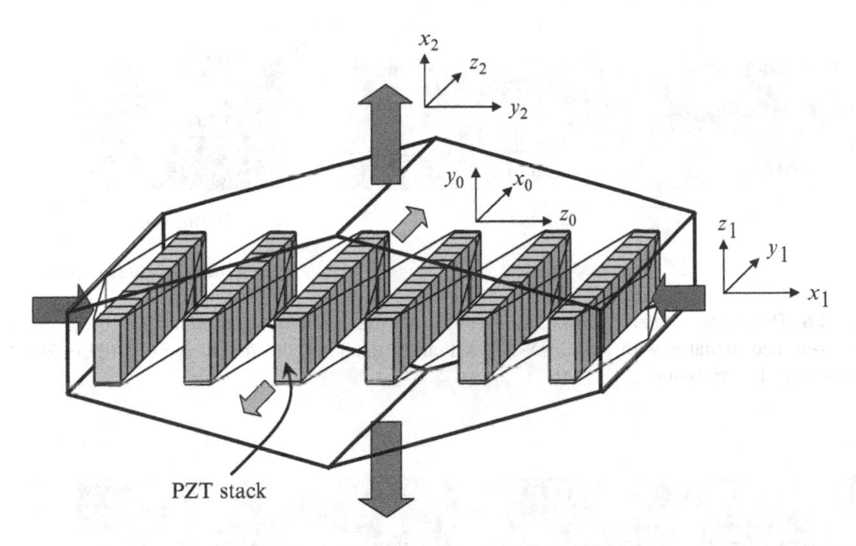

Fig. 2.5 Three-dimensional nesting for 20% strain (© IEEE 2010), reprinted with permission

2.2.2 Prototype Actuator Unit

A prototype nested actuator with over 20% effective strain is designed based on the structural compliance analysis. Over 20% of effective strain can be obtained by a two-layer mechanism; $K = 2$ and $\alpha = 15$. Figure 2.6 shows the design of the second layer structure and assembled actuator unit. Phosphor bronze (C54400, H08) is used for the material. The APA50XS "Moonie" piezoelectric actuators developed by Cedrat Inc. [1] are adopted for the first layer. By stacking six of APA50XS actuators for the first layer, this large strain may be achieved with a proper design of the second layer. The length of the assembled unit in actuation direction is 12 mm, and the width is 30 mm.

Figure 2.7 shows the maximum free-load displacement where two of the prototype units are connected in series. All the nested PZT stack actuators are ON by applying 150 V actuation voltage. The displacement was measured by using a laser displacement sensor (Micro-Epsilon optoNCDT 1401). One of the units produced a displacement of 2.53 mm that is equivalent to 21.1% effective strain (i.e., $2.53/12 = 0.211$). The maximum blocking force measured by using a compact load cell (Transducer Techniques MLP) was 1.69 N. The device can also produce fast movements (i.e., rise-time within 30 ms; see Sect. 2.4). A large control bandwidth is advantageous to achieve dextrous manipulation.

The development of the prototype device confirmed that a large amplification gain on the order of several hundreds can be obtained. The nesting approach realizes a large strain very compactly, making it ideal for the cellular actuator concept. In addition, the strain amplification mechanisms make the actuator unit compliant [23, 33–35]. The analysis of the structural compliance will be discussed in next section.

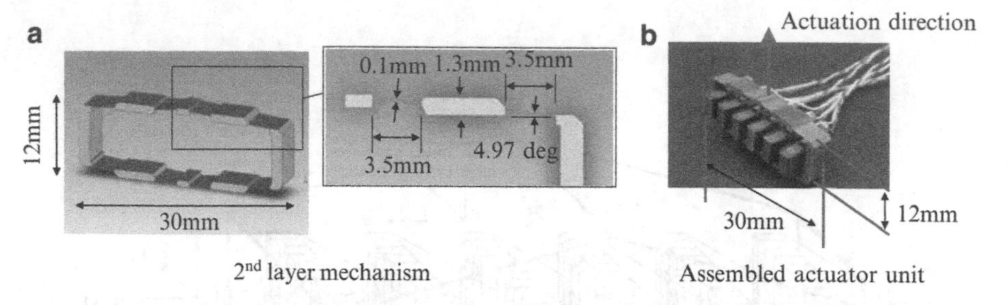

Fig. 2.6 Prototype actuator unit: (**a**) Design of the second layer rhombus mechanism; (**b**) assembled actuator with six CEDRAT actuators used for the first layer (© IEEE 2010), reprinted with permission

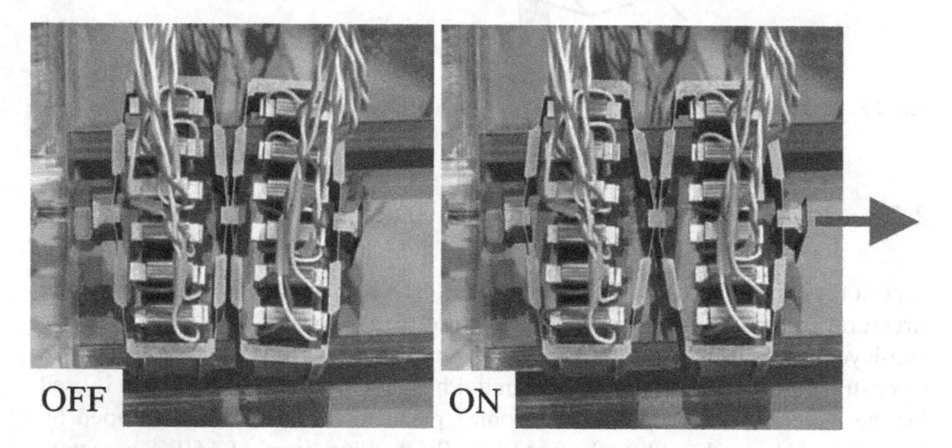

Fig. 2.7 Snapshots of free-load displacement: two nested rhombus mechanisms are connected in series. Each unit generates approximately 21% effective strain compared with its original length (© ASME 2008), reprinted with permission

2.2.3 *Micromanipulator Design*

Two types of miniature manipulators have been developed. Figure 2.8 shows a design of a small gripper. A piece of metal that acts as a robotic "finger" was attached to each of the ends of the outer rhombus mechanism as shown in Fig. 2.8a. The displacement of the tip is approximately 2.5 mm that is sufficient to pinch a small-outline integrated circuit as shown in Fig. 2.8b. Figure 2.9 shows another design of a manipulator. A relatively long rod was attached to one of the ends of the actuator unit that performs a pushing operation of small objects.

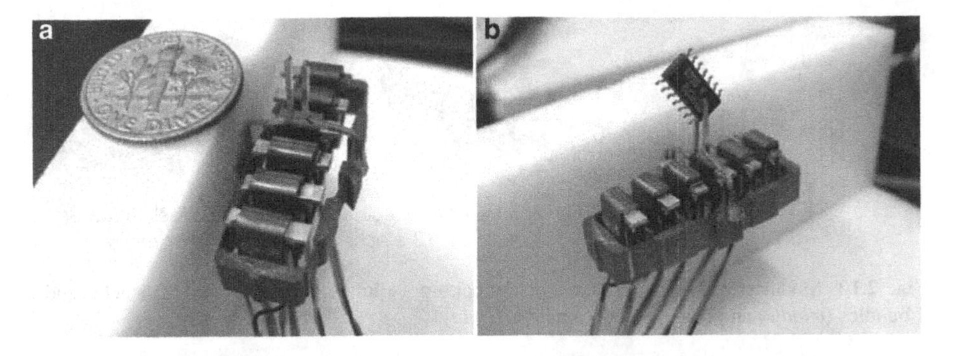

Fig. 2.8 Microgripper

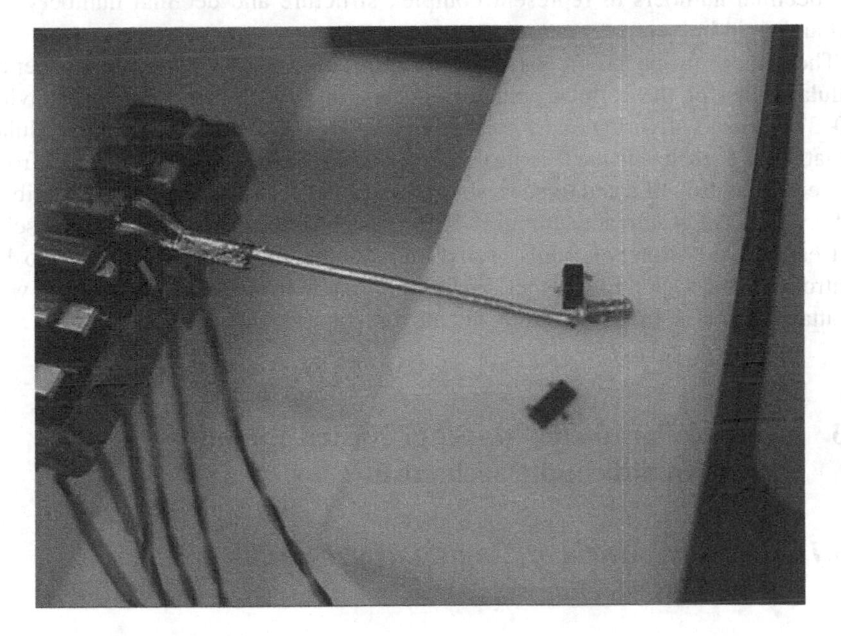

Fig. 2.9 Micromanipulator

2.2.4 Modular Design of Cellular Actuators

A wide variety of sizes and shapes is configurable using the designed actuator as a building-block as shown in Fig. 2.10. Actuator units can be arranged into complex topologies giving a wide range of strength, displacement, and robustness characteristics. Figure 2.10 shows two example configurations where cells are connected in series (stack) and in parallel (bundle). For example, the configuration shown in the right is expected to produce 11.8 N (1.67 N × 7 bundles) of blocking force and 15.2 mm (2.53 mm × 6 stacks) of free displacement if the developed units are applied. Actuator arrays can be represented simply and

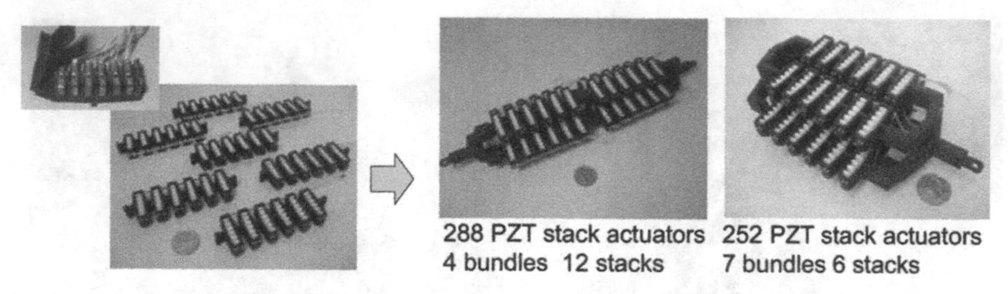

Fig. 2.10 Modularity of cellular actuators: mock-up cellular actuators with 12 stacks and 4 bundles (*middle*) and 6 stacks and 7 bundles (*right*)

compactly using a layer-based description, or "fingerprint," [15] which uses hexidecimal numbers to represent complex structure and decimal numbers for cell and rigid link connections.

The author's group has proposed a new control method to control vast number of cellular units for the cellular actuator concept inspired by the muscle behavior [30–32]. Instead of wiring many control lines to each individual cell, each cellular actuator has a stochastic local control unit that receives the broadcasted signal from the central control unit, and turns its state in a simple ON–OFF manner as described in Sect. 2.4. The macro actuator array in the figure corresponds to a single muscle, and each of the prototype units (cells) corresponds to Sarcomere known to be controlled in an ON–OFF manner. For example, a pair of the actuator arrays will be attached to a link mechanism in an antagonistic arrangement.

2.3 Lumped Parameter Model of Nested Rhombus Strain AmplificationMechanism

2.3.1 *Two-Port Model of Single-Layer Flexible Rhombus Mechanisms*

A single piezoelectric cellular actuator unit, or a combination of a PZT stack actuator and a strain amplification mechanism, can be modeled by a lumped model with three springs representing compliance and one rigid member representing strain amplification [33–35]. One of the advantages of this model is to easily gain physical insights as to which elements degrade actuator performance and how to improve it through design. To improve performance with respect to output force and displacement, the stiffness of the spring connected in parallel with the PZT stack actuator (the admissible motion space) must be minimized, while the one connected in series (in the constrained space) must be maximized, leading to the idealized rhombus mechanism made up of rigid beams and free joints.

Consider the case shown in Fig. 2.11a where a rhombus mechanism, including Moonies, is connected to a spring load. k_{load} is an elastic modulus of the load, and

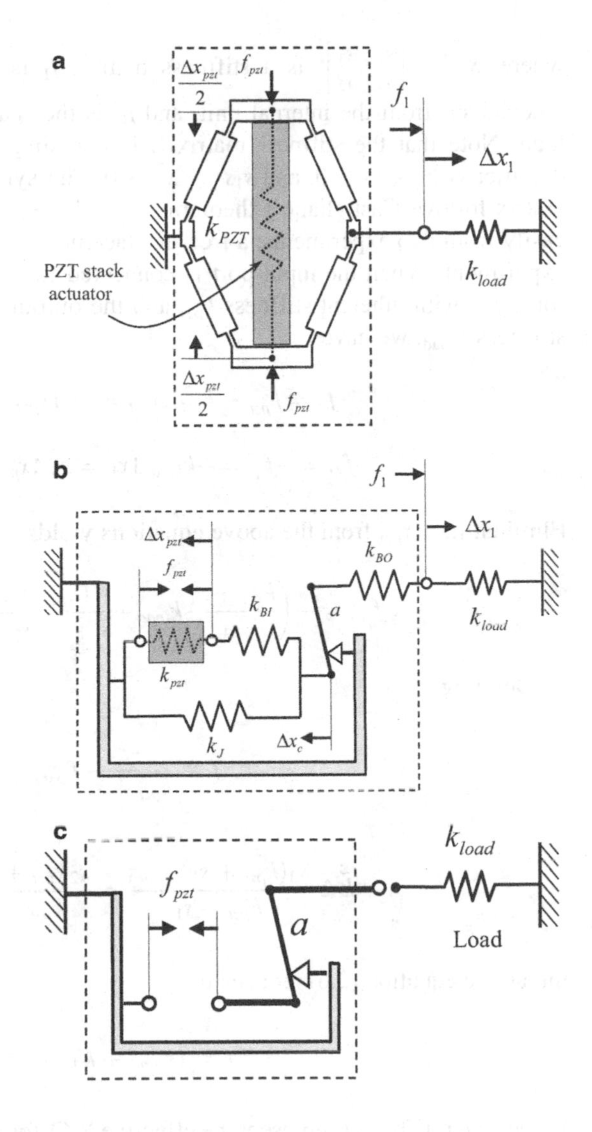

Fig. 2.11 Model of rhombus strain amplification mechanism (© IEEE 2010), reprinted with permission. (**a**) Rhombus mechanism with structural flexibility. (**b**) Proposed lumped parameter model. (**c**) Model of idealized rhombus (k_{BI}, $k_{BO} \to \infty$, $k_J \to 0$)

k_{pzt} is an elastic modulus of the internal unit such as a PZT stack actuator. Δx_{pzt} is the displacement of the internal unit, and f_{pzt} is the force applied to the amplification mechanism from the internal unit. f_1 is the force applied to the load from the actuator, and Δx_1 is the displacement of the load. In this figure, we assume that the internal unit is contractive for later convenience.

The rhombus strain amplification mechanism is a two-port compliance element, whose constitutive law is defined by a 2×2 stiffness matrix:

$$\begin{bmatrix} f_I \\ f_O \end{bmatrix} = S \begin{bmatrix} \Delta x_{pzt} \\ \Delta x_1 \end{bmatrix}, \tag{2.3}$$

where $S^{2\times2} = \begin{bmatrix} s_1 & s_3 \\ s_3 & s_2 \end{bmatrix}$ is a stiffness matrix. $f_{\rm I}$ is the net force applied to the mechanism from the internal unit, and $f_{\rm O}$ is the reaction force from the external load. Note that the stiffness matrix S is non-singular, symmetric, and positive-definite; $s_1 > 0$, $s_2 > 0$, and $s_1s_2 - s_3^2 > 0$. The symmetric nature of the stiffness matrix follows Castigliano's theorems. The elements of matrix **S** can be identified easily from two experiments; a free-displacement experiment and a blocking force experiment. When the input port is connected to a PZT stack actuator producing force $f_{\rm pzt}$ with inherent stiffness $k_{\rm pzt}$ and the output port is connected to a load of stiffness $k_{\rm load}$, we have

$$f_I = f_{pzt} - k_{pzt}\Delta x_{pzt} = s_1\Delta x_{pzt} + s_3\Delta x_1, \tag{2.4}$$

$$f_O = -f_1 = -k_{load}\Delta x_1 = s_3\Delta x_{pzt} + s_2\Delta x_1. \tag{2.5}$$

Eliminating $\Delta x_{\rm pzt}$ from the above equations yields

$$f_{pzt} = -\left(\frac{k_{pzt} + s_1}{s_3}k_{load} + \frac{s_2(k_{pzt} + s_1) - s_3^2}{s_3}\right)\Delta x_1. \tag{2.6}$$

Defining

$$\tilde{f} \triangleq \frac{-s_3}{k_{pzt} + s_1} f_{pzt}, \tag{2.7}$$

$$\tilde{k} \triangleq \frac{s_2(k_{pzt} + s_1) - s_3^2}{k_{pzt} + s_1} = \frac{s_2 k_{pzt} + \det S}{k_{pzt} + s_1} > 0, \tag{2.8}$$

the above equation (2.6) reduces to

$$\tilde{f} = (k_{load} + \tilde{k})\Delta x_1. \tag{2.9}$$

Force $\tilde{f}$ and stiffness $\tilde{k}$ represent the effective PZT force and the resultant stiffness of the PZT stack all viewed from the output port of the amplification mechanism.

2.3.2 Lumped Parameter Model

A drawback with the aforementioned two-port model representation is that it is hard to gain physical insights as to which elements degrade actuator performance and how to improve it through design. In the previous section two distinct compliances were introduced, one in the admissible motion space and the other in the constrained space. To improve performance with respect to output force and displacement, the stiffness in the admissible motion space must be minimized, while the one in the constrained space must be maximized. To manifest these structural compliances, we propose a lumped parameter model shown in Fig. 2.11b with three spring elements, k_J, k_{BI}, and k_{BO}, and one amplification leverage a. As the spring constants, k_{BI} and k_{BO}, tend to infinity, the system reduces to the one consisting of all rigid links, where the output Δx_1 is directly proportional to the input displacement Δx_{pzt}. Stiffness k_J impedes this rigid body motion, representing the stiffness in the admissible motion space. Elastic deformation at k_{BI} and k_{BO} represent deviation from the rigid body motion.

From Fig. 2.11b,

$$f_{pzt} + k_{BI}(\Delta x_c - \Delta x_{pzt}) - k_{pzt}\Delta x_{pzt} = 0, \tag{2.10}$$

$$ak_{BO}(a\Delta x_c - \Delta x_1) + k_J\Delta x_c + k_{BI}(\Delta x_c - \Delta x_{pzt}) = 0, \tag{2.11}$$

$$f_1 = k_{load}\Delta x_1 = k_{BO}(a\Delta x_c - \Delta x_1), \tag{2.12}$$

where Δx_c is the displacement at the connecting point between the leverage and springs; however, this point is virtual and Δx_c does not correspond to a physical displacement. This model is applicable to a wide variety of "rhombus-type" amplification mechanisms including Moonies. See [35] for more detail about the validation of the model and parameter calibration.

Consider the blocking force when the PZT stack actuator generates its maximum force, f_{pztmax}, given as follows:

$$f_1^{block} = \frac{ak_{BI}k_{BO}}{(a^2k_{BI}k_{BO} + k_{BI}k_J) + k_{pzt}(a^2k_{BO} + k_J + k_{BI})} f_{pzt\,max}. \tag{2.13}$$

Similarly, the free-load displacement for this rhombus mechanism, where $k_{load} \to 0$, is given by

$$\Delta x_1^{free} = \frac{ak_{BI}}{k_{pzt}(k_{BI} + k_J) + k_J k_{BI}} f_{pzt\,max}. \tag{2.14}$$

As addressed above, these equations imply that the blocking force will be maximized by k_{BI}, $k_{BO} \to \infty$. Similarly, $k_J \to 0$ maximizes Δx_1^{free}.

Another advantage is that the three-spring model is able to represent the ideal rhombus shown in Fig. 2.2b as a special case as shown in Fig. 2.11c by letting k_{BI}, $k_{BO} \to \infty$ and $k_J \to 0$. Note that the stiffness matrix S cannot be defined for the ideal rhombus. The number of unknown parameters becomes 4 as the rigid amplification leverage is explicitly included, which makes the calibration problem ill-posed; however, this amplification leverage is necessary to include the ideal case. In addition, three lumped springs are considered minimum to satisfy the input–output bidirectionality, which is a basic requirement of Castigliano's theorems.

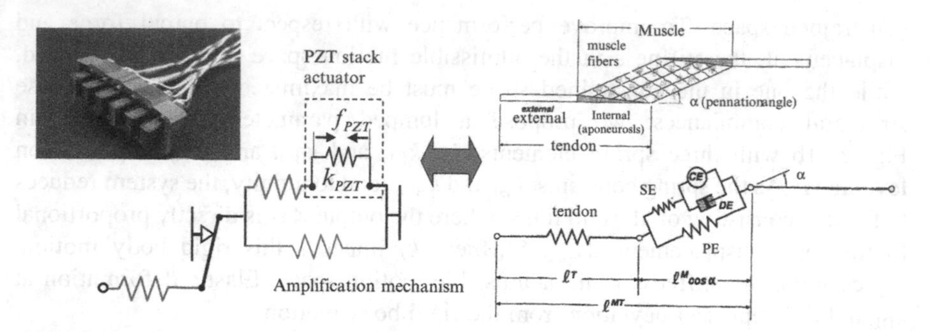

Fig. 2.12 Muscle-like compliance of the proposed actuator unit: (*left*) lumped parameter model of the actuator; (*right*) Hill's muscle model [36] (© Copyright 2001 Springer Science & Business), reprinted with permission

2.3.3 *Muscle-Like Compliance*

The proposed strain amplification mechanisms introduce compliance [23, 33–35] which is particularly important for rehabilitation robotics where a robotic device directly interacts with humans. PZT ceramics itself is not compliant against human body; however, it should be noted that the mechanical model described above for the proposed lumped parameter model is very similar to a well-known muscle model called Hill-type model as shown in Fig. 2.12. Introducing compliance similar to biological muscles is crucial in the engineering point of view, and the material of the actuator itself is not important. In other words, the use of biomaterials is not stipulated to obtain "muscle-like" compliance; PZT stack actuators perform equally well, in fact surpassing biomaterials in terms of the reliability and the speed of response, provided the amplification structures are designed in an appropriate manner.

2.4 Control of Compliant Actuators by Minimum Switching Discrete Switching Law

Simple ON–OFF controls will suffice to drive individual actuator units, since the aggregate outputs will be smooth and approximately continuous if a large number of modules are involved. Expensive analog drive amplifiers are not required. Moreover, ON–OFF controls are effective in overcoming prominent hysteresis and nonlinearity of actuator materials in terms of static actuator control. As shown in Fig. 2.13, bistable ON–OFF control does not depend on these complex nonlinearities, as long as the state of the material is pushed towards either ON or OFF state. In Fig. 2.13, the input is voltage and the output is displacement if we take PZT as an example.

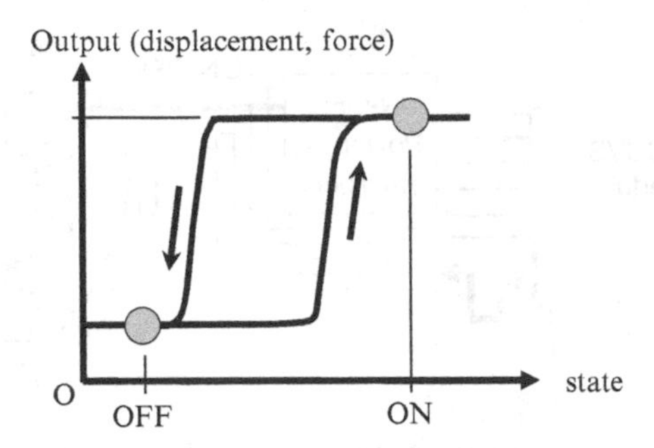

Fig. 2.13 Bi-stable ON–OFF control

Fig. 2.14 Principal vibration modes of cellular actuator: Mode 1 at 76 Hz (*left*), Mode 2 at 347 Hz (*right*) (© IEEE 2009), reprinted with permission

In terms of dynamic control, however, the vibration modes in the mechanical amplifier create residual oscillations. For example, the developed actuator unit has two principal vibration modes as shown in Fig. 2.14. The natural frequencies in general tend to increase when the size of a structure decreases, which makes the vibration suppression more difficult. Applying on–off commands to the cellular actuator actually causes vibration at its own natural frequencies, since the actuator itself has a low damping ratio. Commercially viable actuator systems require more sophisticated, low-cost design and control methods. This means minimizing the amount of costly controllers, such as a model-based controller that would result in high computational loads. An advanced feedback controller might require a high-resolution linear amplifier. It is advantageous to apply a discrete approach for size and cost reduction of the entire control system.

Command shaping techniques [5, 9, 14, 20, 26] may be applicable for this actuator to move to any number of the available discrete positions without vibration, simply by altering the phase at which the command is applied to the individual redundant inputs (PZT stacks). However, most of these conventional command shaping techniques have a single actuator per degree of freedom and therefore each impulse's amplitude is restricted to a value of +1 or − 1. We call this All On/ All Off control. When multiple discrete inputs are available for a single degree of

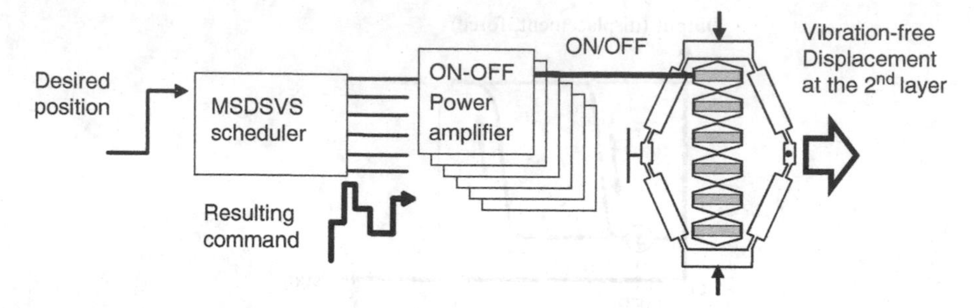

Fig. 2.15 Schematic of piezoelectric cellular actuator with discrete switching command to suppress vibration

freedom, All On/All Off control is actually a subset of a larger set of vibration suppression commands, which we call discrete switching vibration suppression (DSVS). One that is of particular interest is the case where the set of inputs experiences the minimum number of changes in state, or minimum switching discrete switching vibration suppression (MSDSVS) [22]. In certain cases, the minimum switching discrete control solution is monotonic in nature, resembling a staircase. This is desirable because no additional switches (control effort) are necessary to reduce vibration beyond those required to reach the goal position and the oscillation due to unmodeled resonances will not be amplified. The MSDSVS has several advantages over All On/All Off control, providing a satisfactory solution can be obtained. It reduces the amount of heat generated by the switching transistors for a given move, which may allow them to be smaller. It reduces the number and amplitude of loading/unloading cycles on the first and second layers, which could result in longer actuator life. This can be further exploited by imposing an algorithm that distributes the control effort evenly over the first layer actuators over all moves in time. In general, for actuators of this type, the number of changes of state corresponds to the control effort, so MSDSVS is a form of energy saving control.

Figure 2.15 shows a schematic of the piezoelectric cellular actuator system. A desired position is provided to the command scheduler that determines times at which to turn the piezoelectric stacks on and off in such a way so as not to excite the vibration modes of the system. These digital signals are fed to a switching network, which provides voltage to the first layer actuators at the times selected by the command scheduler. The force generated by the first layer actuators produces a small displacement in the first layer actuators, which results in a larger displacement of the second layer in the transverse direction. A discrete command configuration allows the actuator to take five intermediate positions between the extremes of its operating range, corresponding to the number of first layer actuators turned on.

An algorithm that numerically solves the MSDSVS problem has been implemented [22]. The response of a PZT actuator unit consisting of 6 piezoelectric stack actuators was tested for a step input, All On/All Off control, and the proposed MSDSVS. The implemented algorithm found optimal timings for switching piezo-stack actuators in an ON–OFF manner. The number of possible commands grows as

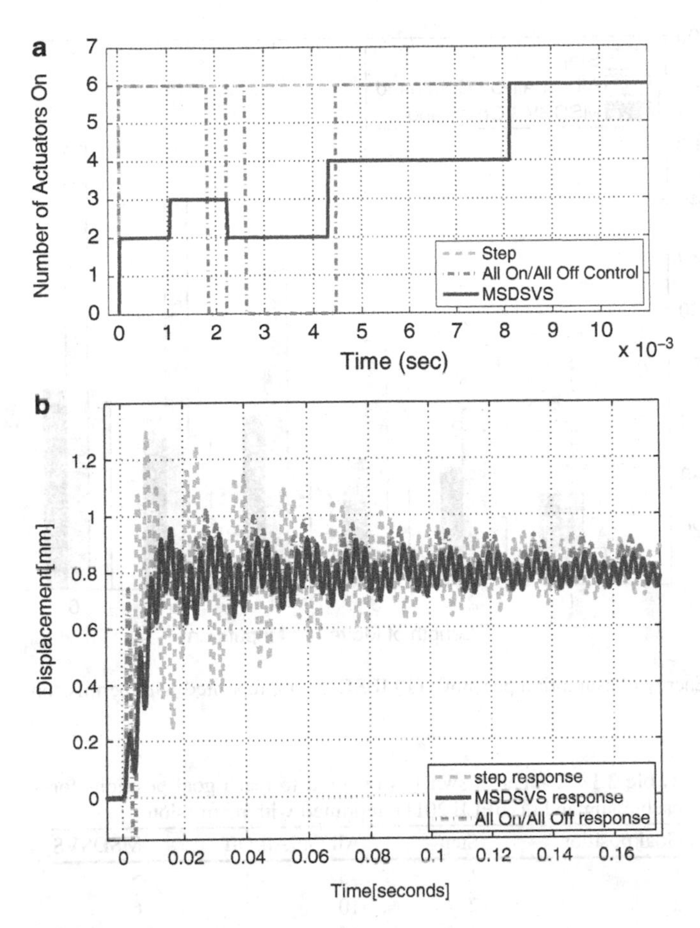

Fig. 2.16 Vibration suppression by MSDSVS (© IEEE 2011), reprinted with permission. (**a**) Step, All On/All Off and MSDS commands to PZT actuator. (**b**) Dynamic response of step, All On/All Off and MSDSVS commands

the total number of impulses increases. To reduce the computational load, a minimum switching candidate, i.e., the monotonic command input, will be given as an initial solution to the algorithm. If the solver is unsuccessful using the monotonic command, the next most optimal will be tried by incrementally introducing additional ON–OFF pulses until a satisfactory solution is obtained.

Figure 2.16 shows the response and command for a goal position of six actuators on. Note that a monotonically increasing command was not found for this set of natural frequencies. The response to a step (all six actuators transitioned at the same time) and the response to both vibration suppression commands. The step response shows considerable oscillation ($>50\%$ overshoot), due to the lightly damped nature of the system. The various commands are completed by 10 ms, so the remaining period of time shows the decay of any residual oscillation once the command is

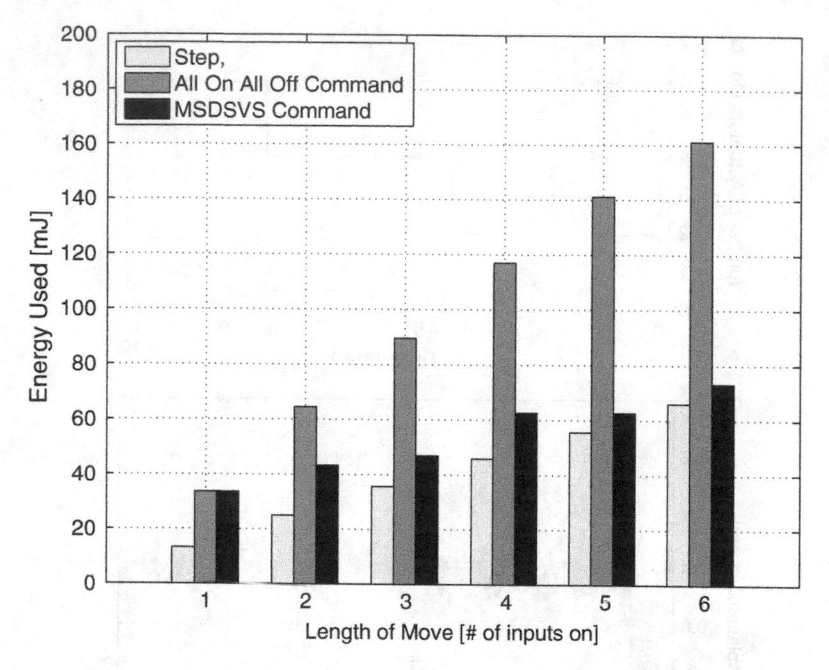

Fig. 2.17 Energy consumption per move (© IEEE 2011), reprinted with permission

Table 2.1 Number of switches required to reach goal positions for each command (© IEEE 2011), reprinted with permission

Goal position	Step	All On/All Off	MSDSVS
1	1	5	5
2	2	10	6
3	3	15	5
4	4	20	8
5	5	25	7
6	6	30	8

completed. The response to All On/All Off command has the largest transient during the command. Despite the use of vibration suppression commands, one still observes some oscillation remaining in the response after such a command is completed. This is likely due to a combination of factors. Both All On/All Off and MSDSVS methods are based on linear analysis, and they suppress oscillation in an actuator that is described accurately by a linear model. In actuality, some non-negligible nonlinear effects are present. The natural frequency of the cellular actuator changes slightly as it extends. Figure 2.17 shows how the difference in energy dissipated grows quickly with an increase in the number of inputs. Table 2.1 shows the number of switches required to reach a given position, which directly correlates with the results shown in Fig. 2.17.

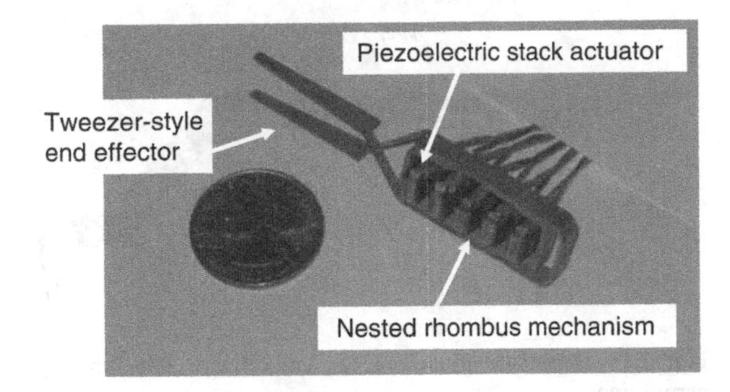

Fig. 2.18 Assembled end-effector nesting actuator module

2.5 Tweezer-Style Piezoelectric End-Effector

2.5.1 Piezoelectric End-Effector for Robotic Surgery and Intervention in MRI

A prototype of a tweezer-style piezoelectric end-effector with nested-rhombus multilayer stain amplification mechanisms has been designed and developed [13]. Figure 2.18 shows an overview of the developed device. This robotic end-effector is a proof-of-concept prototype for a telerobotic system for surgery and intervention guided by magnetic resonance imaging (MRI). PZT piezoelectric actuators (APA35XS, CEDRAT, Inc.) were adopted. Five of the PZT actuators were connected in series and nested in a larger amplification mechanism made by phosphor bronze. A similar actuator unit with two amplification layers as shown in Fig. 2.7 was developed. Then, this unit was further nested in a tweezer-shape structure that acts as another (i.e., third) amplification mechanism.

Note that this particular prototype device contained small pieces of ferromagnetic metal in the first layer mechanisms and was not completely MRI compatible. This issue can easily be resolved by using a non-magnetic model of APA35XS that is also available from CEDRAT, Inc. The development of a refined device with MRI compatibility is in progress.

2.5.2 Modeling and Design

The desired specifications of the robotic end-effector were determined based on a standard surgical clip. The goal is to achieve 1.0 N of force and 10 mm of displacement at the tip. The rhombus mechanism with PZT actuators is nested into a tweezer-shape structure with a reverse action mechanism. In this section, a Bernoulli–Euler beam model of the tweezer-shape structure is considered. A model of the structure is shown in Fig. 2.19. The displacement at Point A, δ_A, where the

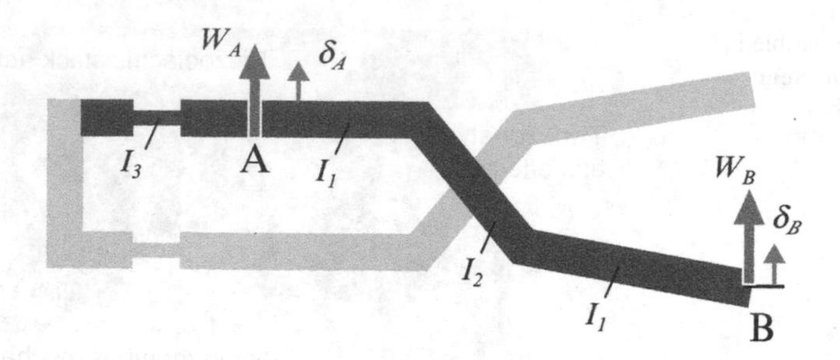

Fig. 2.19 Schematic model of the end-effector structure (© JSME 2010), reprinted with permission

force from the actuator module is applied, and the displacement at the tip δ_B can be written as

$$\delta_A = \left(\frac{C_{A2}}{2EI_1} + \frac{C_{A4}}{2EI_3}\right) W_A + \left(\frac{C_{A1}}{2EI_1} + \frac{C_{A3}}{2EI_3}\right) W_B, \tag{2.15}$$

$$\delta_B = \left(\frac{C_{B2}}{2EI_1} + \frac{C_{B5}}{2EI_3}\right) W_A + \left(\frac{C_{B1}}{2EI_1} + \frac{C_{B3}}{2EI_2} + \frac{C_{B4}}{2EI_3}\right) W_B, \tag{2.16}$$

where W_A, W_B are the forces at Points A and B. W_B is a reaction force from a handling object. E is Young's modulus of the phosphorus bronze, and I_1, I_2, I_3 are the second moment of area of the sections shown in Fig. 2.19. $C_{A1} \sim C_{A4}$ and $C_{B1} \sim C_{B5}$ are structural coefficients obtained from Castigliano's theorem by assuming a Bernoulli–Euler beam model of the tweezer-shape structure (see Appendix). Equations (2.15) and (2.16) are used to determine the dimensions of the end-effector to achieve desired performances. Let $\hat{W}_A$ be the force at the point A that needs to be applied to achieve the desired tip displacement when the tip of the end-effector is free (i.e., $W_B = 0$). From (2.15) we have

$$\hat{W}_A = \left(\frac{C_{B2}}{2EI_1} + \frac{C_{B2}}{2EI_3}\right)^{-1} \hat{\delta}_A. \tag{2.17}$$

The dimensions of the end-effector need to be determined such that $\hat{W}_A$ and $\hat{\delta}_A$ do not exceed the maximum force and displacement of the actuator unit nested in the tweezer-shape structure.

Figure 2.20 shows the final dimensions of the structure determined by a trial-and-error basis by using finite element software. The length of the final assembly is 70 mm in the longitudinal direction and the width is 14 mm. The simulation result achieved 1.1 N of force and 9.0 mm of displacement at the tip that are sufficiently close to the desired values. The stress analysis of the designed end-effector was conducted. The maximum von Mises stress when the actuator module exerts the maximum force was 274 MPa. The designed end-effector has a sufficient strength since the yield stress of phosphorus bronze is 528 MPa.

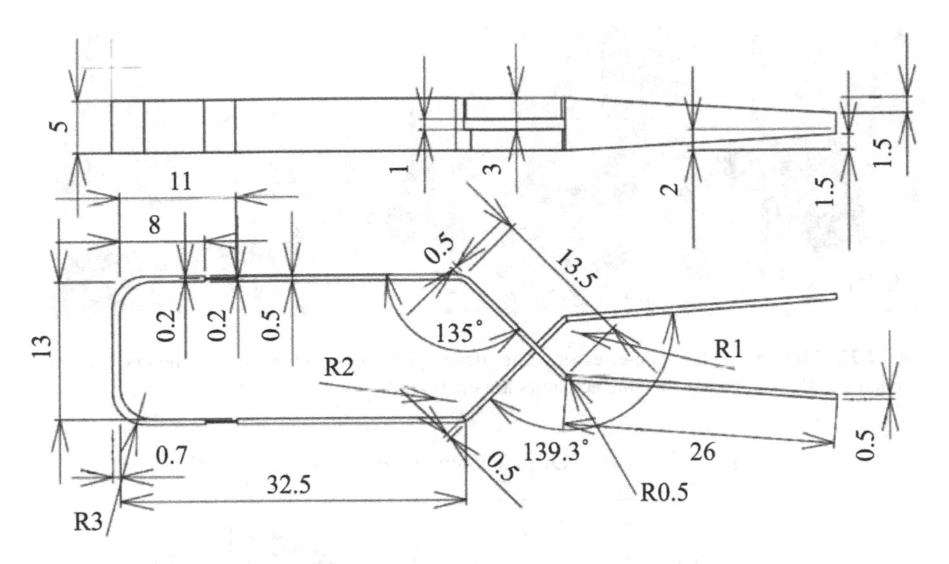

Fig. 2.20 Drawings of the tweezer-style end-effector (© JSME 2010), reprinted with permission

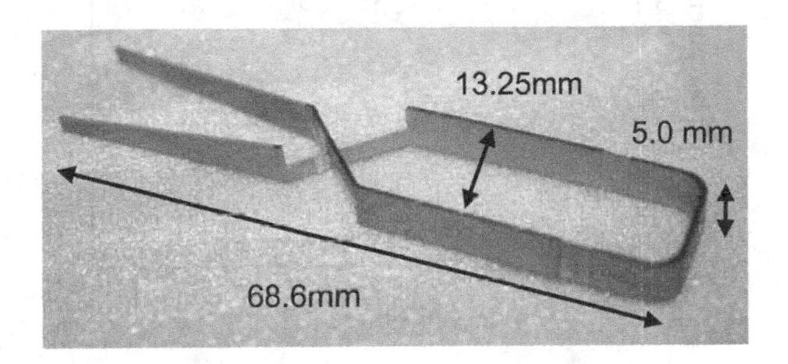

Fig. 2.21 Fabricated end-effector

2.5.3 Fabrication and Performance Test

The fabricated tweezer-shape structure made by phosphorus bronze is shown in Fig. 2.21. The length of the structure is 68.6 mm, the height is 5.0 mm, and the width is 13.25 mm, respectively. Figure 2.22 shows the free displacement of the assembled end-effector. The prototype end-effector produced 8.8 mm of displacement and 1.0 N of static pinching force by applying 150 V input voltage to each of the five PZT actuators. The performance of the assembled end-effector was evaluated by using a force sensor and a laser displacement sensor. Figure 2.23 shows the displacement and force profile when the input voltage changes from 0 to 150 V and 150 to 0 V. Table 2.2 summarizes the results of the simulation and experiment.

Fig. 2.22 Motion of the end-effector. The developed end-effector has a reverse action mechanism; the tips close when the actuators are energized

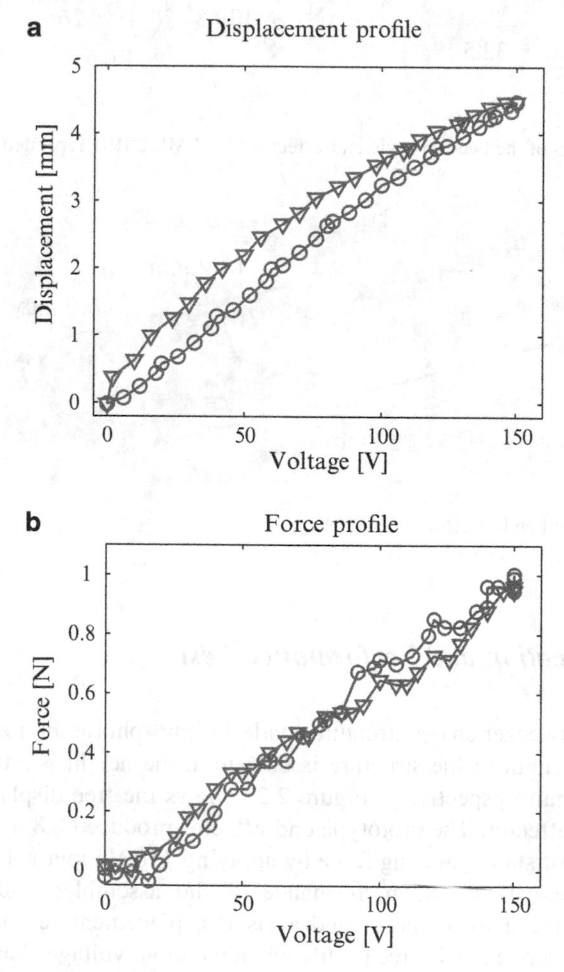

Fig. 2.23 Displacement and force performance. *Circles* are forward (from 0 to 150 V) and *inverse-triangles* are backward (from 150 to 0 V) directions: note that (**a**) shows the absolute displacement of one of the end-points. The total displacement is twice of this measurement. (**b**) Force profile

Table 2.2 Performance of the assembled end-effector

	Displacement (mm)	Force (N)
Simulation	9.0	1.1
Experiment	8.8	1.0

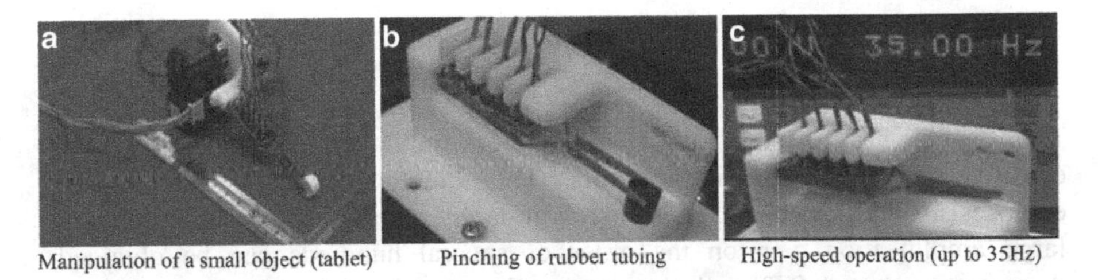

a
b
c
35.00 Hz
Manipulation of a small object (tablet)
Pinching of rubber tubing
High-speed operation (up to 35Hz)

Fig. 2.24 Manipulation using the tweezer-style robotic end-effector

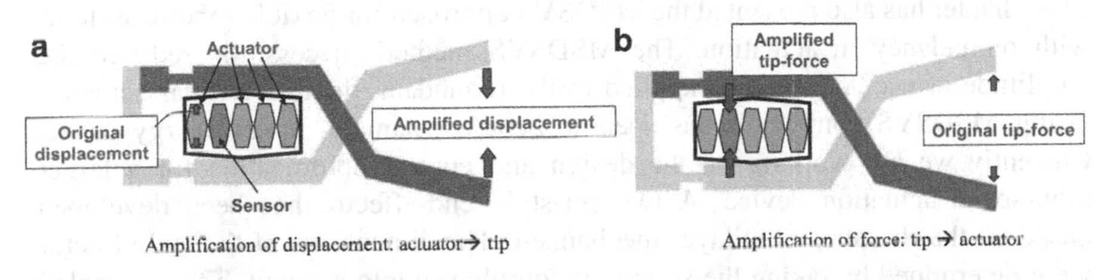

Fig. 2.25 Concept of force sensing using strain amplification mechanisms

In Fig. 2.24a, the developed end-effector was attached on a rotary motor and manipulated a small object. Figure 2.24b shows a pinching of a rubber tube. The natural frequency of the fabricated end-effector was 36 Hz; uncompensated bandwidth of actuation up to 35 Hz was confirmed as shown in Fig. 2.24c. This actuation bandwidth is greatly wider than existing robotic micro grippers driven by tendons or MRI-compatible manipulators driven by fluid actuators. The bandwidth could be further improved if a vibration compensation controller is implemented.

2.5.4 *Force Sensing*

The salient feature of this end-effector is that one of the PZT actuators can be used not only for actuation but also for sensing, by simply switching from a circuit for actuation to the one for sensing. This usage would not cause major loss of function as an end-effector as the proposed mechanism encloses multiple PZT units in series. Figure 2.25a shows the concept of the displacement amplification as described in previous sections. Contrary, as shown in Fig. 2.25b, the tip-force can be measured by measuring an induced electrical potential across one of the nested PZT actuators. Since the force acting on the PZT actuator is the amplified tip-force on the order of hundreds through the strain amplification mechanism, even a small tip-force can be magnified and accurately measured. This high sensitivity of force measurement is extremely important for medical applications that deal with delicate tissues and organs. The measured tip-force can be sent back to the operator's side and used for haptic feedback.

2.6 Conclusion

This chapter has presented a nested rhombus multilayer mechanism for large effective-strain piezoelectric actuators. This hierarchical nested architecture encloses smaller flextensional actuators with larger amplifying structures so that a large amplification gain on the order of several hundreds can be obtained. A prototype nested PZT cellular actuator that weighs only 15 g has produced 21% effective strain (2.53 mm displacement from 12 mm actuator length and 30 mm width) and 1.69 N blocking force. A lumped parameter model has been proposed to represent the mechanical compliance of the nested strain amplifier. This chapter has also presented the MSDSVS approach for flexible robotic systems with redundancy in actuation. The MSDSVS method successfully reduced the amplitude of oscillation when applied to the redundant, flexible cellular actuator. Using MSDSVS commands has specific benefits, namely, lower energy usage. Currently we are working on the design and control optimization for a larger high-speed actuation device. A tweezer-style end-effector has been developed based on the rhombus multilayer mechanism. The dimensions of the end-effector were determined by taking the structural compliance into account. The assembled robotic end-effector produced 1.0 N of force and 8.8 mm of displacement at the tip.

Acknowledgements This research was partially supported by National Science Foundation grant, Cyber-Physical Systems, ECCS-0932208. The author expresses his gratitude to Mr. Fuyuki Sugihara, formerly with Nara Institute of Science and Technology, Japan, for the design and fabrication of the tweezer-type end-effector.

Appendix

Structural Analysis of Tweezer-Style End-Effector

Figure 2.26 shows a schematic diagram of the tweezer-style structure for the end-effector. Here we assume the end-effector is fixed at point C and the force generated from the actuator unit is applied at Point A. From Castigliano's theorem assuming a Bernoulli–Euler beam model, $C_{\mathrm{A1}} \sim C_{\mathrm{A4}}$ and $C_{\mathrm{B1}} \sim C_{\mathrm{B5}}$ in (2.15) and (2.16) can be written as follows:

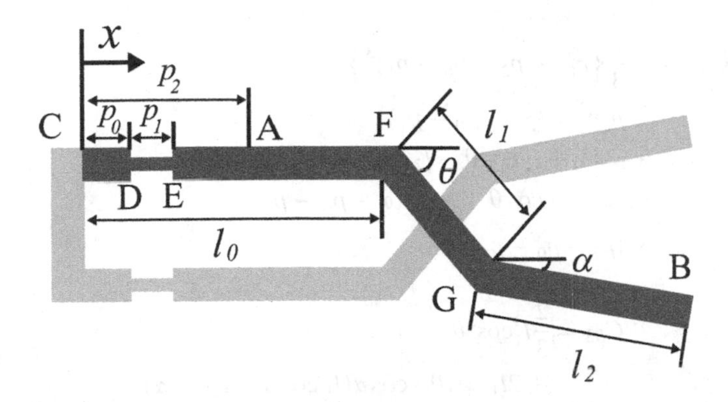

Fig. 2.26 Schematic of the end-effector (© JSME 2010), reprinted with permission

$$
\begin{aligned}
C_{A1} = & -\frac{2}{3}\left\{p_0^3 + (p_2 - p_0 - p_1)^3\right\} \\
& + (l_0 + l_1 \cos\theta + l_2 \cos\alpha)p_0(p_0 - 2p_2) + p_2 p_0^2 \\
& - (l_0 + l_1 \cos\theta + l_2 \cos\alpha - p_0 - p_1)(p_2 - p_0 - p_1)^2 \\
& + (p_2 - p_0 - p_1)^3,
\end{aligned} \tag{2.18}
$$

$$
\begin{aligned}
C_{A2} = & \frac{2}{3}\left\{p_0^3 + (p_2 - p_0 - p_1)^3\right\} \\
& + 2p_2 p_0(p_2 - p_0),
\end{aligned} \tag{2.19}
$$

$$
\begin{aligned}
C_{A3} = & -\frac{2}{3}p_1^3 + (p_2 - p_0)p_1^2 \\
& + (l_0 + l_1 \cos\theta + l_2 \cos\alpha - p_0)p_1(p_1 - 2p_2 + 2p_0),
\end{aligned} \tag{2.20}
$$

$$
C_{A4} = \frac{2}{3}p_1^3 + 2(p_2 - p_0)p_1(a - p_0 - p_1), \tag{2.21}
$$

$$
\begin{aligned}
C_{B1} = & \frac{2}{3}\left\{p_0^3 + (p_2 - p_0 - p_1)^3\right\} \\
& + \frac{2}{3}(l_0 - p_2)^3 + \frac{2}{3}l_2^3 \cos^3\alpha \\
& + 2p_0(l_0 + l_1 \cos\theta + l_2 \cos\alpha) \\
& \times (l_0 + l_1 \cos\theta + l_2 \cos\alpha - p_0) \\
& + 2(l_0 + l_1 \cos\theta + l_2 \cos\alpha - p_0 - p_1)(p_2 - p_0 - p_1) \\
& \times (l_0 + l_1 \cos\theta + l_2 \cos\alpha - p_2) \\
& + 2(l_0 - p_2)(l_0 + l_1 \cos\theta + l_2 \cos\alpha - p_2) \\
& \times (l_1 \cos\theta + l_2 \cos\alpha),
\end{aligned} \tag{2.22}
$$

$$
\begin{aligned}
C_{B2} = & -\frac{2}{3}\left\{p_0^3 + (p_2 - p_0 - p_1)^3\right\} \\
& + p_2 p_0^2 + (p_2 - p_0 - p_1)^3 \\
& + p_0(l_0 + l_1 \cos\theta + l_2 \cos\alpha)(p_0 - 2p_2) \\
& - (l_0 + l_1 \cos\theta + l_2 \cos\alpha - p_0 - p_1) \\
& \times (p_2 - p_0 - p_1)^2,
\end{aligned} \tag{2.23}
$$

$$
\begin{aligned}
C_{B3} = & \frac{2}{3} l_1^3 \cos^3\theta \\
& + 2l_1 \cos\theta l_2 \cos\alpha (l_1 \cos\theta + l_2 \cos\alpha),
\end{aligned} \tag{2.24}
$$

$$
\begin{aligned}
C_{B4} = & \frac{2}{3} p_1^3 + 2p_1(l_0 + l_1 \cos\theta + l_2 \cos\alpha - p_0) \\
& \times (l_0 + l_1 \cos\theta + l_2 \cos\alpha - p_0 - p_1),
\end{aligned} \tag{2.25}
$$

$$
\begin{aligned}
C_{B5} = & -\frac{2}{3} p_1^3 + (p_2 - p_0) p_1^2 \\
& + p_1(l_0 + l_1 \cos\theta + l_2 \cos\alpha - p_0)(p_1 - 2p_2 + 2p_0),
\end{aligned} \tag{2.26}
$$

where p_0, p_1, p_2 are the lengths between C and D, D and E, C and A, respectively, l_0, l_1, l_2 are the lengths between C and F, F and G, G and B, respectively, and θ, α are the angles shown in Fig. 2.26.

References

1. CEDRAT Inc. http://www.cedrat.com/
2. Fukui I, Yano T, Kamatsuki T (1984) Lever actuator comprising a longitudinal-effect electroexpansive transducer. US Patent 4,435,666
3. MacGregor R (2003) Shape memory alloy actuators and control methods. US Patent 6,574,958
4. Conway N, Traina Z, Kim S (2007) A strain amplifying piezoelectric MEMS actuator. J Micromech Microeng 17(4):781–787
5. Diaz IM, Pereira E, Feliu V, Cela JJL (2009) Concurrent design of multimode input shapers and link dynamics for flexible manipulators. IEEE/ASME Trans Mechatron 15(4):646–651. DOI 10.1109/TMECH.2009.2031434
6. Dogan A, Uchino K, Newnham R (1997) Composite piezoelectric transducer with truncated conical endcaps "cymbal". IEEE Trans Ultrason Ferroelectrics Freq Contr 44(3):597–605 DOI 10.1109/58.658312
7. Dogan A, Xu Q, Onitsuka K, Yoshikawa S, Uchino K, Newnham R (1994) High displacement ceramic metal composite actuators (moonies). Ferroelectrics 156(1):1–6
8. Ervin J, Brei D (1998) Recurve piezoelectric-strain-amplifying actuator architecture. IEEE/ASME Trans Mechatron 3(4):293–301
9. Fiene J, Niemeyer G (2006) Toward switching motor control. IEEE/ASME Trans Mechatron 11(1): 27–34. DOI 10.1109/TMECH.2005.863368

10. Haertling G (1994) Rainbow ceramics—a new type of ultra-high-displacement actuator. Am Ceram Soc Bull 73(1):93–96
11. Janker P, Christmann M, Hermle F, Lorkowski T, Storm S (1999) Mechatronics using piezoelectric actuators. J Eur Ceram Soc 19(6):1127–1131 (1999)
12. Kostyukov AI (1998) Muscle hysteresis and movement control: a theoretical study. Neuroscience 83(1):303–320
13. Kurita Y, Sugihara F, Ueda J, Ogasawara T (2010) MRI compatible robot gripper using large-strain piezoelectric actuators. Trans Jpn Soc Mech Eng C 76(761):132–141
14. Lim S, Stevens H, How JP (1999) Input shaping for multi-input flexible systems. ASME J Dyn Syst Meas Contr 121:443–447
15. MacNair D, Ueda J (2009) Modeling & characterizing stochastic actuator arrays. In: IEEE/RSJ international conference on intelligent robots and systems, 2009. IROS 2009, St. Louis, USA, October 11–15, pp 3232–3237
16. Moskalik A, Brei D (1997) Quasi-static behavior of individual C-block piezoelectric actuators. J Intell Mater Syst Struct 8(7):571–587
17. Newnham R, Dogan A, Xu Q, Onitsuka K, Tressler J, Yoshikawa S (1993) Flextensional moonie actuators. In: 1993 I.E. proceedings on ultrasonics symposium, vol 1, pp 509–513. DOI 10.1109/ULTSYM.1993.339557
18. Niezrecki C, Brei D, Balakrishnan S, Moskalik A (2001) Piezoelectric actuation: state of the art. Shock Vib Digest 33(4):269–280. DOI 10.1177/058310240103300401
19. Nurung S, Magsino KC, Nilkhamhang I (2009) Force estimation using piezoelectric actuator with adaptive control. In: Proceedings of international conference on electrical engineering/electronics, computer, telecommunications and information technology, pp 350–353
20. Pao LY (1996) Input shaping design for flexible systems with multiple actuators. In: Proceedings of the 13th world congress of the international federation of automatic control, San Francisco
21. Ronkanen P, Kallio P, Koivo HN (2007) Simultaneous actuation and force estimation using piezoelectric actuators. In: Proceedings of international conference on mechatronics and automation, Harbin, China, August 5–8, pp 3261–3265
22. Schultz J, Ueda J (2009) Discrete switching vibration suppression for flexible systems with redundant actuation. In: IEEE/ASME international conference on advanced intelligent mechatronics, 2009. AIM 2009, Singapore, July 14–17, pp 544–549
23. Secord T, Ueda J, Asada H (2008) Dynamic analysis of a high-bandwidth, large-strain, pzt cellular muscle actuator with layered strain amplification. In: Proceedings of 2008 I.E. international conference on robotics and automation (ICRA 2008), Pasadena, CA, USA, May 19–23, pp 761–766
24. Seffen K, Toews E (2004) Hyperhelical actuators: coils and coiled-coils. In: 45th AIAA/ASME/ASCE/AHS/ASC structures, structural dynamics and materials conference, Palm Springs California, April 19–22, pp 19–22
25. Shimizu T, Shikida M, Sato K, Itoigawa K, Hasegawa Y (2002) Micromachined active tactile sensor for detecting contact force and hardness of an object. In: Proceedings of international symposium on micromechatronics and human science, Nagoya, Japan, Oct. 20–23, pp 67–71
26. Singer NC, Seering WP (1990) Preshaping command inputs to reduce system vibration. ASME J Dyn Syst Meas Contr 112:76–82
27. Stanley W, Jacob MD (1982) Structure and function in man, 5th edn. W B Saunders Co, Philadelphia. http://amazon.com/o/ASIN/0721650945/
28. Tansocka J, Williamsa C (1992) Force measurement with a piezoelectric cantilever in a scanning force microscope. Ultramicroscopy 42–44(2):1464–1469
29. Uchino K (1997) Piezoelectric actuators and ultrasonic motors. Kluwer Academic Publishers, Boston
30. Ueda J, Odhnar L, Asada H (2006) A broadcast-probability approach to the control of vast dof cellular actuators. In: Proceedings of 2006 I.E. international conference on robotics and automation (ICRA '06), Orlando, Florida, May 15–19, pp 1456–1461

31. Ueda J, Odhner L, Asada HH (2007) Broadcast feedback for stochastic cellular actuator systems consisting of nonuniform actuator units. In: Proceedings of 2007 I.E. international conference on robotics and automation (ICRA '07), pp 642–647. DOI 10.1109/ROBOT.2007.363059
32. Ueda J, Odhner L, Asada HH (2007) Broadcast feedback of stochastic cellular actuators inspired by biological muscle control. Int J Robot Res 26(11–12):1251–1265. DOI 10.1177/0278364907082443
33. Ueda J, Secord T, Asada H (2008) Piezoelectric cellular actuators using nested rhombus multilayer mechanisms. In: First annual dynamic systems and control conference (DSCC 2008), Ann Arbor, Michigan, October 20–22
34. Ueda J, Secord T, Asada H (2008) Static lumped parameter model for nested PZT cellular actuators with exponential strain amplification mechanisms. In: IEEE international conference on robotics and automation, 2008. ICRA 2008, Pasadena, CA, USA, May 19–23, pp 3582–3587
35. Ueda J, Secord T, Asada H (2010) Large effective-strain piezoelectric actuators using nested cellular architecture with exponential strain amplification mechanisms. IEEE/ASME Trans Mechatron 15:770–782
36. Yamaguchi G (2001) Dynamic modeling of musculoskeletal motion. Kluwer Academic Publishers, Boston

Chapter 3
Autocalibration of MEMS Accelerometers

Iuri Frosio, Federico Pedersini, and N. Alberto Borghese

Abstract In this chapter, we analyze the critical aspects of the widely diffused calibration and autocalibration procedures for MEMS accelerometers. After providing a review of the main applications of this kind of sensors, we introduce the different sensor models proposed in literature, highlighting the role of the axis misalignments in the sensor sensitivity matrix. We derive a principled noise model and discuss how noise affects the norm of the measured acceleration vector. Since autocalibration procedures are based on the assumption that the norm of the measured acceleration vector, in static condition, equals the gravity acceleration, we introduce the international gravity formula, which provides a reliable estimate of the gravity acceleration as a function of the local latitude and altitude. We derive then the autocalibration procedure in the context of maximum likelihood estimate and we provide examples of calibrations. For each calibrated sensor, we also illustrate how to derive the accuracy on the estimated parameters through the covariance analysis and how to compute the angles between the sensing axes of the sensor. In the conclusion, we summarize the main aspects involved in the autocalibration of MEMS accelerometers.

3.1 Introduction

The advent of microelectromechanical system (MEMS) technology has allowed miniaturized, high performance and cheap accelerometers to be built using a variety of different approaches [1–3]. These sensors were initially used to detect sudden, critical events like in airbag control [4], but nowadays applications cover a wide range of fields, briefly described in the following. A common factor for many applications is that an accurate measurement of the local acceleration is needed, but MEMS sensors

I. Frosio (✉) • F. Pedersini • N. Alberto Borghese
Computer Science Department, University of Milan, Via Comelico 39/41, Milano 20135, Italy
e-mail: iuri.frosio@unimi.it; federico.pedersini@unimi.it; alberto.borghese@unimi.it

D. Zhang (ed.), *Advanced Mechatronics and MEMS Devices*, Microsystems,
DOI 10.1007/978-1-4419-9985-6_3, © Springer Science+Business Media New York 2013

are only imprecisely (or not at all) calibrated after production; therefore, a practical and precise sensor calibration procedure is necessary to get satisfying accuracy.

MEMS accelerometers are used, in static or quasi-static conditions, as tilt sensors [5] and to reconstruct the movements of human body segments [6–10]. A surveillance application is for instance described in [9], where the analysis of the signal measured by an accelerometer on the trunk of the subject permits to classify human activities and posture transitions with an accuracy higher than 90 %. In [6], a wearable device containing several accelerometers and gyroscopes is used to track and analyze the movements of body segments; the sensor calibration permits to limit the effect of the drift: the experimental results reported in the paper revealed positioning errors smaller than 6 mm and angular errors smaller than 0.2° during the acquisition of a typical sit-to-stand movement; such accuracy is comparable to that of a traditional motion capture system, which is however more costly and cumbersome. An algorithm for the on-line, automatic recalibration of the accelerometer is proposed in [10], where another medical application of the accelerometers is described; calibration is in this case necessary to guarantee that the accuracy of the sensor remains constant over the time, as its output changes significantly, for instance, with respect to the external temperature or other physical quantities.

MEMS accelerometers are also widely employed in Inertial Measurements Units (IMUs), together with gyroscopes and magnetometers, to track the motion of a pedestrian [11], terrestrial [12–14] or even aerial vehicles [15]. In [11], the authors suggest using a wearable IMU comprising accelerometers and magnetometers to track a pedestrian; an accurate calibration procedure of the IMU sensors is essential for this application; in particular, the misalignment of the sensor axes has to be explicitly taken into account to guarantee a satisfying accuracy of the tracking procedure. In [13], an IMU is used to track the position of a car when the GPS signal is lost; if the sensor is not calibrated before its use, a position error larger than 2.5 km can be accumulated during a 60 s loss of the GPS signal; a calibrated sensor limits the position drift to less than 400 m. The same authors describe in [14] an integrated IMU/GPS system, where the drifts of the offsets and sensitivities of the sensor with respect to the temperature are taken into account by a proper thermal variation model. A similar IMU system is considered in [15] to track the position and orientation of a small helicopter; also in this case, a model of the accelerometers including non-orthogonal axes and the corresponding calibration procedure are described.

In a slightly different scenario, MEMS accelerometers are also used to integrate traditional tracking systems based on vision technology [16]. In fact, vision based tracking systems are generally capable of tracking objects, without introducing drifts, at low frequencies (a camera acquires typically images at a rate of 60 frames per second); accelerometer based procedures generate data at a significantly higher frequency; they are used in [16] to track the movement of an object between the acquisition of two consequent frames, therefore increasing the tracking frequency to hundreds of Hertz.

Other quasi-static applications of the accelerometers requiring an accurate measurements of the local acceleration vectors include, but are not limited to, handwriting recording [17], real-time character recognition [18], quantification of the physical activity of a subject [19], analysis of sport equipment during movements [20], and so on.

A common factor to all these applications is that the sensors are used in quasi-static condition; therefore, they measure an acceleration which is supposed to be slowly varying with time. Modeling the frequency response of the sensor is generally not necessary for this kind of applications; however, an accurate calibration is needed to avoid drifts that may turn the measured acceleration into an untrustworthy position measurement in a short time.

Other applications of the accelerometers include a dynamic usage, for instance controlling damages of mechanical structures [21, 22] or seismography [23]. In this case, the output of the accelerometer has to be characterized over the entire frequency domains and specific calibration procedures have been proposed to this aim: this goes beyond the scope of this chapter.

The chapter is organized as follows: after this brief introduction, where different applications of the MEMS accelerometers have been reviewed, the sensor models adopted by various authors in literature are briefly described. In the same section, it is also illustrated the matrix decomposition procedure that allows passing from a symmetric to a left triangular sensitivity matrix. The next section introduces a novel, principled noise model for MEMS accelerometers and clearly illustrates the effect of the noise on the measurements of small accelerations. Afterwards, the importance of correcting for local gravity variations is highlighted. The autocalibration procedure is described in the subsequent section, where a novel derivation of the procedure in the context of maximum likelihood estimate is derived. The autocalibration results obtained on different accelerometers are illustrated and discussed, and a brief review of the main concepts closes the chapter.

3.2 Sensor Models

Different sensor models have been proposed in literature, to describe the relation between the acceleration applied to the sensor and its output. These models differ for their complexity (i.e., number of model parameters); generally speaking, the more complex is the model, the more accurate is the description of the physics of the sensor; on the other hand, complex models also require complex (i.e., costly, time-consuming, or requiring specific hardware) calibration procedures, and they suffer the risk of overfitting. Therefore, depending on the specific application, the best compromise between the model accuracy and its complexity has to be found.

For instance, in [9] it is assumed that the three axes of the accelerometer produce the same output voltage when they are solicited by the same force; such simple model is described by the following equation:

$$\mathbf{v} = V_{CC}\left(\begin{bmatrix} s & 0 & 0 \\ 0 & s & 0 \\ 0 & 0 & s \end{bmatrix}\mathbf{a}' + \begin{bmatrix} o \\ o \\ o \end{bmatrix}\right), \tag{3.1}$$

where $\mathbf{v}$ is a 3×1 vector representing the three voltages output by the sensor, $\mathbf{a}'$ is the acceleration vector in the reference system of the sensor,[1] whereas s and o represent the sensitivity and offset of each channel. The parameter V_{CC} represents the power supply voltage, and it is introduced here to take into account that the major part of the MEMS accelerometers is ratiometric [4]. Notice that, although this model is quite simplistic, it is sufficient to recognize with high accuracy the posture of a human being [9]; however, a more realistic model is needed for applications requiring a higher accuracy.

In [23] and [10], a six-parameter model is considered, representing a sensor with different offset and sensitivity for each channel. This is described by:

$$\mathbf{v} = V_{CC} \left(\begin{bmatrix} s_X & 0 & 0 \\ 0 & s_Y & 0 \\ 0 & 0 & s_Z \end{bmatrix} \mathbf{a}' + \begin{bmatrix} o_X \\ o_Y \\ o_Z \end{bmatrix} \right), \tag{3.2}$$

where s_k and o_k represent respectively the sensitivity and the offset of the kth sensor channel. Equation (3.2) can take into account, for instance, the different weight of the proof masses used by the three channels of the accelerometer, or the different stiffness of their supporting arms, introduced during the fabrication of the sensor.

This model, however, cannot give account of the misalignments of the three axes of the sensor. Small misalignments are in fact introduced by a not perfectly controlled fabrication process; even more important, large deviations from set of three orthogonal axes can be measured when a triaxial accelerometer is obtained putting together mono or biaxial accelerometers, like for instance in [5].

The sensor model which includes the axes misalignment requires nine parameters, and it is the most widely adopted for the description of the accelerometers [4, 11, 13, 15, 24] and, more in general, of any triaxial sensor [25]. To derive the model, we have to consider that the vector $\mathbf{a}'$ in (3.1) and (3.2) represents the three projections of the local acceleration vector, $\mathbf{a}$, onto the three non orthogonal sensing axes of the sensor. To put in relation $\mathbf{a}$ and $\mathbf{a}'$, we have to consider a set of rotation matrices that define the relationship of the misaligned axes, $\{\mathbf{X}', \mathbf{Y}', \mathbf{Z}'\}$ to those of the perfectly orthogonal reference system, $\{\mathbf{X}, \mathbf{Y}, \mathbf{Z}\}$ (Fig. 3.1). To this aim let us first assume, without loss of generality, that axes $\mathbf{X}$ and $\mathbf{X}'$ are coincident and that, in the sensor reference system, they are both equal to $[1\ 0\ 0]^T$. The $\mathbf{Y}'$ axis is obtained rotating $\mathbf{Y}$ around the Z axis by an angle α_{ZY} (see Fig. 3.1); therefore, its coordinates, in the orthogonal reference system defined by $\{\mathbf{X}, \mathbf{Y}, \mathbf{Z}\}$, are given by:

$$\mathbf{Y}' = \begin{bmatrix} c(\alpha_{ZY}) & -s(\alpha_{ZY}) & 0 \\ s(\alpha_{ZY}) & c(\alpha_{ZY}) & 0 \\ 0 & 0 & 1 \end{bmatrix} \begin{bmatrix} 0 \\ 1 \\ 0 \end{bmatrix} = \begin{bmatrix} -s(\alpha_{ZY}) \\ c(\alpha_{ZY}) \\ 0 \end{bmatrix}, \tag{3.3}$$

[1] Each component of $\mathbf{a}'$ represents the projection of the local acceleration vector onto one of the sensing axes of the sensor; these are not necessarily orthogonal to each other.

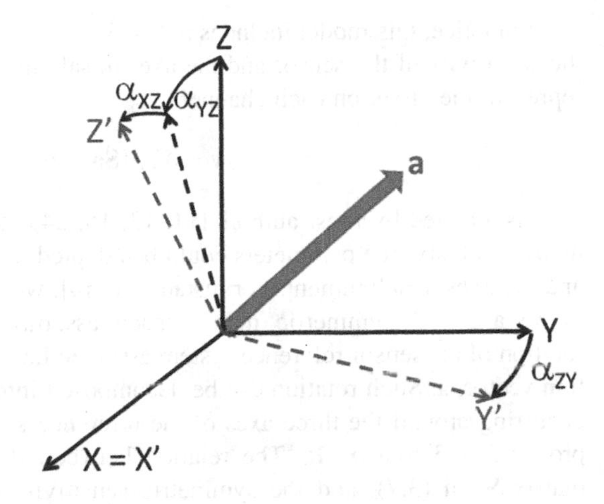

Fig. 3.1 The geometrical model adopted to describe the misalignments of the sensor axes

where $c(t)$ and $s(t)$ indicate respectively the $\cos(t)$ and $\sin(t)$ functions. The $\mathbf{Z}'$ axis is obtained rotating $\mathbf{Z}$ around the Y axis, by an angle α_{YZ}, and then around the X axis, by an angle α_{XZ}. Its coordinates are therefore given by:

$$\mathbf{Z}' = \begin{bmatrix} 1 & 0 & 0 \\ 0 & c(\alpha_{XZ}) & -s(\alpha_{XZ}) \\ 0 & s(\alpha_{XZ}) & c(\alpha_{XZ}) \end{bmatrix} \begin{bmatrix} c(\alpha_{YZ}) & 0 & s(\alpha_{YZ}) \\ 0 & 1 & 0 \\ -s(\alpha_{YZ}) & 0 & c(\alpha_{YZ}) \end{bmatrix} \begin{bmatrix} 0 \\ 0 \\ 1 \end{bmatrix}$$

$$= \begin{bmatrix} -s(\alpha_{YZ}) \\ -s(\alpha_{XZ})c(\alpha_{YZ}) \\ c(\alpha_{XZ})c(\alpha_{YZ}) \end{bmatrix} \tag{3.4}$$

The components of the vector $\mathbf{a}'$ are obtained projecting $\mathbf{a}$ onto the three axes X', Y', Z'; such process is described by:

$$\mathbf{a}' = \begin{bmatrix} 1 & 0 & 0 \\ -s(\alpha_{ZY}) & c(\alpha_{ZY}) & 0 \\ -s(\alpha_{YZ}) & -s(\alpha_{XZ})c(\alpha_{YZ}) & c(\alpha_{XZ})c(\alpha_{YZ}) \end{bmatrix} \mathbf{a}. \tag{3.5}$$

By substituting (3.5) into (3.2), we finally obtain the model of the accelerometer including three sensitivities, three offsets, and three angles (associated to axes misalignments), which is given by:

$$\mathbf{v} = V_{CC}\left(\begin{bmatrix} s_X & 0 & 0 \\ -s_Y s(\alpha_{ZY}) & s_Y c(\alpha_{ZY}) & 0 \\ -s_Z s(\alpha_{YZ}) & -s_Z s(\alpha_{XZ})c(\alpha_{YZ}) & s_Z c(\alpha_{XZ})c(\alpha_{YZ}) \end{bmatrix} \mathbf{a} + \begin{bmatrix} o_X \\ o_Y \\ o_Z \end{bmatrix} \right). \tag{3.6}$$

In practice, this model includes a 3×3 lower triangular matrix, **S**, to represent the sensitivity of the sensor and the axes misalignments, and a 3×1 vector, **o**, to represent the offset on each channel:

$$\mathbf{v} = V_{CC}(\mathbf{Sa} + \mathbf{o}). \tag{3.7}$$

It is adopted by most authors [11, 13, 15, 24, 25]. More generally, any 3×3 matrix with six free parameters could be adopted to describe the sensor sensitivity and the axes misalignment. For instance, in [4], we assumed the sensitivity matrix to be a 3×3 symmetric matrix; such assumption simply corresponds to a rotation of the sensor reference system assumed here to describe the local acceleration vector, **a**. Such rotation can be decomposed into a sequence of three rotations occurring around the three axes of the reference system, and it is described by a proper 3×3 matrix, **R**. The relation between the lower triangular sensitivity matrix **S**, in (3.7), and the symmetric sensitivity matrix $\mathbf{S}'$, adopted in [4], is therefore: $\mathbf{S} = \mathbf{RS}'$; the matrix **R** can be obtained from $\mathbf{S}'$ extracting the sequence of three Givens rotations that transform $\mathbf{S}'$ into **S** [26].

An example of such matrix decomposition is illustrated in the following. In particular, to illustrate how the decomposition works, let us consider the following sensitivity matrix:

$$\mathbf{S}' = \begin{bmatrix} 0.0188 & -0.0004 & -0.0006 \\ -0.0004 & 0.0185 & -0.0002 \\ -0.0006 & -0.0002 & 0.0207 \end{bmatrix}. \tag{3.8}$$

This matrix was computed for a typical triaxial MEMS accelerometer (the LIS3L02AL by ST Microelectronics [27]) with the autocalibration procedure described in this chapter and, more specifically, this is the matrix associated to the sensor #3 in Tables 3.6, 3.7 and 3.8.

The first two Givens rotations have to delete the first and second elements of the third column of $\mathbf{S}'$. Notice that the three columns of $\mathbf{S}'$ can be interpreted as a set of three non orthogonal axes. The first two terms of the third column can then be zeroed by a proper rotation around the X axis, followed by a proper rotation around the Y axis. In particular, from the explicit form of two rotation matrices around the X and Y axes applied to $\mathbf{S}'$, we obtain:

$$\mathbf{S^{XY}} = \mathbf{R_Y R_X S}' = \begin{bmatrix} \cos(R_Y) & 0 & -\sin(R_Y) \\ 0 & 1 & 0 \\ \sin(R_Y) & 0 & \cos(R_Y) \end{bmatrix} \begin{bmatrix} 1 & 0 & 0 \\ 0 & \cos(R_X) & \sin(R_X) \\ 0 & -\sin(R_X) & \cos(R_X) \end{bmatrix} \mathbf{S}',$$

$$S_{YZ}^{XY} = \cos(R_X)S'_{YZ} + \sin(R_X)S'_{ZZ} = 0 \Rightarrow R_X = -\tan^{-1}\left(\frac{S'_{YZ}}{S'_{ZZ}}\right),$$

$$S_{XZ}^{XY} = \cos(R_Y)S'_{XZ} - \sin(R_Y)\cos(R_X)S'_{ZZ} + \sin(R_Y)\sin(R_X)S'_{YZ} = 0,$$

$$\Rightarrow R_Y = \tan^{-1}\left[\frac{-S'_{XZ}}{-\cos(R_X)S'_{ZZ} + \sin(R_X)S'_{YZ}}\right], \tag{3.9}$$

where $\mathbf{S^{XY}}$ is the sensitivity matrix $\mathbf{S}'$ rotated by $\mathbf{R_X}$ and $\mathbf{R_Y}$; R_X and R_Y are the angles of rotation around the X and Y axes; $S^{XY}{}_{YZ}$ and $S^{XY}{}_{XZ}$ are respectively the element on the second row, third column and on the first row, third column of $\mathbf{S^{XY}}$; S'_{ij} represents the elements in position ij of $\mathbf{S}'$. For the matrix $\mathbf{S}'$ in (3.8), we obtain $R_X = 0.5903°$ and $R_Y = -1.5431°$ and the resulting matrix $\mathbf{S^{XY}}$ is:

$$\mathbf{S^{XY}} = \mathbf{R_Y R_X S} = \begin{bmatrix} 0.0188 & -0.0004 & 0 \\ -0.0004 & 0.0185 & 0 \\ -0.0011 & -0.0004 & 0.0207 \end{bmatrix}. \tag{3.10}$$

To get a lower triangular matrix from (3.10), a further rotation matrix representing a rotation around the Z axis has to be applied to $\mathbf{S^{XY}}$. Notice, in fact, that the third column of $\mathbf{S^{XY}}$ is parallel to the Z axis, and it is therefore not affected by a rotation around it; as a consequence, after the application of such rotation, the first and second element of the third column of $\mathbf{S^{XY}}$ remain zeros. The third Givens rotation is therefore obtained from the explicit form of the matrix $\mathbf{R_Z}$, representing a rotation of an angle R_Z around the Z axis, applied to $\mathbf{S^{XY}}$; this gives:

$$\mathbf{S} = \mathbf{R_Z S^{XY}} = \begin{bmatrix} \cos(R_Z) & \sin(R_Z) & 0 \\ \sin(R_Z) & \cos(R_Z) & 0 \\ 0 & 0 & 1 \end{bmatrix} \mathbf{S^{XY}},$$

$$S_{XY} = \cos(R_Z) S^{XY}_{XY} + \sin(R_Z) S^{XY}_{YY} = 0 \Rightarrow R_Z = -\tan^{-1}\left(\frac{S^{XY}_{XY}}{S^{XY}_{YY}}\right). \tag{3.11}$$

The rotation angle around the Z axis for $\mathbf{S}$ in (3.11) results to be $R_Z = 1.3683°$; after application of $\mathbf{R_Z}$ to $\mathbf{S^{XY}}$, we obtain the decomposition of the symmetric matrix $\mathbf{S}'$ into a rotation matrix $\mathbf{R}$ and a triangular matrix $\mathbf{S}$ as:

$$\mathbf{S} = \mathbf{R_Z R_Y R_X S}' = \mathbf{R S}',$$

$$\mathbf{R} = \begin{bmatrix} 0.9994 & 0.0236 & 0.0272 \\ -0.0239 & 0.9997 & 0.0097 \\ -0.0269 & -0.0103 & 0.9996 \end{bmatrix},$$

$$\mathbf{S} = \begin{bmatrix} 0.0188 & 0 & 0 \\ -0.0009 & 0.0185 & 0 \\ -0.0011 & -0.0004 & 0.0207 \end{bmatrix}. \tag{3.12}$$

The lower triangular matrix $\mathbf{S}$ in (3.12) represents therefore the sensor sensitivity matrix as in (3.6); the three rows represent three vectors whose magnitude is equal to the sensitivity of the associated channel; in particular, we obtain in this case: $s_X = 0.0188$ V/(m/s^2), $s_Y = 0.0186$ V/(m/s^2), $s_Z = 0.0208$ V/(m/s^2). Moreover, these vectors are aligned with the axes of the sensor; we can therefore estimate angle between the axes. For instance, the angle between the X and Y axis for the matrix $\mathbf{S}$ in (3.12) is obtained as:

$$XY = \cos^{-1}\left(\frac{[0.0188 \quad 0 \quad 0][-0.0009 \quad 0.0185 \quad 0]^T}{\sqrt{0.0188^2+0^2+0^2}\sqrt{(-0.0009)^2+0.0185^2+0^2}}\right) = 92.737^\circ. \tag{3.13}$$

The angles between the other axes are obtained in a similar manner.

To summarize, (3.7) represents the commonly adopted model to describe the relation between the acceleration vector in the sensor reference system, $\mathbf{a}$, and the voltage output of the sensor, $\mathbf{v}$. The 3×3 sensitivity matrix, $\mathbf{S}$, contains six free parameters that have to be estimated during the calibration procedure. $\mathbf{S}$ is generally assumed to be left triangular or symmetric; decomposition of a symmetric matrix through a set of three Givens rotations allows computing the corresponding left triangular matrix and the rotation matrix which transform one into the other. From the triangular matrix, the angles between the sensor axes can finally be derived.

Many experimental results, reported for instance in [4, 16], suggest that models of quasi static accelerometers with number of parameters higher than nine are generally not justifiable, as the additional parameters cannot be reliably estimated by the existing calibration procedures. In [4], we considered a sensitivity matrix with 12 free parameters; the experimental results demonstrate that such model does not provide better results than the nine parameters model, and the additional parameters are estimated with low accuracy. In [16], a model including the sensor nonlinearity is analyzed, but the authors conclude that the parameters associated to the nonlinearity cannot be reliably estimated and the linear model performs better.

Other sensor models, tailored to specific applications, may include the dependency of $\mathbf{S}$ and $\mathbf{o}$ from the temperature or other physical parameters [14]. However, a sensor model including these parameters require a costly and time consuming calibration procedure, where data are acquired at a set of different, controlled temperatures. When possible, it is preferable to use a nine model parameters and calibrate the sensor on the field and to track in real time the changes of $\mathbf{o}$ and $\mathbf{S}$, as done, for instance, in [9].

Finally, we notice that also sensor models representing the frequency response of the sensor have been proposed for applications like seismography [23] or mechanical testing [21, 22]. In this case, the output of the accelerometer has to be characterized over the entire frequency domain and specific calibration procedures have been proposed to this aim. As already noted in the former chapter, this goes beyond the scope of this chapter; the sensor model and the autocalibration procedure described here could however be applied also in these cases, to accurately characterize the response of the sensor at zero frequency.

3.3 Noise Model

Although the major part of the authors spent a lot of time to describe the physical models of the sensor, only a few of them (see for instance [25]) spent some words about the inclusion of the noise into the sensor model; actually this constitutes a critical point for many of the calibration procedures in the literature, as the cost function associated to the calibration procedure should be strictly related to the characteristics of the noise present on the calibration data. Here we introduce a novel, reasonable noise model and we analyze how this affects the measured acceleration vector and the calibration procedures.

In the following, we assume that additive, white noise with zero mean and Gaussian distribution is present on each output channel of the accelerometer. In particular, let us indicate with $G_M(\sigma^2)$ a column vector with M components where each component is a Gaussian random variable with zero mean and variance σ^2 and let us define the noise vector $\mathbf{n} \sim G_3(\sigma^2)$. The noisy sensor output is therefore defined as:

$$\mathbf{v_n} = V_{CC}(\mathbf{Sa} + \mathbf{o}) + \mathbf{n}. \tag{3.14}$$

Inverting (3.14), we obtain the estimated acceleration vector, $\mathbf{a_n}$, from the output of the accelerometer, which is:

$$\mathbf{a_n} = \mathbf{S}^{-1}\left(\frac{\mathbf{v_n}}{V_{CC}} - \mathbf{o}\right) = \mathbf{S}^{-1}\left(\frac{\mathbf{v}}{V_{CC}} - \mathbf{o}\right) - \mathbf{S}^{-1}\frac{\mathbf{n}}{V_{CC}}. \tag{3.15}$$

The first term in (3.15), $\mathbf{S}^{-1}(\mathbf{v}/V_{CC} - \mathbf{o})$, represents the real acceleration vector, whereas the term $-\mathbf{S}^{-1}\mathbf{n}/V_{CC}$ is the noise contribution. Here we will analyze the effect of the noise term on the length of $\mathbf{a_n}$ and we will show how this can affect a sensor calibration procedure. To this aim, let us consider the expected value of the squared norm of $\mathbf{a_n}$, which is given by:

$$\begin{aligned}
E\left[\mathbf{a_n}^{\mathrm{T}}\mathbf{a_n}\right] &= E\left[\left\{\mathbf{S}^{-1}\left(\frac{\mathbf{v_n}}{V_{CC}} - \mathbf{o}\right)\right\}^{\mathrm{T}} \mathbf{S}^{-1}\left(\frac{\mathbf{v_n}}{V_{CC}} - \mathbf{o}\right)\right], \\
&= E\left[\left\{\mathbf{S}^{-1}\left(\frac{\mathbf{v}}{V_{CC}} - \mathbf{o}\right) - \mathbf{S}^{-1}\frac{\mathbf{n}}{V_{CC}}\right\}^{\mathrm{T}} \left\{\mathbf{S}^{-1}\left(\frac{\mathbf{v}}{V_{CC}} - \mathbf{o}\right) - \mathbf{S}^{-1}\frac{\mathbf{n}}{V_{CC}}\right\}\right], \\
&= \left(\frac{\mathbf{v}}{V_{CC}} - \mathbf{o}\right)^{\mathrm{T}} \mathbf{S}^{-\mathrm{T}}\mathbf{S}^{-1}\left(\frac{\mathbf{v}}{V_{CC}} - \mathbf{o}\right) - 2E\left[\left(\mathbf{S}^{-1}\frac{\mathbf{n}}{V_{CC}}\right)^{\mathrm{T}}\right]\mathbf{S}^{-1}\left(\frac{\mathbf{v}}{V_{CC}} - \mathbf{o}\right) \\
&\quad + \frac{E\left[(\mathbf{S}^{-1}\mathbf{n})^{\mathrm{T}}\mathbf{S}^{-1}\mathbf{n}\right]}{V_{CC}^2},
\end{aligned} \tag{3.16}$$

where $E[x]$ indicates the expected value of x. The first term in (3.16) is associated to the true acceleration vector, whereas the second and third terms are noise contributes.

The second term is zero as the expected value of **n** is zero for each component; in fact:

$$E\left[\left(\mathbf{S}^{-1}\frac{\mathbf{n}}{V_{CC}}\right)^{\mathrm{T}}\right] = E\left[\mathbf{n}^{\mathrm{T}}\right]\frac{\mathbf{S}^{-\mathrm{T}}}{V_{CC}} = 0. \tag{3.17}$$

To characterize the effect of the third term in (3.16), we approximate **S** with a diagonal matrix, with diagonal elements equal to s^*; also $\mathbf{S}^{-1}$ is in this case diagonal, with diagonal elements equal to $1/s^*$. Such approximation corresponds to the assumptions that sensor axes are perfectly orthogonal (it is therefore reasonable considering the typical small misalignments registered in the real accelerometers) and that the sensitivity of the three axes is the same. Under these assumptions, the third term in (3.16) can be rewritten as:

$$\begin{aligned}\frac{E\left[\left(\mathbf{S}^{-1}\mathbf{n}\right)^{\mathrm{T}}\mathbf{S}^{-1}\mathbf{n}\right]}{V_{CC}^2} &\approx \frac{E\left[n_X^2/s^{*2} + n_Y^2/s^{*2} + n_Z^2/s^{*2}\right]}{V_{CC}^2} \\ &= \frac{E\left[n_X^2\right] + E\left[n_Y^2\right] + E\left[n_Z^2\right]}{s^{*2}V_{CC}^2}.\end{aligned} \tag{3.18}$$

Each component of **n** is a Gaussian random variable with zero mean and variance σ^2: n_X, n_Y, $n_Z \sim \sigma \cdot G_1(1)$. The sum of k squared normal random variables is a χ^2 random variable with k degrees of freedom, indicated by χ_k^2; its mean is k and its variance is $2k$. Therefore, we can write:

$$\frac{E\left[\left(\mathbf{S}^{-1}\mathbf{n}\right)^{\mathrm{T}}\mathbf{S}^{-1}\mathbf{n}\right]}{V_{CC}^2} \approx \frac{\sigma^2}{s^{*2}V_{CC}^2}\left(E\left[\chi_3^2\right]\right) = 3\frac{\sigma^2}{s^{*2}V_{CC}^2}. \tag{3.19}$$

Substituting (3.17) and (3.19) into (3.16) we finally obtain:

$$\begin{aligned}E\left[\mathbf{a_n}^{\mathrm{T}}\mathbf{a_n}\right] &\approx \left(\frac{\mathbf{v}}{V_{CC}} - \mathbf{o}\right)^{\mathrm{T}}\mathbf{S}^{-\mathrm{T}}\mathbf{S}^{-1}\left(\frac{\mathbf{v}}{V_{CC}} - \mathbf{o}\right) + 3\frac{\sigma^2}{s^{*2}V_{CC}^2} \\ &= \mathbf{a}^{\mathrm{T}}\mathbf{a} + 3\frac{\sigma^2}{s^{*2}V_{CC}^2}.\end{aligned} \tag{3.20}$$

This equation highlights that, in presence of noise on the measured sensor output, the squared length of the measured acceleration vector is biased by a factor which increases with the inverse of the squared sensor sensitivity, $1/s^{*2}$, and with the variance of the noise, σ^2.

This observation has some consequences for any application that requires the norm of the estimated acceleration vector to be considered, as the bias produced by noise has to be eliminated to avoid drifts. An example of this situation is illustrated in the following on a simple, didactic problem.

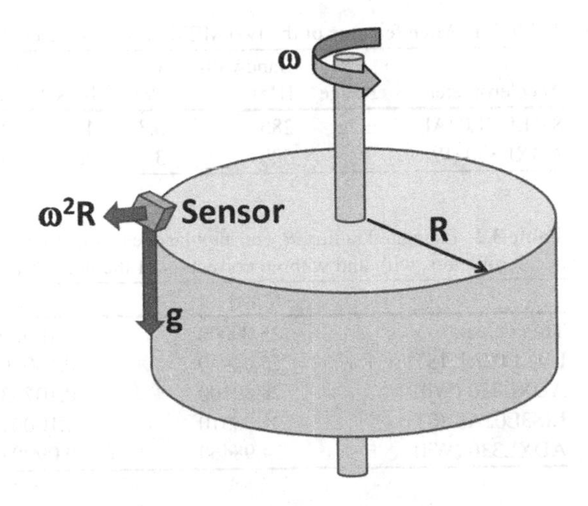

Fig. 3.2 Experimental setup for estimating the radius R or the angular speed ω of a planar rotating disk, using an accelerometer

Let us suppose that a triaxial accelerometer is positioned at the border of a disk of radius R, rotating at a constant angular speed ω (Fig. 3.2). Let us also suppose that the sensor orientation is unknown. The sensor is subjected to the local gravity force, $\mathbf{g}$, in vertical direction, and to the centripetal force induced by the disk rotation, which is radially oriented and has modulus equal to $\omega^2 R$. Since these two forces are orthogonal to each other, the norm of the resultant acceleration measured by the sensor should be equal to:

$$\|\mathbf{a}\| = \sqrt{g^2 + [\omega^2 R]^2}. \tag{3.21}$$

From (3.21), the radius R of the circle can be estimated from the module of the acceleration measured by the accelerometer and ω; alternatively, the disk angular speed ω may be derived from $\|\mathbf{a}\|$ and R, as:

$$\begin{cases} \omega = \sqrt{\dfrac{\sqrt{\|\mathbf{a}\|^2 - g^2}}{R}}, \\ R = \dfrac{\sqrt{\|\mathbf{a}\|^2 - g^2}}{\omega^2}. \end{cases} \tag{3.22}$$

It is evident, from (3.22), that to estimate R or ω, one has to reliably estimate $\|\mathbf{a}\|^2$. However, in presence of noise, averaging the squared norm of the measured acceleration vectors gives $\|\mathbf{a}\|^2 + 3\sigma^2/(s^{*2} V_{CC}^2)$ in spite of $\|\mathbf{a}\|^2$; as a consequence, ω and R will be both overestimated.

To give a more quantitative idea of this phenomenon, let us consider a disk of radius $R = 25$ m, rotating at an angular speed $\omega = 0.01$ rad/s, corresponding at about one rotation every 10 min. The resulting centripetal acceleration is equal to $\omega^2 R = 0.0025$ m/s^2, orthogonal to the gravity force; assuming $g = 9.80665$ m/s^2, the norm of the acceleration experimented by the sensor is equal to 9.80678 m/s^2.

Table 3.1 Main features of the two MEMS accelerometers considered here

Accelerometer	Range	Bandwidth [Hz]	V_{CC} [V]	$1/s^*$ [m/s^2]	o	σ^2 [V^2]	$3\sigma^2/[s^{*2}V_{CC}^2]$ [m^2/s^4]
ST LIS3L02AL	±2g	285	3.3	49.03	0.5	4.9658E – 07	0.000329
ADXL330 (Wii)	±3g	500	3	98.07	0.5	6.4800E – 06	0.020773

Table 3.2 Estimated radius, R, and angular speed, ω, of a rotating planar disk, using two different accelerometers, with and without correction of the noise bias

	R [m]	ω [rad/s]	Noise bias correction
True value	25.00000	0.10000	–
LIS3L02AL (ST)	25.04940	0.10010	No
ADXL330 (Wii)	28.80100	0.10733	No
LIS3L02AL (ST)	25.00510	0.10001	Yes
ADXL330 (Wii)	24.98460	0.09997	Yes

Let us consider then two typical triaxial MEMS accelerometers: the first one is the LIS3L02AL by ST Microelectronics [27], already analyzed in [4]; this is a high performance accelerometer for low frequency applications, characterized by low noise. The second sensor is the ADXL330 by Analog Devices [28], which is currently employed in the Wiimote controller by Nintendo [29]. The main characteristics of these sensors, taken from their datasheets and including the noise variance σ^2 and the corresponding noise bias, are summarized in Table 3.1.

Table 3.2 reports the values of R and ω estimated from 10,000,000 different values of $||\mathbf{a}||^2$ obtained through simulation. In particular, the correct value of $\mathbf{v}$ was generated for each sensor, and white Gaussian noise with variance σ^2 equal to that reported in Table 3.1 was added to the output channels of the accelerometers. The value of $||\mathbf{a}||^2$ in (3.16) was then estimated averaging 10,000,000 values of the squared norm of the acceleration. Notice that the values of R and ω are overestimated by both accelerometers; the second sensor, which is noisier than the first one, overestimates the disk radius by more than 10 %. However, if the bias introduced by the noise term is taken into account, R and ω are properly estimated.

Such results demonstrate that the bias introduced by noise can dramatically affect the estimates, especially when small accelerations are considered like in this case [30]. A typical application where such situation occurs in practice is breath controlling through accelerometers [31, 32]: in this case, an accelerometers is positioned onto the stomach of the subject and it has to measure the small accelerations induced by the breath process.

3.4 Local Gravity Variations

As already noted, the major part of the calibration procedures are generally based on the fact that, in static condition, the module of the acceleration vector measured by the sensor must equal the local acceleration gravity. This is generally assumed to

Table 3.3 This table shows the maximum drift induced by a 1 min integration of the accelerometer output, calibrated without considering the local gravity variations due to latitude, altitude, gravity anomaly and tidal effect. The drift effects induced by the noise bias are also reported

	Induced Δg, % (9.80665 m/s^2 = 100 %)	Induced Δg [m/s^2]	Position drift after 1 min [m]
Latitude	0.5300	0.05198	93.56
Altitude	0.2800	0.02746	49.43
Gravity anomaly	0.0100	0.00098	1.77
Tidal effect	0.0030	0.00029	0.53
$3\sigma^2/(s^{*2}V_{CC}^2)$, ST LIS3L02AL	0.0002	0.00002	0.03
$3\sigma^2/(s^{*2}V_{CC}^2)$, ADXL300 (Wii)	0.0108	0.00106	1.91

be $g = 9.80665$ m/s^2, that represents a mean value of the acceleration induced by the Earth gravity field. However, as noticed in [33, 34] or [5], and clearly illustrated in [35], different factors may contribute to modify the local value of g. These are the following:

1. *Latitude*—Because of the centrifugal force resulting from the rotation of the Earth, which is null at the poles and maximum at the equator, and also because of the bulge of the Earth at the equator, the value of the acceleration of gravity varies from 9.780 m/s^2 at the equator to 9.832 m/s^2 at the poles: a difference of 0.53 %.
2. *Altitude*—Since the gravitational force is proportional to the inverse of the square of the distance from the center of mass, the value of the acceleration of gravity varies with the altitude. Comparison of the acceleration of gravity measured at sea level and on the top of mount Everest shows a difference of about 0.28 %.
3. *Gravity anomaly*—The local gravity field is locally affected by the presence of mountains, canyons or concentration or low/high density rocks in the mantel. Typical variations are generally included in the interval ±0.001 m/s^2, corresponding to a variation of 0.01 % of the nominal value of the acceleration gravity [36].
4. *Tidal effect*—Depending on the geometrical configuration of the Earth/moon/sun system, the sun and moon gravitational mass attraction may induce temporary variations as large as 0.003 % of the local gravity force.

Most authors do not consider these factors during the calibration procedure. As a consequence, the estimated sensor sensitivity is over or underestimated, and drifts are introduced when the output of the accelerometers in an IMU is integrated over time to derive the position information. Table 3.3 shows the maximum drifts induced by these factors when the sensor output is integrated over 60s, and an average acceleration close to 10 m/s^2 is registered for such interval of time. The same table shows the drifts introduced by the noise bias (3.20) for the two types of sensor considered here.

This table dramatically demonstrates the need for correcting the value of g during calibration, at least for the variations induced by latitude and altitude. In fact, without such correction, the position drift can easily exceed 50 m after 1 min. The same table also demonstrates that gravity anomaly, tidal effect and bias introduced by noise are one or two orders of magnitude smaller and, at least in first approximation, they can be disregarded.

Fortunately, the influence of the first two main factors can be easily taken into account to get a reliable estimate of the local acceleration vector, as a function of the latitude and altitude. In particular, considering the radius of the Earth at the equator and at the poles (and assuming its shape as an ellipsoid), the mass of the planet and the centripetal force due to Earth rotation and the ellipsoidal shape of our planet, the *International Gravity Formula* has been derived [37]:

$$g(\theta) = 9.780327\left[1 + 5.3024 \times 10^{-3}\sin^2(\theta) - 5.8 \times 10^{-6}\sin^2(\theta)\right] \text{ m/s}^2, \quad (3.23)$$

which gives the value of the acceleration gravity, $g(\theta)$, as a function of the latitude, θ. For instance, at the equator (3.17) gives $g(0°) \approx 9.78$ m/s^2; at the poles, g $(\pm 90°) \approx 9.83$ m/s^2; whereas at 45° latitude, it gives $g(\pm 45°) \approx 9.81$ m/s^2.

Altitude can be taken into account considering that the mass attraction decreases with the square of the distance from the mass center. Linearization of this relation around the sea level gives raise to the *Free Air Correction* formula; the gravity force decreases approximately of 0.000003086 m/s^2 per meter height; once included into (3.23), this gives:

$$\begin{aligned} g(\theta, h) = 9.780327&\left[1 + 5.3024 \times 10^{-3}\sin^2(\theta) - 5.8 \times 10^{-6}\sin^2(\theta)\right] \\ &- 3.086 \times 10^{-6}\text{h m/s}^2, \end{aligned} \quad (3.24)$$

where h is the height (in meters) with respect to the sea level.

Equation (3.24) provides a simple way to compute the local value of the acceleration of gravity depending on the latitude and altitude; such computation should always be performed before starting any sensor calibration procedure, as an inaccurate estimation of the gravity acceleration corrupts the estimate of the sensor sensitivity and, as a consequence, may introduce drifts over the integrated signal which easily overcomes 50 m after 1 min. An alternative to use of (3.24) is measuring the local gravity field through a gravimeter. However, this solution is time consuming and costly; it is preferable for most applications to approximate it with (3.24).

3.5 Calibration Procedures

Calibration procedures are aimed at estimating the offset, sensitivity and axis misalignments of the sensor. More precisely, the inverse sensitivity matrix, $\mathbf{S}^{-1}$, and the offset vector, $\mathbf{o}$, in (3.15), are generally estimated, as these permit to directly transform the output of the sensor, in *V*, into an acceleration vector, given in m/s^2, as in (3.15).

Most of the calibration procedures are based on the fact that, in static conditions, the module of the acceleration vector measured by the sensor must equal the local acceleration gravity, g. As noted before, most authors assume $g = 9.80665$ m/s^2 independently from latitude and altitude. However, to avoid a biased estimate of the sensitivity parameters, it is reasonable to use (3.24) to get an accurate prediction of g, after that the calibration procedure can be performed.

The accelerometer calibration procedures are divided into different families, that are briefly analyzed in the following [13]: the *six-positions method*, the *extended six-positions method* and the *autocalibration* procedures; a brief description of some alternative calibration procedures is also provided at the end of the paragraph. More space is dedicated to the description of the autocalibration procedure that constitutes the most cost effective and reliable solution, and it is therefore widely adopted.

3.5.1 Six-Positions Method

This kind of calibration procedure is based on six orthogonal measurements of the gravity force vector. It is therefore capable to estimate the parameters of a sensor model with at most six unknowns that represent the sensor offsets and sensitivities along the three axes. On the other hand, axis misalignment cannot be estimated.

For each measurement of gravity, the sensor should be ideally oriented such that one of its axes is aligned with respect to the gravity force. For each axis, the offset and sensitivity are then derived from (3.2); in fact, for the *X* axis, we can for instance write the following equations:

$$\begin{cases} v_{X,1} = V_{CC}(s_X g + o_X), \\ v_{X,2} = V_{CC}(-s_X g + o_X), \end{cases} \tag{3.25}$$

where $v_{X,1}$ and $v_{X,2}$ are the measured outputs of the *X* channel when the sensor's *X* axis is respectively parallel and antiparallel with respect to the gravity force. Solving (3.25) leads to the estimate of s_X, o_X.

This calibration procedure is characterized by the simplicity of its implementation: each axis can be calibrated independently from the others and this only requires solving a linear system with two equations and two unknowns. Nevertheless, the procedure has some critical drawbacks. First of all, it requires to use a costly machinery to accurately position the sensor into six different orientations, each orthogonal to the others; however, since the sensor package hides the internal structure, it is not possible to guarantee that the sensor is properly oriented with respect to the calibration bank; therefore, a global misalignment of the sensor can result. Moreover, the procedure is time consuming and does not allow estimating the misalignment between the axes.

Calibrating an accelerometer with the six-position method is therefore limited to low-accuracy applications, like for instance in [23], where a low precision seismograph is described, or in [19], where the only norm of the acceleration vector is used to quantify the physical activity of a human subject. For lower precision applications, like in the case of the Wiimote controller [29], the sensor calibration can even be performed without resorting to a costly rotating bank, but simply rotating by hand the sensor into six "almost" orthogonal positions; in this case, the six-position method results in a simple, fast and poorly accurate calibration method.

3.5.2 Extended Six-Positions Method

Many authors perform a sequence of at least nine measurements, with controlled sensor orientations, to obtain the calibration data [16, 30]. This allows estimating not only the offset and sensitivity of each axis, but also the axis misalignments and, in some cases, the sensor nonlinearity [16] and the noise properties [30].

Different authors use the controlled calibration data in different manners to obtain different calibration strategies. In [30], a sensor model including offsets, sensitivities, axes misalignments and non linearities is considered; moreover, wavelets are used to analyze the output signal of the accelerometers in the time-frequency domain; this allows for estimating the typical noise characteristics and developing a soft thresholding method which is used to eliminate the wavelet components associated to the noise from the signal.

In [16], a model including offsets, sensitivities, axes misalignments and nonlinearity is considered; calibration is performed measuring the acceleration at different orientations with respect to gravity, while one sensor axis is aligned horizontally; a turn-table is used to this aim. The sensor parameters are then estimated minimizing a cost function which is composed of the sum of two terms: the first term represents the fact that the sensor output must be zero for the horizontal sensing axis; the second term takes into account that the norm of the measured acceleration must equal g; a scalar parameter is also included to assign different weights to the two costs. Based on their results, the authors suggest avoiding the use of non linear models, as the parameters describing the non linearities cannot be reliably estimated.

Overall, the extended six-positions method presents some advantages with respect to the six-position method, as it allows adopting more complex sensor models. However, this kind of calibration can only be performed in a controlled environment, using costly machinery and a time consuming procedure.

3.5.3 Autocalibration Methods

Autocalibration methods are the most used [4, 9–11, 13, 15, 25, 38]; the only assumption required is that, in static conditions, the module of the acceleration vector measured by the sensor equals the local acceleration gravity, g. From this and from a sequence of acquisitions performed at random orientations, the parameters of the sensor model are estimated. Simple sensor models with only one offset and sensitivity parameters as well as complex models including nonlinearity can be adopted in autocalibration. However, the most diffused choice is represented by the nine parameters model including three offsets, three sensitivities and three cross axis terms.

More precisely, autocalibration is based on the minimization of a cost function, representing the sum of the differences between the norm of the measured accelerations and g, as a function of the model parameters. Slightly different cost functions are however adopted by different authors, leading to more or less complex minimization strategies. In [13], a weighted least squares scheme is adopted to compare the squared norm of the measured accelerations with g^2; the linearization of the cost function leads to a constrained minimization which is generally achieved by a proper iterative algorithm in less than 30 iterations. On the other hand, in [9] a very simple sensor model including only one offset and one sensitivity parameter is adopted. This allows minimizing the cost function (and therefore estimating the model parameters) in closed form. In [15], as in [4, 11, 25] a cost function in the form:

$$E(\theta) = \sum_{i=1}^{N} \left(\left\| \mathbf{a}_{i,\theta} \right\|^2 - g^2 \right)^2 \tag{3.26}$$

is adopted, where $\boldsymbol{\theta}$ is the vector of the model parameters and N is the number of measured acceleration vectors. Such cost function is characterized by the fact that only the squared norm of the measured accelerations appears in it; therefore, it can be minimized through efficient iterative algorithms like the Newton's method [4].

The success of autocalibration procedures is explained by the fact that they do not require any special machinery to be performed, and they are therefore extremely cheap and easy to perform; they only require the sensor to be randomly oriented a sufficient number of times, covering the whole range of orientations and measuring each time the module of the static acceleration, and the minimization of a cost function. For this reason, autocalibration can even be performed on-line: for instance, in [9], each time a period of sensor immobility is detected, the modulus of the gravity force is measured and the sensor parameters are updated. In this manner, changes of the sensor sensitivity due, for instance, to changes of temperature, can easily be managed and corrected in real time.

In spite of the large literature on autocalibration methods, only one author [25], between the ones cited here, has spent some words about the adequacy of the cost function with respect to the sensor noise model. Here we therefore propose an original derivation of the cost function in the context of the maximum likelihood estimate, starting from the sensor noise model described by (3.15). Coherently with the other autocalibration procedures, we will assume that, in static condition, the sensor is subjected only to the force of gravity, whose local intensity g can be computed through (3.24). To derive a simpler cost function, similarly to what done in [4, 11, 15, 25], we will finally consider the squared norm of the measured acceleration vector (like in (3.26)) instead of the simple norm.

Let us consider therefore a set of N acceleration vectors, measured in static condition. The squared norm of the noisy gravity acceleration vector, $\mathbf{g_n}$, measured by the sensor, is given by:

$$\begin{aligned}
\mathbf{g}_\mathbf{n}^\mathrm{T}\mathbf{g}_\mathbf{n} &= \left\{\mathbf{S}^{-1}\left(\frac{\mathbf{v}-\mathbf{n}}{V_{CC}}-\mathbf{o}\right)\right\}^\mathrm{T}\mathbf{S}^{-1}\left(\frac{\mathbf{v}-\mathbf{n}}{V_{CC}}-\mathbf{o}\right),\\
&= \left(\frac{\mathbf{v}}{V_{CC}}-\mathbf{o}\right)^\mathrm{T}\mathbf{S}^{-\mathrm{T}}\mathbf{S}^{-1}\left(\frac{\mathbf{v}}{V_{CC}}-\mathbf{o}\right) - 2\mathbf{n}^\mathrm{T}\frac{\mathbf{S}^{-\mathrm{T}}}{V_{CC}}\mathbf{S}^{-1}\left(\frac{\mathbf{v}}{V_{CC}}-\mathbf{o}\right) + \frac{\mathbf{n}^\mathrm{T}\mathbf{S}^{-\mathrm{T}}\mathbf{S}^{-1}\mathbf{n}}{V_{CC}^2},\\
&= \mathbf{g}^\mathrm{T}\mathbf{g} - 2\mathbf{n}^\mathrm{T}\frac{\mathbf{S}^{-\mathrm{T}}}{V_{CC}}\mathbf{g} + \frac{\mathbf{n}^\mathrm{T}\mathbf{S}^{-\mathrm{T}}\mathbf{S}^{-1}\mathbf{n}}{V_{CC}^2}.
\end{aligned} \tag{3.27}$$

The first term in (3.27) is constant and equal to g^2. To analyze the second term in (3.27), let us consider a sensor with perfectly aligned axes with equal sensitivity and offset; let us assume that $1/s^*$ is the inverse sensitivity of each axis. When this ideal sensor is randomly oriented in the space and measures an acceleration of intensity g, this term reduces to:

$$2\mathbf{n}^\mathrm{T}\frac{\mathbf{S}^{-\mathrm{T}}}{V_{CC}}\mathbf{g} \rightarrow 2\frac{g}{s * V_{CC}}\mathbf{n}^\mathrm{T}\mathbf{m}, \tag{3.28}$$

where $\mathbf{m}$ is a random 3×1 versor. The term $g\mathbf{m}$ represents the gravity acceleration in the sensor reference frame: as the sensor is randomly oriented, also the measured acceleration has a random orientation. Notice now that, since $\mathbf{m}$ is a versor, $\mathbf{n}^\mathrm{T}\mathbf{m}$ is a linear combination of three Gaussian random variables with variance σ^2, and the squared mixing coefficients sum to one ($\mathbf{m}^\mathrm{T}\mathbf{m} = 1$); therefore, $\mathbf{n}^\mathrm{T}\mathbf{m}$ itself is a Gaussian random variable with variance σ^2. Overall, the second term in (3.27) results to be a Gaussian random variable with zero mean and variance equal to $[2g\sigma/(s^*V_{CC})]^2$.

For the same ideal sensor, the third term in (3.27) is a random variable with approximate distribution $[\sigma^1/(s^*V_{CC})]^2\chi_3^{\,2}$; this has a mean value of $3[\sigma/(s^*V_{CC})]^2$ and variance equal to $6[\sigma/(s^*V_{CC})]^4$.

Table 3.4 show the average and variance of the three terms in (3.27), for the two accelerometers considered here. Since the variance of the third term is typically very small with respect to the variance of the second term, we assume that the third term contributes only with its mean value to $\mathbf{g_n}^\mathrm{T}\mathbf{g_n}$. As a consequence, we can approximate $\mathbf{g_n}^\mathrm{T}\mathbf{g_n}$ with a Gaussian random variable, that is $\mathbf{g_n}^\mathrm{T}\mathbf{g_n} \sim g^2 + 3[\sigma/(s^*V_{CC})]^2 + G_1(3[2g\sigma/(s^*V_{CC})]^2)$. The maximum likelihood

Table 3.4 The average and the variance for the three terms of $g^{\mathrm{T}}g$ in (3.27) are reported here for two typical MEMS triaxial accelerometers

		LIS3L02AL (ST)	ADXL330 (Wii)
$(\mathbf{v}/V_{CC}-\mathbf{o})^{\mathrm{T}}\mathbf{S}^{-\mathrm{T}}\mathbf{S}^{-1}(\mathbf{v}/V_{CC}-\mathbf{o})$	Average [m/s^2]	9.80665	9.80665
	Variance [m^2/s^4]	0	0
$2\mathbf{n}^{\mathrm{T}}/V_{CC}\mathbf{S}^{-\mathrm{T}}\mathbf{S}^{-1}(\mathbf{v}/V_{CC}-\mathbf{o})$	Average [m/s^2]	0	0
	Variance [m^2/s^4]	4.2174E − 02	2.66364
$\mathbf{n}^{\mathrm{T}}\mathbf{S}^{-\mathrm{T}}\mathbf{S}^{-1}\mathbf{n}/V_{CC}{}^2$	Average [m/s^2]	3.2890E − 04	2.0772E − 02
	Variance [m^2/s^4]	7.21180E − 08	2.8767E − 04

estimate of the sensor model parameters $\boldsymbol{\theta}$ from a set of N measurements corrupted by Gaussian noise is obtained by the least squares estimate; the cost function to be minimized (corresponding to the negative log likelihood) is therefore given by:

$$E(\theta) = \sum_{i=1}^{N}\left[\left\|\mathbf{a}_{i,\theta}\right\|^2 - \left(g^2 + 3\frac{\sigma^2}{s^{*2}V_{CC}^2}\right)\right]^2. \tag{3.29}$$

Such reasoning line justifies the choice of a quadratic cost function operated, for instance, in [4, 11, 25]. With respect to these, however, we have introduced in (3.29) the term $3[\sigma/(s^*V_{CC})]^2$, which takes into account the bias introduced by the quadratic noise term.

Choosing the traditional sensor model with nine parameters, including three offsets, three sensitivity and the axes misalignments, (3.29) is rewritten as:

$$E\left(\mathbf{o},\mathbf{S}^{-1}\right) = \sum_{i=1}^{N}\left[\left(\frac{\mathbf{v_{n,i}}}{V_{CC}}-\mathbf{o}\right)^{\mathrm{T}}\mathbf{S}^{-\mathrm{T}}\mathbf{S}^{-1}\left(\frac{\mathbf{v_{n,i}}}{V_{CC}}-\mathbf{o}\right) - \left(g^2+3\frac{\sigma^2}{s^{*2}V_{CC}^2}\right)\right]^2, \tag{3.30}$$

where $\mathbf{v_{n,i}}$ is the noisy sensor output measured at the ith orientation, and $\mathbf{S}$ (and therefore also $\mathbf{S}^{-1}$) is assumed to be symmetric as in [4]. As already noticed, the autocalibration procedure does not estimate $\mathbf{S}$ directly, but it estimates $\mathbf{S}^{-1}$; from this, $\mathbf{S}$ is then easily derived by matrix inversion. Notice, however, that (3.30) is not linear with respect to the elements of $\mathbf{S}^{-1}$ and $\mathbf{o}$. A closed solution for the minimization of $E(\mathbf{o}, \mathbf{S}^{-1})$ is not available: an iterative minimization algorithm has to be employed to this aim. To the scope, we have adopted the Newton's method, which is an iterative optimization procedure that guarantees quadratic convergence towards the solution [39]. Starting from an initial guess of the sensor parameters, which is reasonably the one provided by the sensor manufacturer in the datasheet, the solution is iteratively updated as:

$$\boldsymbol{\theta}^{\mathrm{t+1}} = \boldsymbol{\theta}^{\mathrm{t}} - \alpha\mathbf{H}^{-1}\left(\boldsymbol{\theta}^{\mathrm{t}}\right)\cdot\mathbf{J}\left(\boldsymbol{\theta}^{\mathrm{t}}\right), \tag{3.31}$$

where $\boldsymbol{\theta}^{\mathbf{t}} = [\theta_1, \theta_1, \ldots, \theta_9] = [o_X, o_Y, o_Z, S^{-1}{}_{XX}, S^{-1}{}_{YY}\, S^{-1}{}_{ZZ}\, S^{-1}{}_{XY}\, S^{-1}{}_{XZ}\, S^{-1}{}_{YZ}]$ is the unknown vector parameters at the tth iteration, containing the offset vector $\mathbf{o}$ and the six independent elements of $\mathbf{S}^{-1}$. The terms $\mathbf{J}(\boldsymbol{\theta}^{\mathbf{t}})$ and $\mathbf{H}(\boldsymbol{\theta}^{\mathbf{t}})$ in (3.31) are the Jacobian vector and the Hessian matrix of the cost function $E(\boldsymbol{\theta})$, defined respectively as follows:

$$\mathbf{J}(\theta^{\mathbf{t}}) = \left[\frac{\partial E}{\partial \theta_1}, \frac{\partial E}{\partial \theta_2}, \ldots, \frac{\partial E}{\partial \theta_9}\right], \quad \mathbf{H}(\theta^{\mathbf{t}}) = \left\{h_{ij} = \frac{\partial^2 E}{\partial \theta_i \partial \theta_j}\right\}_{i,j=1\ldots9}, \tag{3.32}$$

α is a damping parameter ($0 < \alpha < 1$) and it is computed at each iteration by means of a line search procedure [39].

Notice that, since the cost function (3.30) is a sum of squared terms, the derivatives in (3.31) and (3.32) can be easily computed from the following definitions:

$$\begin{cases} \boldsymbol{\theta} = (\mathbf{o}, \mathbf{S}^{-1}), \\ \varepsilon_i(\boldsymbol{\theta}) = \left(\frac{\mathbf{V_{n,i}}}{V_{CC}} - \mathbf{o}\right)^{\mathrm{T}} \mathbf{S}^{-\mathrm{T}}\mathbf{S}^{-1}\left(\frac{\mathbf{V_{n,i}}}{V_{CC}} - \mathbf{o}\right) - \left(g^2 + 3\frac{s_{-1}^2\sigma^2}{V_{CC}^2}\right), \\ E(\boldsymbol{\theta}) = \sum_{i=1}^{N} \varepsilon_i{}^2(\boldsymbol{\theta}) = \varepsilon(\boldsymbol{\theta})^{\mathbf{T}}\varepsilon(\boldsymbol{\theta}), \\ \frac{\partial E}{\partial \theta_j} = 2\sum_{i=1}^{N} \varepsilon_i(\boldsymbol{\theta}) \frac{\partial \varepsilon_i(\boldsymbol{\theta})}{\partial \theta_j}, \\ \frac{\partial^2 E}{\partial \theta_j \partial \theta_k} = \frac{\partial^2 E}{\partial \theta_k \partial \theta_j} = 2\sum_{i=1}^{N} \left(\frac{\partial \varepsilon_i(\boldsymbol{\theta})}{\partial \theta_j}\frac{\partial \varepsilon_i(\boldsymbol{\theta})}{\partial \theta_k} + \varepsilon_i(\boldsymbol{\theta})\frac{\partial^2 \varepsilon_i(\boldsymbol{\theta})}{\partial \theta_j \partial \theta_k}\right).. \end{cases} \tag{3.33}$$

Iterations are stopped when the following convergence criterion is satisfied:

$$\max_{i=1\ldots9}\left\{\left|\frac{\theta_i^t - \theta_i^{t-1}}{(\theta_i^t + \theta_i^{t-1})/2}\right|\right\} < \delta, \tag{3.34}$$

where δ is a threshold, which has empirically been set equal to 1.5×10^{-6} to guarantee that the estimated parameters assume stable values. Such criterion expresses the fact that the maximum change of a model parameter between two consecutive iterations does not exceed 0.00015 %. Despite this strong convergence criterion, and thanks to the high convergence rate achieved by the Newton's method, less than ten iterations are generally sufficient to converge. Two typical calibration results obtained with this autocalibration procedure are described in the next pages, where the problem of estimating the accuracy of the computed model parameters is also assessed.

3.5.4 *Other Calibration Methods*

Several other calibration methods, although less diffused, have been proposed in literature. Some of them, like those in [40, 41], avoid explicit measurements of the sensor output, since they are based only on electrical tests, and they generally permit to estimate the full frequency response of the sensor. They can however be performed only in laboratory, using the proper electrical equipment. Other calibration methods, like that proposed in [14], include the dependency of the sensor parameters from temperature; also these require special equipments and can be performed only in laboratory.

3.6 Results

In this section we illustrate the typical results of an autocalibration procedure, based on the minimization of the cost function (3.30). We also present a methodology to evaluate the sensor accuracy after calibration, based on comparing the MEMS measurements with those obtained by a commercial motion capture system (SMART3D™ [42]). This also allows comparing the six parameters sensor model (3.2) with the nine parameters model, and to demonstrate the superiority of the last one. It also serves to highlight the dramatic increase of accuracy registered after the sensor is calibrated, and therefore the need for calibration.

The results are referred to the LIS3L02AL triaxial accelerometer by ST, whose main characteristics are reported in Tables 3.1 and 3.5. Four different accelerometers of this kind were considered in this study. The output of the MEMS accelerometers were acquired by a host computer at a sampling rate of 960 Hz through a NI-DAQ board. An analog low-pass RC filter was added to the accelerometer output to filter high frequency noise and avoid aliasing, which would have been introduced by sampling. The bandwidth of the filter was set to 285 Hz, which is large enough, for instance, for tracking the human finest movements [4, 43].

The sensors were calibrated using a different number of sensor random orientations, ranging from $N = 72$ for the sensor #1, to $N = 42$ for the sensor #2 and $N = 35$ for sensors #3 and #4. The parameters of the bias vector, $\mathbf{o}$, and of the inverse sensitivity matrix, $\mathbf{S}^{-1}$, are reported in Tables 3.6, 3.7 and 3.8 for each of the four accelerometers.

Table 3.5 Some of the most significant parameters of the LIS3L02AL MEMS accelerometer

Parameter	Value	Unit
Zero-*g* level	$V_{CC}/2 \pm 6\,\%$	V
Bias drift vs. temperature	$\pm 1.47 \times 10^{-2}$	m/(s^2 °C)
Acceleration range	± 20	m/s^2
Sensitivity	$V_{CC}/49 \pm 10\,\%$	V/(m/s^2)
Cross-axis	± 2	% of sensitivity
Acceleration noise density	5×10^{-4}	m/(s^2 $\sqrt{\text{Hz}}$)

Table 3.6 Estimated values of the components of the bias vector, with estimated standard deviation. The nominal value is 0.5 for both the LISL02AL and ADXL330 sensors

		o_X	o_Y	o_Z
LIS3L02AL (ST) #1	Estimated value	0.49153	0.48619	0.54819
	Std	1.31E − 06	2.87E − 06	1.50E − 06
	Std %	0.00027 %	0.00059 %	0.00027 %
LIS3L02AL (ST) #2	Estimated value	0.52071	0.49948	0.51016
	Std	3.54E − 06	1.02E − 05	5.25E − 06
	Std %	0.00068 %	0.00204 %	0.00103 %
LIS3L02AL (ST) #3	Estimated value	0.42085	0.49741	0.49981
	Std	2.05E − 05	9.91E − 06	4.59E − 06
	Std %	0.00488 %	0.00199 %	0.00092 %
LIS3L02AL (ST) #4	Estimated value	0.45424	0.52032	0.45312
	Std	2.50E − 05	9.84E − 06	5.09E − 06
	Std %	0.00549 %	0.00189 %	0.00112 %
ADXL330 (AD—Wii)	Estimated value	0.49248	0.48815	0.49163
	Std	2.59E − 05	1.79E − 05	1.82E − 05
	Std %	0.00525 %	0.00366 %	0.00371 %

Table 3.7 Estimated values of the diagonal elements of the inverse sensitivity matrix, with estimated standard deviation. The nominal value is 49.05 m/s^2 for the LISL02AL sensor, and 98.1 m/s^2 for the ADXL330 sensor

		$S^{-1}{}_{XX}$ (m/s^2)	$S^{-1}{}_{YY}$ (m/s^2)	$S^{-1}{}_{ZZ}$ (m/s^2)
LIS3L02AL (ST) #1	Estimated value	50.02726	50.47767	51.38764
	Std	3.71E − 04	9.57E − 04	3.63E − 04
	Std %	0.00074 %	0.00190 %	0.00071 %
LIS3L02AL (ST) #2	Estimated value	49.50062	50.94925	49.61606
	Std	8.52E − 04	3.33E − 03	1.15E − 03
	Std %	0.00172 %	0.00654 %	0.00232 %
LIS3L02AL (ST) #3	Estimated value	53.24319	53.99703	48.29837
	Std	6.62E − 03	2.58E − 03	1.23E − 03
	Std %	0.01244 %	0.00478 %	0.00255 %
LIS3L02AL (ST) #4	Estimated value	51.21930	51.96574	47.13613
	Std	7.63E − 03	2.50E − 03	1.46E − 03
	Std %	0.01490 %	0.00482 %	0.00309 %
ADXL330 (AD—Wii)	Estimated value	97.74048	97.21622	99.36618
	Std	3.39E − 02	2.22E − 02	2.24E − 02
	Std %	0.03469 %	0.02281 %	0.02253 %

Notice that the computed values may significantly differ from the nominal ones: for instance, the bias of the *X* channel of the accelerometer #3 is 16 % smaller than its nominal value (Table 3.6), whereas the inverse sensitivity of the *Y* channel of the same accelerometer is 10 % larger than its nominal value (Table 3.7).

Table 3.8 Estimated values of the off-diagonal elements of the inverse sensitivity matrix, with standard deviation. The nominal value is 0 for both the LISL02AL and ADXL330 sensors

		$S^{-1}{}_{XY}$ (m/s^2)	$S^{-1}{}_{XZ}$ (m/s^2)	$S^{-1}{}_{YZ}$ (m/s^2)
LIS3L02AL (ST) #1	Estimated value	−0.38794	0.22297	0.01980
	Std	3.56E − 04	3.26E − 04	4.58E − 04
	Std %	0.09186 %	0.14610 %	2.31562 %
LIS3L02AL (ST) #2	Estimated value	−0.52038	−0.17683	0.07647
	Std	1.01E − 03	6.88E − 04	1.49E − 03
	Std %	0.19354 %	0.38900 %	1.94430 %
LIS3L02AL (ST) #3	Estimated value	1.25742	1.44731	0.59025
	Std	2.27E − 03	1.15E − 03	8.70E − 04
	Std %	0.18082 %	0.07972 %	0.14744 %
LIS3L02AL (ST) #4	Estimated value	−0.58641	−0.35658	−0.24576
	Std	2.43E − 03	1.22E − 03	8.70E − 04
	Std %	0.41446 %	0.34336 %	0.35410 %
ADXL330 (AD—Wii)	Estimated value	−1.37789	0.02840	0.06594
	Std	1.92E − 02	2.04E − 02	1.96E − 02
	Std %	1.39585 %	71.79472 %	29.74861 %

3.6.1 Accuracy of the Estimated Sensor Parameters

Notice that, in Tables 3.6, 3.7 and 3.8, beyond the estimated parameter values, we also reported the estimated standard deviation for each parameter. This information is crucial to determine the reliability of the estimated parameters. As a matter of fact, a large standard deviation would indicate that the parameter cannot be accurately estimated, and therefore, it should be eliminated from the sensor model. To assess the reliability of the parameter estimate, a widely used solution in the statistical domain is employed to derive an estimate of the variability of the parameters from the estimated variance of the measurements. The accuracy on the estimated parameters is therefore evaluated through the covariance analysis [44], carried out on the linearized version of the cost function (3.30) around the final value of the parameters. More in details, we will illustrate in the following how this analysis can be performed, starting from the case of a linear least squares system, and extending then the theory to the case of a nonlinear system.

Let us suppose that measurements are affected by additive Gaussian noise, zero mean, and with variance, ${\sigma_0}^2$, that is often called sample standard deviation. Let us also suppose that noise is independent on the different samples. This information is often available from the knowledge of the measurement process.

In our case, in particular, the sensor noise is provided in the sensor datasheet; the expected noise on the norm of the squared acceleration vector can be derived from the computed sensor parameters and (3.27); more in detail, we have demonstrated that such noise can be reasonably assumed to be Gaussian, with variance ${\sigma_0}^2$ approximately equal to $[2g\sigma/(s^*V_{CC})]^2$.

In this case any linear model that relates a vector of parameters, $\mathbf{x}$, to a vector of measurements, $\mathbf{b}$, can be written in matrix form as:

$$\mathbf{Ax} = \mathbf{b} + \boldsymbol{\nu}, \tag{3.35}$$

where $\mathbf{A}$ is the design matrix and $\boldsymbol{\nu}$ is a vector containing the measurement error for each sample.

System (3.35) is solved in the least squares sense; this produces the maximum likelihood estimate of the parameters as:

$$\mathbf{x} = \left(\mathbf{A}^{\mathrm{T}}\mathbf{A}\right)^{-1}\mathbf{A}^{\mathrm{T}}\mathbf{b}. \tag{3.36}$$

We are interested in having an estimate of the reliability in the estimate of $\mathbf{x}$ and the correlation between a pair of parameters. To this purpose, let us analyze the effect of the residual $\boldsymbol{\nu}$ on the estimated parameters $\mathbf{x}$:

$$\mathbf{x} + \mathbf{u} = \mathbf{CA}^{\mathrm{T}}(\mathbf{b} + \boldsymbol{\nu}), \tag{3.37}$$

where $\mathbf{C} = (\mathbf{A}^{\mathrm{T}}\mathbf{A})^{-1}$ is called the covariance matrix and $\mathbf{u}$ is the error on the estimated parameters introduced by a measurement error $\boldsymbol{\nu}$. We therefore easily derive:

$$\mathbf{u} = \mathbf{CA}^{\mathrm{T}}\boldsymbol{\nu}. \tag{3.38}$$

Given the hypothesis on $\boldsymbol{\nu}$, the vector $\mathbf{u}$ is zero mean; in fact:

$$E[\mathbf{u}] = E\left[\mathbf{CA}^{\mathrm{T}}\boldsymbol{\nu}\right] = \mathbf{CA}^{\mathrm{T}}E[\boldsymbol{\nu}] = \mathbf{0}. \tag{3.39}$$

This means that the estimate of $\mathbf{x}$ is not biased. We can now compute the variability of $\mathbf{u}$, by evaluating:

$$E\left[\mathbf{uu}^{\mathrm{T}}\right] = E\left[\mathbf{CA}^{\mathrm{T}}\boldsymbol{\nu}\left(\mathbf{CA}^{\mathrm{T}}\boldsymbol{\nu}\right)^{\mathrm{T}}\right] = E\left[\mathbf{CA}^{\mathrm{T}}\boldsymbol{\nu}\boldsymbol{\nu}^{\mathrm{T}}\mathbf{AC}^{\mathrm{T}}\right], \tag{3.40}$$

where the only term that has a statistical distribution is $\boldsymbol{\nu}$ and therefore (3.39) can be rewritten as:

$$E\left[\mathbf{uu}^{\mathrm{T}}\right] = \mathbf{CA}^{\mathrm{T}}E\left[\boldsymbol{\nu}\boldsymbol{\nu}^{\mathrm{T}}\right]\mathbf{AC}^{\mathrm{T}}. \tag{3.41}$$

Given the hypothesis of independence and Gaussianity of $\boldsymbol{\nu}$, $E[\boldsymbol{\nu}\ \boldsymbol{\nu}'] = \sigma_0{}^2\mathbf{I}$ and we obtain:

$$E\left[\mathbf{uu}^{\mathrm{T}}\right] = \mathbf{CA}^{\mathrm{T}}\mathbf{I}\sigma_0^2\mathbf{AC}^{\mathrm{T}} = \sigma_0^2\mathbf{CA}^{\mathrm{T}}\mathbf{AC}^{\mathrm{T}} = \sigma_0^2\mathbf{CC}^{-1}\mathbf{C}^{\mathrm{T}} = \sigma_0^2\mathbf{C}^{\mathrm{T}}. \tag{3.42}$$

The variance on the estimated parameters is therefore proportional to $\sigma_0{}^2$ and depends exclusively on the covariance matrix **C** (note that $\mathbf{C}^{\mathrm{T}} = \mathbf{C}$):

$$\mathrm{Var}[u_i] = \sigma_0^2 C_{ii}. \tag{3.43}$$

More generally, it can be demonstrated that the covariance between the ith and the jth elements of **u** is given by C_{ij}. The relationship between two parameters can then be expressed through the correlation index r_{ij}, which is computed as:

$$-1 \leq r_{ij} = \frac{E\left[u_i u_j\right]}{\sqrt{E[u_i u_i]E\left[u_j u_j\right]}} = \frac{C_{ij}}{\sqrt{C_{ii}C_{jj}}} \leq 1. \tag{3.44}$$

In the present situation, the relationship between the measurements (the voltage output vector **v**) and the model parameters ($\mathbf{S}^{-1}$ and **o**) is obtained through the nonlinear cost function (3.30). However, the here above derivation can still be applied linearizing (3.30) around the actual value of the parameters. In this case, the coefficient matrix **A** in (3.35) corresponds to the Jacobian matrix of $E(\mathbf{o}, \mathbf{S}^{-1})$ in (3.30), that is $\partial\boldsymbol{\varepsilon}(\boldsymbol{\theta})/\partial\boldsymbol{\theta}$ [see (3.33)]; the matrix **A** describes how a change of the estimated parameter vector, $\boldsymbol{\theta}$ [corresponding to **x** in (3.35)] influences the error vector, $\boldsymbol{\varepsilon}(\boldsymbol{\theta})$. More precisely, such relation is given by $\partial\boldsymbol{\varepsilon}(\boldsymbol{\theta})/\partial\boldsymbol{\theta}\cdot\boldsymbol{\Delta\theta} \approx \boldsymbol{\varepsilon}(\boldsymbol{\theta} - \boldsymbol{\Delta\theta}) - \boldsymbol{\varepsilon}(\boldsymbol{\theta})$, where the right term of the equation corresponds to $\mathbf{b} + \boldsymbol{\nu}$ in (3.35), and it represents a vector where each component has a Gaussian distribution with variance equal to $\sigma_0{}^2 = [2g\sigma/(s{*}V_{CC})]^2$. Therefore, (3.43) still provides an estimate of the error in the computation of the parameters.

As it can be appreciated in Table 3.6, the estimation uncertainty for the bias vector is always lower than 0.006 % for the ST accelerometer; this means that **o** is estimated with great precision by the calibration procedure. For the diagonal elements of $\mathbf{S}^{-1}$, the standard deviation never exceeds 0.015 %: also these parameters are accurately estimated. The standard deviation for the off-diagonal elements of $\mathbf{S}^{-1}$ reaches 2.32 % of its nominal value in the worst case: this term are estimated with a less, but still reasonable, accuracy.

3.6.2 Evaluation of the Accuracy of the Calibrated Sensor Used as a Tilt Sensor

Once the autocalibration procedure has been completed, the accelerometer is ready to be used. We first observe that, when used as a tilt sensor, the orientation of the sensor in 3D space can be defined by only two angles (ρ and φ in Fig. 3.3), which represent the orientation of the device with respect to the gravity vector, **g**: the rotation around an axis parallel to **g** cannot be observed since the sensor output is invariant for rotations around such axis.

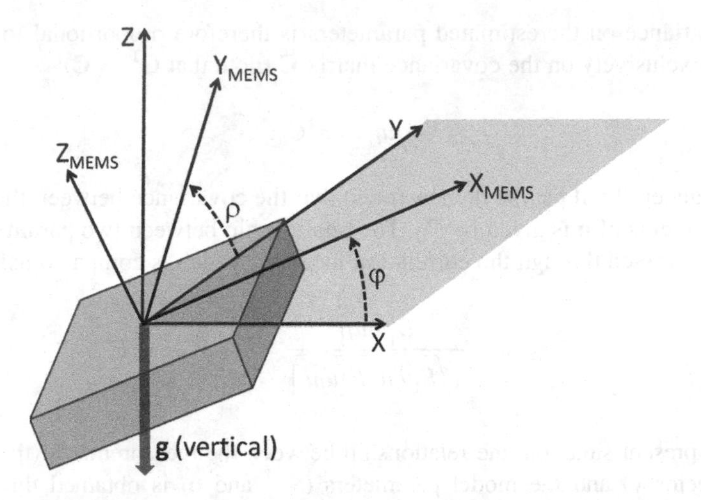

Fig. 3.3 The angles φ and ρ, which describe the orientation of the sensor accelerometer with respect to an absolute reference system having the Z axis oriented parallel to the gravity vector. The local reference system is indicated by the axes X_{MEMS}, Y_{MEMS}, Z_{MEMS} (© 2009 IEEE, reprinted with permission)

Let us define an absolute reference frame with the Z axis parallel to **g**. Let us indicate as (φ, ρ) the angles between the X and Y axes of the sensor and the horizontal plane (Fig. 3.3). From the acceleration vector obtained inverting (3.7), the angles φ and ρ could be computed by means of the following equations:

$$\begin{cases} \phi = \arcsin(a_X), \\ \rho = \arcsin(a_Y). \end{cases} \tag{3.45}$$

These equations are frequently used with biaxial accelerometers, but they suffer from a critical drawback: the sensitivity on the estimated values of φ and ρ depends on the value of φ and ρ itself, as shown in Fig. 3.4 and already highlighted by some but not all authors (see for instance [4, 5, 33]). To overcome this problem, the following trigonometric equations can be used to compute φ and ρ:

$$\begin{cases} \phi = \arctan\left(\dfrac{a_x}{\sqrt{a_y^2 + a_z^2}}\right), \\ \\ \rho = \arctan\left(\dfrac{a_y}{\sqrt{a_x^2 + a_z^2}}\right). \end{cases} \tag{3.46}$$

These equations guarantee that the accuracy is almost constant inside the whole range of values which φ and ρ can assume, as shown by the dotted and dashed lines in Fig. 3.4.

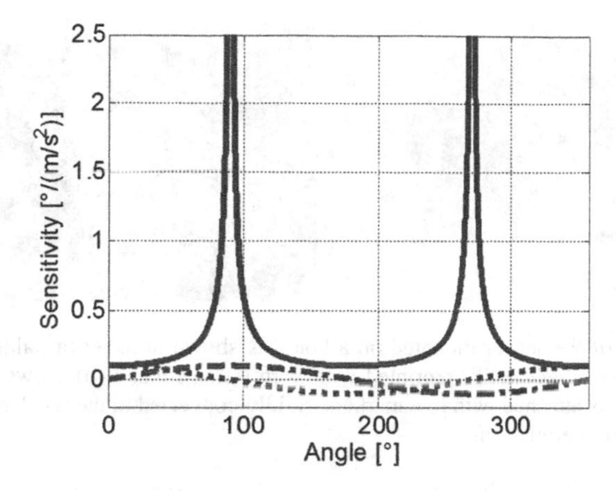

Fig. 3.4 This figure shows the sensitivity of the measured acceleration as a function of the tilt angle. The continuous line refers to (3.45) and it expresses the sensitivity of φ with respect to a_X. The dotted and dashed lines refer to (3.46); the first expresses the sensitivity of φ with respect to a_X and the second one as a function of a_Z. For clarity, these two functions are computed for $a_Y = 0$ (© 2009 IEEE, reprinted with permission)

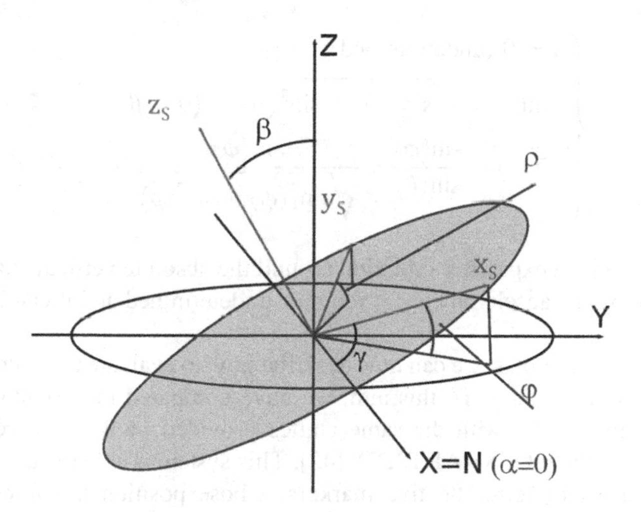

Fig. 3.5 The orientation of the MEMS accelerometer expressed both with the orientation angles ϕ and ρ, and with the Euler angles (α, β, γ). Because the value of α is irrelevant, it is arbitrarily set to 0; consequently, the nodes' axis N coincides with the X axis (© 2009 IEEE, reprinted with permission)

The two angles defined in (3.46) represent two independent orientation parameters in the sense that any error on the estimate on φ does not influence the estimate on ρ and vice versa (Fig. 3.5). Another possible set of angles used to describe orientation is through the Euler's angles. Since these are defined as a sequence of rotations, they do not enjoy the same property of independence, as for the angle set (φ, r).

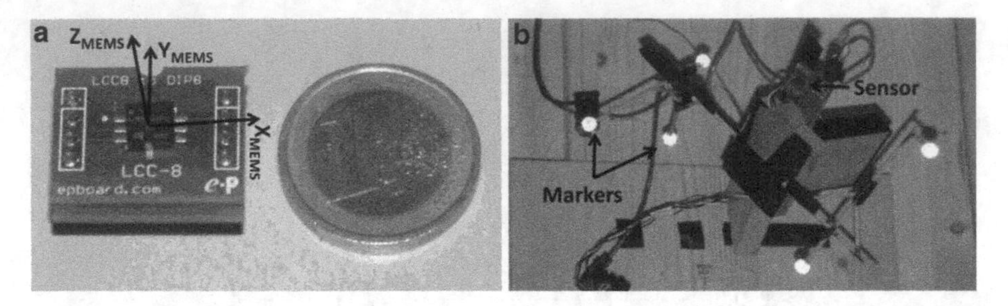

Fig. 3.6 A zoom of the sensor mounted on a board is shown in panel (**a**), along with its local reference system (© 2009 IEEE, reprinted with permission). Panel (**b**) shows the structure on which the sensor was attached with five markers, rigidly connected to the accelerometer (© 2009 IEEE, reprinted with permission)

In order to express the sensor orientation in terms of the Euler angles (α, β, γ), it is possible to derive the relationship between (φ, ρ) and (α, β, γ); this is for instance reported in Appendix A (A3 and A8) of [4]. The following relationships are obtained:

$$\begin{cases} \alpha = 0 \text{ (undetermined)}, \\ \sin(\beta) = \sqrt{\sin^2(\phi) + \sin^2(\rho)}, \quad (0 \le \beta \le \pi), \\ \sin(\gamma) = \dfrac{\sin(\phi)}{\sin(\beta)} = \dfrac{\sin(\phi)}{\sqrt{\sin^2(\phi) + \sin^2(\rho)}}. \end{cases} \tag{3.47}$$

Since the angle α expresses a rotation around the absolute vertical axis, it cannot be measured by the accelerometer; it remains undetermined and it can be arbitrary set to zero.

Given these preambles, we can now describe how to evaluate the accuracy of the autocalibration procedure. To this aim, we have compared the orientation angles computed using (3.46), with the same angles provided by a commercial motion capture system, the BTS-SMART3D™ [42]. This system is able to compute the 3D position of a set of retroreflective markers, whose position is surveyed by six cameras. The working volume of the motion capture system was approximately 500 mm × 500 mm × 500 mm, which allowed accommodating different sensor orientations with optimal marker visibility. With this working volume, the markers are localized with an accuracy of 0.1 mm (rms error).

The markers and the accelerometer, rigidly connected to each other, were fixed on a frame that could be oriented in any direction, as shown in Fig. 3.6b. Five markers were located, noncoplanarly, in the vicinity of the accelerometer. The orientation of the supporting structure was then computed as the mean rotational component of the rigid motion undergone by the markers. The angular accuracy in the measurement of this orientation can be derived from the spatial accuracy in the localization of the markers. This can be done by determining the sensitivity of the angular displacement with respect to a spatial displacement, as described in [45]. For the adopted setup (five markers, spatial accuracy of 0.1 mm (rms), and a minimum distance of 100 mm from the accelerometer) the angular accuracy in the measurement of the orientation of the MEMS results to be better than 0.025° (rms).

The vertical direction of the motion capture reference system was carefully established to guarantee that it is parallel to gravity, by surveying two markers, fixed onto the wire of a plumb line, held along the vertical.

Once the vertical direction was determined, and the 3D position of the five markers on the accelerometer board was known for each orientation, the rotation matrix and the corresponding Euler angles (α, β, γ) for each orientation were computed. This has been done by using quaternions [46] that allow determining the rotation by solving a linear system, thereby guaranteeing the orthonormality of the obtained rotation matrix. From the Euler angles, we computed the angles φ_{REF} and ρ_{REF}, which define the reference orientation of the structure, and consequently of the accelerometer, in the motion capture reference system, by exploiting the relationships of (3.47). At the same time, for each sampled orientation, the angles φ_{MEMS} and ρ_{MEMS} were also estimated by processing the MEMS sensor output through (3.46).

The high accuracy of the motion capture system allowed taking its orientation measurements as the ground truth. For this reason, we adopted a comparative approach to evaluate the accuracy of the orientation angles φ and ρ, comparing the value output by the accelerometer to those computed through the motion capture system data.

The estimated orientation angles along with their standard deviation are reported in Table 3.9, which describes the metrological performance of the sensor in the attitude estimation. In particular, we compared the output of the accelerometer before calibration ("Factory" column), calibrated with a six parameter model (excluding therefore the axes misalignments) and with a nine parameters model. The error in the estimated angles ranges from −23.10° to +6.15° when factory calibration data are used. This bias decreases to the range from −1.54° to +1.15°, when the six-parameters model is used, and it further decreases to the range from −0.26° to +0.26° when the nine-parameters model is adopted. Such results clearly demonstrate the need for calibrating the sensor and, moreover, the need for including the axes misalignment terms in the sensor model.

To investigate the spatial distribution of the errors we defined the error angles, $\Delta\varphi$ and $\Delta\rho$, as:

$$\begin{cases} \Delta\phi = \phi_{\mathrm{MEMS}} - \phi_{\mathrm{REF}} \\ \Delta\rho = \rho_{\mathrm{MEMS}} - \rho_{\mathrm{REF}} \end{cases} \tag{3.48}$$

and plotted them in Fig. 3.7 for one of the sensors (#1). As it can be seen, errors do not show any particular dependence from the sensor orientation: by using factory calibration data (panel (a)) errors are distributed approximately inside a circle centered in [0, 0]. The dimension of this circle is greatly reduced when auto calibration is carried out, and, in particular, when the nine-parameters model is adopted (panel (b)). Panels (c) and (d) show that, thanks to the particular trigonometric formulation of (3.46), the error is isotropic, as it does not depend on

Table 3.9 Mean error and standard deviation of the measured orientation, expressed by angles φ and ρ (first and second section of the table) and with the Euler angles β and γ (third and fourth sections), for the four ST accelerometers calibrated with the factory calibration data (column "Factory"), with the six-parameters model (column "Model 6") and with the nine-parameters model (column "Model 9"). Errors are in degrees

Sensor #	# Samples (N)	Factory	Model 6	Model 9
φ: *mean error* $\pm$ *SD* (°)				
#1	72	-2.23 ± 6.84	$+0.24 \pm 0.53$	$+0.18 \pm 0.47$
#2	42	$+3.38 \pm 2.77$	$+0.15 \pm 0.60$	-0.21 ± 0.59
#3	35	-23.10 ± 3.37	-1.54 ± 1.81	-0.26 ± 1.44
#4	35	-14.90 ± 4.86	$+0.58 \pm 0.91$	-0.20 ± 0.74
ρ: *mean error* $\pm$ *SD* (°)				
#1	72	-0.05 ± 6.40	-0.09 ± 0.65	$+0.11 \pm 0.59$
#2	42	$+0.80 \pm 2.52$	$+0.41 \pm 0.51$	$+0.03 \pm 0.34$
#3	35	$+0.77 \pm 10.38$	$+1.15 \pm 1.66$	$+0.26 \pm 1.11$
#4	35	$+6.15 \pm 8.12$	-0.42 ± 0.68	$+0.00 \pm 0.50$
β: *mean error* $\pm$ *SD* (°)				
#1	72	$+1.47 \pm 11.17$	-0.01 ± 0.59	$+0.11 \pm 0.59$
#2	42	-0.34 ± 3.18	-0.09 ± 0.56	$+0.03 \pm 0.56$
#3	35	-8.34 ± 10.73	-0.57 ± 1.34	-0.31 ± 1.11
#4	35	-8.62 ± 12.60	$+0.31 \pm 0.56$	-0.03 ± 0.47
γ: *mean error* $\pm$ *SD* (°)				
#1	72	-1.99 ± 4.37	$+0.31 \pm 1.02$	$+0.23 \pm 0.81$
#2	42	$+5.28 \pm 7.22$	$+0.24 \pm 0.61$	-0.10 ± 0.37
#3	35	-27.11 ± 13.38	-0.94 ± 6.48	-0.46 ± 1.76
#4	35	-18.81 ± 22.34	$+0.99 \pm 1.81$	-0.12 ± 0.45

the values assumed by φ and ρ. Similar results are obtained for the other calibrated accelerometers.

The results reported in the column "Model 6" of Table 3.9 show a residual error of the order of few degrees on the orientation measurements, while measurement error was as large as 20° with factory calibration data (sensor #3). However, accuracy can be further improved considering also the misalignment term in the sensitivity matrix, leading to the nine-parameters model (column "Model 9"). With this model, the accuracy was reduced to less than one degree (±0.26°). The improvement with respect to the six-parameters model is therefore almost of one order of magnitude and consistent in all the calibration experiments. This fact, together with the very low uncertainty in the parameter estimate (see Tables 3.6, 3.7 and 3.8), can be reasonably interpreted as a better model-fitting capability of the nine-parameters model, with respect to the six-parameters model. This allows drawing the conclusion that the axes misalignment parameters in the model allow a better fitting of the physical sensor behavior.

The value of sensitivities and biases obtained through auto calibration is generally close but not equal to that provided by the manufacturer; typical differences are in the order of ±10 % for the bias and ±5 % for the scale factor (cf. Tables 3.6, 3.7 and 3.8).

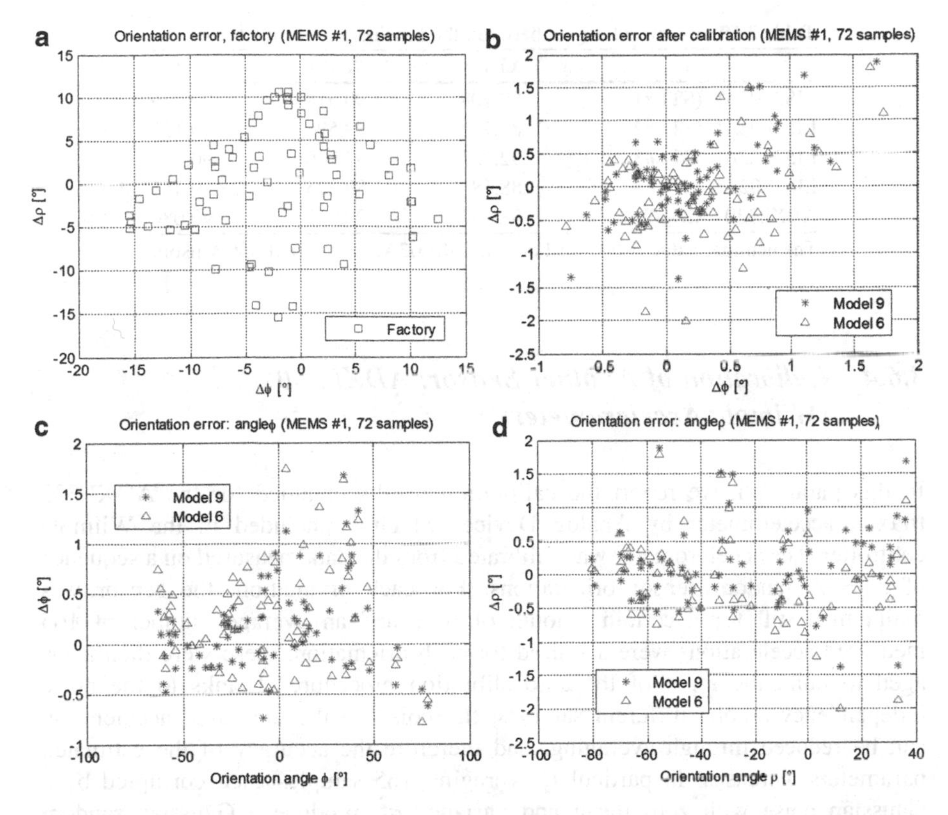

Fig. 3.7 Error in the estimate of φ and ρ, for the 72 orientations measured for sensor #1 (© 2009 IEEE, reprinted with permission)

However, when they are used to compute the sensor orientation, this error is amplified: the error was in some cases larger than 20º (sensor #3 in Table 3.9), with standard deviation exceeding 10°, when the values of **S** and **o** given by the factory (and null axes misalignment) were used. Nevertheless, factory parameters can be used as a reliable initialization point for Newton's optimization and a small number of iterations are sufficient to obtain a reliable and accurate estimate of the sensor parameters.

3.6.3 From the Sensitivity Matrix to the Axes Misalignments

From the inverse sensitivity matrix $\mathbf{S}^{-1}$ computed in calibration, the matrix **S** is easily obtained and the axes misalignments can be computed from the decomposition of **S** into a rotation matrix and a lower triangular matrix [see (3.9)–(3.11)]. This also allows computing the sensitivity of each sensor channel. An example of such decomposition is illustrated at the beginning of the chapter. The estimated angles between the axes of the accelerometers considered here are reported in Table 3.10. For the LIS3L02AL accelerometer, deviations from the orthogonality as large as 3° have been measured.

Table 3.10 Estimated angles between the sensing axes of the sensor

	XY (°)	*XZ* (°)	*YZ* (°)
LIS3L02AL (ST) #1	89.108	90.518	90.059
LIS3L02AL (ST) #2	88.778	89.593	90.155
LIS3L02AL (ST) #3	92.737	92.926	90.943
LIS3L02AL (ST) #4	88.681	89.225	89.453
ADXL330 (AD—Wii)	88.387	90.035	90.080

The nominal value is 90° for both the LISL02AL and ADXL330 sensors

3.6.4 Calibration of Another Sensor: ADXL330 (Wiimote Accelerometer)

In this paragraph, we report the calibration results obtained for the ADXL333 triaxial accelerometer by Analog Device, which is included in the Wiimote controller. The accelerometer was calibrated from the data measured on a sequence of $N = 49$ random sensor orientations. For each orientation, the sensor was maintained still for a certain amount of time and an average number of 165 measured accelerations were acquired for each orientation; these were then averaged to build the input of the autocalibration procedure. Thanks to the noise independency among different samples, the noise on the measured acceleration can be reduced through averaging, and therefore the accuracy of the estimated parameters increases. In particular, averaging 165 samples each corrupted by a Gaussian noise with zero mean and variance σ^2, produce a Gaussian random variable with variance reduced to $\sigma^2/165$. For the sensor considered here, this is indispensable to get a reliable estimate of the axes misalignments. Results of the calibration (estimated values together with their estimated standard deviations) are reported in Tables 3.6, 3.7 and 3.8, whereas in Table 3.10 the estimated angles between the sensor axes are reported.

Notice that the misalignments of the axes of the ADXL330 sensor results to be smaller than that of the LIS3L02AL sensor. This is coherent with what reported in the sensor datasheets [27, 28]: a 2 % cross-axis term is expected for the LIS3L02AL sensor, whereas a 1 % cross-axis term is declared for the ADXL330 accelerometer. Notice, however, that the second sensor is noisier than the first one; as a consequence, the uncertainty of the off-diagonal terms in $\mathbf{S}^{-1}$ is comparable with the estimated values for these terms (see Table 3.8—for instance, the standard deviation of $s^{-1}{}_{XZ}$ is 72 % of its estimated value). This fact suggests that, when a noisy sensor is used, the six parameter model could provide a more accurate estimate of the sensor parameters. In fact, the nine-parameters model can be used even in this case, but the estimate of the cross-axis term would be unreliable. Notice, at last, that without averaging the 165 measured accelerations, the standard deviations reported in Table 3.9 should be about 12 times larger; this would make the estimate of the axes misalignments completely unreliable.

3.7 Conclusion

We summarize here the main aspects involved in the autocalibration of MEMS accelerometers, described in this chapter. Each of this has to be carefully considered before designing a proper autocalibration procedure, taking into account the specific application and the sensor characteristics.

3.7.1 Choice of the Sensor Model

The choice of the most adequate sensor model depends on two main factors: the specific application in which the sensor is involved, and the noise characteristics of the sensor. The simplest model include only one sensitivity and one offset parameters, and it is suited for applications like [9], where the accelerometer is used to roughly detect the posture of a human being; because of the low number of parameters, this model can be adopted also for very noisy sensors. On the opposite, the most refined sensor models may include nonlinearities, hysteresis [16] or even the electrical cross talk between the channels [4]. The analysis of the literature, however, highlights that models with more than nine parameters often lead to an unreliable estimate of the parameters, and therefore do not contribute to increase the overall accuracy of the sensor. As a general rule, we can state that the nine parameter model, including three offsets and sensitivities and the axes misalignments, is generally suited for most applications and produces an accurate estimate of the model parameters. In case of noisy sensor, like the ADXL330 by Analog Device considered here, even this model can be over-parameterized; in this case, a proper noise reduction strategy has to be adopted; otherwise, a simpler six parameters model has to be used.

3.7.2 Modeling the Sensor Noise

Many authors assume the noise on the measured acceleration vector to be distributed as a zero mean, Gaussian vector. This is actually true, as shown in (3.15). However, we have also clearly demonstrated that, starting from the hypothesis of Gaussian vector on the measured sensor output, the squared norm of the measured acceleration vector is affected by a zero mean Gaussian component plus a χ^2 component, whose mean differs from zero and it is proportional to the variance of the sensor noise. In many practical cases, the bias introduced he by this component is little when compared with the static gravity acceleration (see Table 3.1) and it can be neglected during the calibration procedure. However, when small accelerations have to be measured, as for instance illustrated by the example in Fig. 3.2, neglecting this term can strongly bias the computation performed from the measured accelerations (see Table 3.2).

3.7.3 The Right Value of g

All the autocalibration procedures are based on the assumption that, in static conditions, the norm of the measured acceleration vector equals the value of the gravity acceleration, g. However, only a few authors explicitly consider that the value of g strongly depends by the latitude, altitude and other factors. Latitude and altitude, in particular, are responsible for the major changes in the values of g, which can go beyond 0.5 % of its nominal values. Equation (3.24) provides an accurate estimate of g as a function of these parameters. Other phenomena, like tidal or gravity anomalies, are responsible of minor changes in the value of g, and can generally be neglected.

Correcting the value of g before calibration is actually not necessary when the accelerometer is used as a pure tilt sensor; in fact, in this case the only useful information is represented by the orientation of the measured acceleration vector; however, when the sensor is used in quasi static condition in a IMU, the norm of the acceleration vector is taken in consideration: in this case, correcting the value of g is absolutely necessary; otherwise, drifts of hundred of meters can be accumulated in less than 1 min.

3.7.4 Accuracy of the Estimated Parameters

The covariance analysis illustrated in the chapter represents an easy-to-use and powerful tool to determine the effectiveness of the calibration procedure. More generally, such analysis permits to determine whether some parameters of the adopted model cannot be reliably estimated; based on this information, one can decide to reduce the parameters of the sensor model, or to increase the size of the dataset used for calibration. Overall, the covariance analysis should always be performed after calibration to identify possibly critical situations.

3.7.5 Axes Misalignments from Calibration Data

From the calibration data, and in particular from the decomposition of the sensitivity matrix into a rotation and a lower triangular matrix, it is possible to derive the angles between the sensing axes of the accelerometer. This information can be useful for the producer, that has to control the consistency of the produced sensors; it can also be used to verify that the calibration has produced reasonable values of the sensor parameters; at last, it can be used to verify the correct alignment of monoaxial and biaxial accelerometers, that are often manually composed to construct a triaxial sensor.

References

1. Rudolf F, Frosio R, Zwahlen P, Dutoit B (2009) Accelerometer with offset compensation. US Patent 2009/0223276 A1, 10 September 2009
2. Olney A (2010) Evolving MEMS qualification requirements. In: Proceedings of IEEE international reliability physics symposium (IRPS), pp 224–230, Wilmington, MA, USA, May 2010
3. Tadigadapa S, Mateti K (2009) Piezoelectric MEMS sensors: state-of-the-art and perspectives. Meas Sci Technol 20:092001 (30 pp)
4. Frosio I, Pedersini F, Stuani S, Borghese NA (2009) Autocalibration of MEMS accelerometers. IEEE Trans Instrum Meas 58(6):2034–2041
5. Łuczak S, Oleksiuk W, Bodnicki M (2006) Sensing tilt with MEMS accelerometers. IEEE Sensor J 6(6):1669–1675
6. Giansanti D, Maccioni G, Macellari V (2005) The development and test of a device for the reconstruction of 3-D position and orientation by means of a kinematic sensor assembly with rate gyroscopes and accelerometers. IEEE Trans Biomed Eng 52(7):1271–1277
7. Pérez R, Costa Ú, Torrent M, Solana J, Opisso E, Cáceres C, Tormos JM, Medina J, Gómez EJ (2010) Upper limb portable motion analysis system based on inertial technology for neurorehabilitation purposes. Sensors 10(12):10733–10751
8. Roetenberg D, Slycke PJ, Veltink PH (2007) Ambulatory position and orientation tracking fusing magnetic and inertial sensing. IEEE Trans Biomed Eng 54(5):883–890
9. Curone D, Bertolotti GM, Cristiani A, Secco EL, Magenes G (2010) A real-time and self-calibrating algorithm based on triaxial accelerometer signals for the detection of human posture and activity. IEEE Trans Inf Technol Biomed 14(4):1098–1105
10. Utters JC, Schipper J, Veltink PH, Olthuis W, Bergveld P (1998) Procedure for in-use calibration of triaxial accelerometers in medical applications. Sensor Actuator A68:221–228
11. Camps F, Harasse S, Monin A (2009) Numerical calibration for 3-axis accelerometers and magnetometers. In: Proceedings of the IEEE international conference on electro/information technology, pp 217–221, Toulouse, France, June 2009
12. Tan CW, Park S (2005) Design of accelerometer-based inertial navigation systems. IEEE Trans Instrum Meas 54(6):2520–2530
13. Syed ZF, Aggarwal P, Goodall C, Niu X, El-Sheimy N (2007) A new multi-position calibration method for MEMS inertial navigation systems. Meas Sci Technol 18:1897–1907
14. Aggarwal P, Syed Z, Niu X, El-Sheimy N (2006) Cost-effective testing and calibration of low cost MEMS sensors for integrated positioning, navigation and mapping systems. In: Shaping the change XXIII FIG Congress Munich, Germany, 8–13 October 2006
15. Wang J, Liu Y, Fan W (2006) Design and calibration for a smart inertial measurement unit for autonomous helicopters using MEMS sensors. In: Proceedings of the 2006 I.E. international conference on mechatronics and automation, Luoyang, China, 25–28 June 2006
16. Lang P, Pinz A (2005) Calibration of hybrid vision/inertial tracking systems. In: Proceedings of the second InerVis: workshop on integration of vision and inertial senors, Barcelona, Spain, 18 April 2005
17. Dong Z, Zhang G, Luo Y, Tsang CC, Shi G, Kwok SY, Li WJ, Leong PHW, Wong MY (2007) A calibration method for MEMS inertial sensors based on optical tracking. In: Proceedings of the second IEEE international conference on nano/micro engineered and molecular systems, Bangkok, Thailand, 16–19 January 2007
18. Dong Z, Uchechukwu C, Wejinya C, Zhou S, Shan Q, Li WJ (2009) Real-time written-character recognition using MEMS motion sensors: calibration and experimental results. In: Proceedings of the 2008 I.E. international conference on robotics and biomimetics, Bangkok, Thailand, 21–26 February 2009.
19. Freedson P, Pober D, Janz KF (2005) Calibration of accelerometer output for children. Med Sci Sports Exerc 37(11 Suppl):S523–S530
20. King K, Yoon SW, Perkins NC, Najafi K (2008) Wireless MEMS inertial sensor system for golf swing dynamics. Sensor Actuator A 141:619–630

21. Zhao M, Xiong X (2009) A new MEMS accelerometer applied in civil engineering and its calibration test. In: The ninth international conference on electronic measurement & instruments ICEMI'2009, Shanghai, China, 2:122–125, Aug. 2009
22. Ratcliffe C, Heider D, Crane R, Krauthauser C, Keun Yoon M, Gillespie JW Jr (2008) Investigation into the use of low cost MEMS accelerometers for vibration based damage detection. Compos Struct 82:61–70
23. Zanjani PN, Abraham A (2010) A method for calibrating micro electro mechanical systems accelerometer for use as a tilt and seismograph sensor. In: 2010 12th international conference on computer modelling and simulation, pp 637–641, Theran, Iran, March 2010
24. Krohn A, Beigl M, Decker C, Kochendorfer U, Robinson P, Zimmer T (2005) Inexpensive and automatic calibration for acceleration sensors. In: Ubiquitous computing systems, Lecture notes in computer science, vol 3598/2005, pp 245–258, DOI: 10.1007/11526858_19
25. Pylvanainen T (2008) Automatic and adaptive calibration of 3D field sensors. Appl Math Model 32:575–587
26. Golub GH, Van Loan CF (1996) Matrix computations, 3rd edn. Johns Hopkins, Baltimore. ISBN 978-0-8018-5414-9
27. STMicroelectronics (2004) LIS3L02AL 3axis-2g linear accelerometer [Online]. http://www.st.com/stonline/
28. Analog Devices (2006) ADXL330 3-axis-3g linear accelerometer [Online]. http://www.analog.com/static/imported-files/data_sheets/ADXL330.pdf
29. Nintendo Wii console web site [Online]. http://www.nintendo.com/wii
30. Shen SC, Chen CJ, Huang HJ (2010) A new calibration method for MEMS inertial sensor module. In: The 11th IEEE international workshop on advanced motion control, Nagaoka, Japan, 21–24 March 2010
31. Ketelhohn CH (2004) Accelerometer-based infant movement monitoring and alarm device. US Patent 6765489, 20 July 2004
32. Kolen PT (2008) Infant SID monitor based on accelerometer. Publication No. US 2008/0262381 A1, 23 October 2008
33. Luczak S, Oleksiuk W (2007) Increasing the accuracy of tilt measurements. Eng Mech 14(1):143–154
34. Ang WT, Khosla PK, Riviere CN (2007) Nonlinear regression model of a low-g MEMS accelerometer. IEEE Sensor J 7(1):81–87
35. Boynton R (2001) Precise measurement of mass. In: 60th annual conference of the society of allied weight engineers, SAWE Paper no. 3147, Arlington, TX, 21–23 May 2001
36. NASA GRACE mission homepage [Online]. http://www.csr.utexas.edu/grace/
37. Ramprasad T (2007) The earth's gravitational field, refresher course on marine geology and geophysics [Online]. Lecture Notes, pp 191–195. http://drs.nio.org/drs/handle/2264/720
38. Wu ZC, Wang ZF, Ge Y (2002) Gravity based online calibration for monolithic triaxial accelerometers' gain and offset drift. In: Proceedings of the 4th world congress intelligent control and automation, pp 2171–2175, Shanghai, China, June 2002
39. Madsen K, Nielsen HB, Tingleff O (2004) Methods for non-linear least squares problems, 2nd edn. IMM Technical University, Lyngby, Denmark
40. Dumas N, Azaïs F, Mailly F, Nouet P (2009) A method for electrical calibration of MEMS accelerometers through multivariate regression. In: Mixed-signals, sensors, and systems test workshop, 2009. IMS3TW '09. IEEE 15th International, 10–12 June 2009, pp 1–6
41. Dumas N, Azaïs F, Mailly F, Nouet P (2008) Evaluation of a fully electrical test and calibration method for MEMS capacitive accelerometers. In: Proceedings on IEEE mixed-signals, sensors, and systems test workshop, pp 1–6, Vancouver, BC, Canada, June 2008
42. BTS website [Online]. http://www.bts.it
43. Winter D (1990) Biomechanics and motor control of human movement. Wiley, New York
44. Press WH, Teukolsky SA, Vetterling WT, Flannery BP (2003) Numerical recipes: the art of scientific computing, 3rd edn. Cambridge University Press, New York
45. Barford NC (1985) Experimental measurements: precision, error and truth. Wiley, Hoboken, NJ
46. Alonso M, Finn EJ (1974) Fundamental university physics. Fields and waves, vol 2. Inter European Editions, Amsterdam

Chapter 4
Miniaturization of Micromanipulation Tools

Brandon K. Chen and Yu Sun

Abstract To enable the manipulation of objects below micrometers, miniaturization of MEMS tools is crucial. The process of device miniaturization, however, poses several challenges that are yet to be overcome. Due to force scaling, the significant increase in surface forces also demands the development of new manipulation strategies. This chapter provides a summary of the difficulties associated with this miniaturization process as well as an up-to-date review on recent progress.

Abbreviations

CNT	Carbon nanotube
FIB	Focused ion beam
EBID	Electron beam induced deposition
SOI	Silicon on insulator
KOH	Potassium hydroxide
DRIE	Deep reactive ion etching
DEP	Dielectrophoresis
SEM	Scanning electron microscope
SE	Secondary electron

4.1 Introduction

Manipulation at the micrometer scale provides a bridge between human and a world only visible under microscopes. It has the potential to extend the dexterity of a human hand, enable physical interaction for material characterization, in situ

B.K. Chen • Y. Sun (✉)
University of Toronto, 5 King's College Road, Toronto, ON M5S 3G8, Canada
e-mail: brandon.chen@utoronto.ca; sun@mie.utoronto.ca

D. Zhang (ed.), *Advanced Mechatronics and MEMS Devices*, Microsystems,
DOI 10.1007/978-1-4419-9985-6_4, © Springer Science+Business Media New York 2013

sample preparation and manipulation, as well as for the prototyping of novel micro- and nanosystems. A number of MEMS based manipulation tools have been developed over the past three decades, mostly for proof-of-concept purposes with a few targeting specific manipulation applications at the meso- and microscales. Manipulation of submicrometer sized objects, however, poses several challenges that are still subjects of present research. These challenges include both the construction of miniaturized tools and the use of these tools for micronanomanipulation.

4.2 Fabrication

The manipulation of objects at the sub-micrometer scale requires end-effectors comparable in dimensions. Miniaturization of entire MEMS devices has several drawbacks. This includes poor structural aspect ratio that results in undesired out-of-plane bending, reduced actuation and sensing performance from reduced device surface area or volume, and poor structure integrity that makes device handling difficult.

These problems with overall device miniaturization can be overcome by selectively miniaturize only the tool tip alone. Fabrication methods for miniaturizing tool tips are summarized as follows.

4.2.1 Serial Fabrication

Manual assembly and local material deposition are both able to create sub-micrometer sized gripping structures on a larger end-effector. High-aspect-ratio nanomaterials such as carbon nanotubes (CNT) or nanowires can be manually assembled onto end-effectors through the pick-place operation, extending the gripping arm with sub-micrometer sized structures. Kim and Lieber assembled two CNTs onto two micrometer-sized electrodes to create nanotweezers [1] (Fig. 4.1a). Actuation is achieved by applying a potential difference between the two CNTs to generate electrostatic force that closes the tweezer tips. Methods for

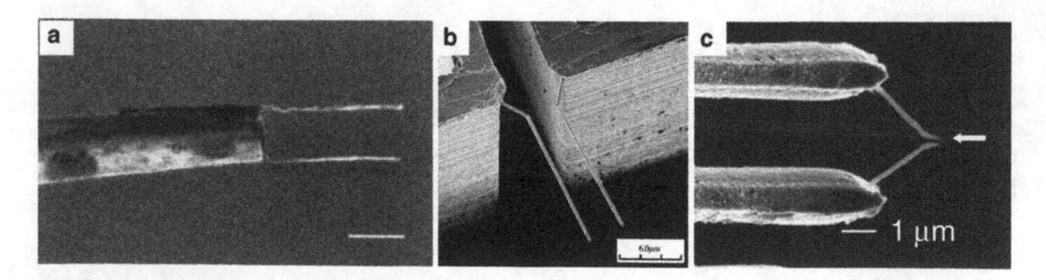

Fig. 4.1 Serially assembled/grown tool tips using: (**a**) Carbon nanotubes (© Science 1999). (**b**) FIB deposition (© IOPP 2009). (**c**) EBID (© IOPP 2001). All reprinted with permissions

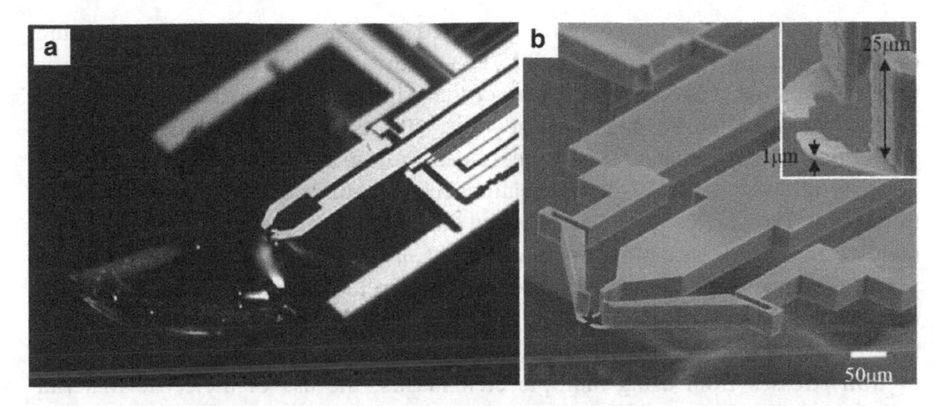

Fig. 4.2 Batch fabricated gripping devices with miniaturized tips. (**a**) Silicon nanotweezer for DNA manipulation (© IEEE 2008). (**b**) Microgripper with fingernail-like gripping tips (© IEEE 2010). All reprinted with permissions

controlling the length of the CNT tweezer arms with a cutting resolution of a few hundreds of nanometers using electrochemical etching was also reported [2].

An alternative serial based approach involves local deposition of materials, such as focus ion beam (FIB) [3] or electron beam induced deposition (EBID) [4] to create miniaturized gripping tips (Fig. 4.1b, c). The approach offers flexibility in terms of gripping tip shape construction since the deposition process can be well controlled by the user.

The challenges with both of these serial based fabrication approaches include long fabrication time, alignment of the tweezer tips, consistency of fabrication, and low force output of the devices due to poor radial stiffness.

4.2.2 Batch Fabrication

Different from serial fabrication approaches that construct the structures of each device one at a time, batch fabrication methods permit more devices to be constructed in parallel and often with better consistency.

A batch fabrication process that utilizes silicon anisotropic etching was proposed for the construction of nanotweezers [5] (Fig. 4.2a). Utilizing silicon-on-insulator (SOI) wafers, the process uses potassium hydroxide (KOH) wet etching to produce sharp tweezer tips and uses deep reactive ion etching (DRIE) for forming all other high aspect ratio structures to achieve satisfactory actuation and sensing performance. The process maintains the advantages of having high aspect ratio structures; furthermore, it is capable of producing nanometer-sized gripping tips for interacting with nanomaterials. The nanotweezer tips of the device were used as electrodes for dielectrophoresis (DEP) trapping of DNA. The fabrication process is complex, and the tweezer tips geometry cannot be readily altered in shape and are not suitable for physical grasping of nano objects.

Another fabrication process was proposed to create fingernail-like structure on microgripping tips [6] (Fig. 4.2b). The feature of this fabrication process is the use of buried oxide layer to construct the gripping tips while using the device silicon layer to create high aspect ratio structures for maximized sensing and actuation performance. The process can be applied to existing microdevice designs, allowing selected device feature be miniaturized to sub-micrometer in thickness. The device has demonstrated the pick and place of 100 nm gold nanospheres inside scanning electron microscope (SEM). The advantage of the process includes high fabrication yield due to the relative simplicity of the fabrication process, and offers flexibility in gripping tip geometry design. The disadvantage of the design includes the need for postprocessing to balance the buildup of film stresses from using multiple etch masks, the use of dielectric material as gripping tips that may causes charge buildup inside the SEM, and the gripping tip dimensions limited by the standard photolithography resolution. Electron-beam lithography or postprocessing using FIB can be used to reduce the gripping tip dimensions. However, these postprocessing steps lengthen the total fabrication time.

4.3 Applications

Besides the fabrication of nanomanipulation tools, the development of manipulation strategies is also an active area of research to work under microscopy and tackle strong adhesion forces at the micro- and nanometer scales.

4.3.1 Imaging Platform

The imaging resolution of an optical microscope is diffraction limited. Thus, electron microscopes have been used as a powerful imaging platform for nanomanipulation. Out of different types of electron microscopes, the SEM is particularly suitable for micro- and nanometer scales manipulation. It offers real-time imaging capability with a resolution down to nanometer scale, isolated high vacuum environment with minimum external disturbances, relatively large vacuum chamber with sufficient space for installing nanomanipulators for precise positioning, and requires minimal sample preparation prior to imaging. The use of SEM as a nanomanipulation platform, however, brings about several challenges.

Electron–solid interaction is the main source of problems inside SEM, especially when the manipulation tool or sample contains dielectric materials. Due to electron irradiation, charges accumulate on the surface of dielectric materials, resulting in either positive or negative charging. This charging effect induces complex electrostatic interactions between objects. In the case of negative charging, incoming electrons are deflected, causing image distortion and poor imaging resolutions.

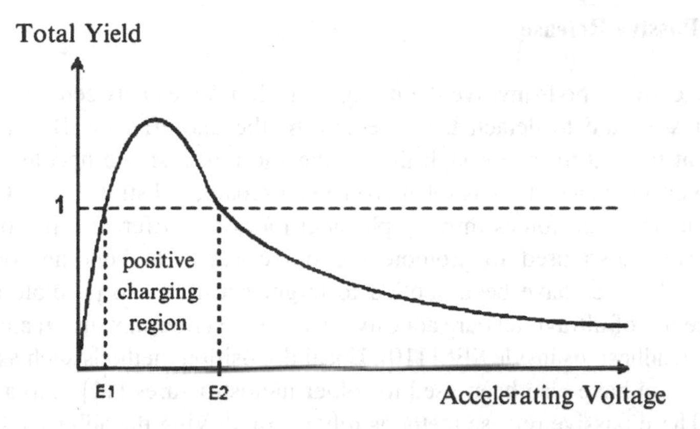

Fig. 4.3 Dielectric sample under electron irradiation can charge up positively or negatively depending upon the accelerating voltage used

To avoid negative charging, it is common practice to coat the specimen with either gold or carbon film, but this is not always possible to perform on samples for manipulation purposes.

To minimize the negative charging effect, choosing a suitable accelerating voltage becomes crucial. Figure 4.3 shows the qualitative relationship between charging and accelerating voltages. When the total electron yield is below one, more electrons are stored in the dielectric material than are scattered. Hence, the sample charges negatively. When the total yield is above one, the dielectric material is charged positively. When the total yield is equal to one, the charge is neutral. The neutral charging of sample occurs at accelerating voltage, E1 and E2, which must be experimentally determined. In practice, the suitable choice of accelerating voltages to minimize charging lies between E1 and E2.

EBID is another concern to account for when working inside SEM. EBID is the break down of gas molecules by electron beam and redeposition of them onto the nearby surfaces. Having a layer of contamination deposited on the tool can affect the tool's electrical conductivity, and also can be act as a soldering technique that bonds objects tighter together, both of which are undesirable for certain applications.

4.3.2 Surface Adhesion

Due to force scaling, surface adhesion forces (e.g., capillary, electrostatics, and van der Waals forces) dominate volumetric forces (e.g., gravity), preventing the rapid and accurate release of micro- and nanometer sized objects. Several strategies have been proposed to deal with this problem, and these strategies can be classified into passive release methods and active release methods.

4.3.2.1 Passive Release

Passive release methods involve the use of adhesion forces between the object and the target substrate to detach the object from the end-effector. By altering the adhesion at the interfaces through the rolling motion, a single needle probe was used for pick and place of microspheres on an Au-coated substrate [7, 8]. Other than altering the adhesion forces through physical motions, different types of fixation methods were also used to promote the object-substrate bonding for release. Chemical adhesives have been applied to target substrates to promote adhesion, such as the use of ultraviolet cure adhesive in ambient environment [9] and electron beam cured adhesives inside SEM [10]. Local deposition methods such as EBID or FIB deposition have also been used to solder nanostructures [11] onto a substrate directly. These passive release methods rely on modifying the adhesive properties of different surfaces through either physical or chemical means, which can be time consuming, poor in repeatability, and difficult to extend to other manipulation applications.

Double-ended tools such as microgrippers can apply grasping motions that significantly ease the pick-up procedure. However, object release remains to be a challenge. It is generally desirable for a gripping tool to have limited adhesive properties to allow object release. To achieve this objective, micropyramids on tool surfaces were used to reduce tool–object contact area and reduce adhesion forces [12]. Similarly, chemical coating can be applied to reduce adhesion forces [13]. However, the effectiveness of gripping tip treatment for release is limited since often the reduced amount of adhesion forces is still significant enough to keep the microobject adhered to a gripping tip.

4.3.2.2 Active Release

Active release methods involve the application of an external force to overcome the object–tool adhesion to facilitate release. For situation where the tool, the object, and the substrate are all electrically conductive, the coulomb interaction induced by applying potential differences between the tool tips and the substrate can be used to help overcome the tool–object adhesion [14, 15]. However, once the released object comes in contact with the substrate, it would reverse in charge and "fly" back towards the tool tips, resulting in the bouncing back and forth of the object between the tool tips and substrate and eventually flying away. It was also reported that a voltage sequence can be applied to minimize this object bouncing effect [16], but only 25 % of released microobjects were found after release [17].

Another active release method utilizes the large bandwidth of piezoelectric actuators to produce mechanical vibration [18, 19]. This approach takes advantage of the inertial effect of both the end-effector and the object to overcome adhesion forces. The release repeatability and accuracy, however, were not quantitatively reported.

Vacuum-based tools are able to create a pressure difference to attach and detach objects from a tool tip. For example, the negative pressure can be applied through a glass micropipette for manipulating micrometer-sized objects [20], or generated from heating and cooling of air in a cavity [21]. The miniaturization and accurate control of vacuum-based tools can be challenging, and its use in a vacuum environment can be limited.

For manipulation inside an aqueous environment, micro-Peltier cooler was used to form ice droplets for pick-place of microobjects [22]. Thawing of the ice droplet effectively released objects. This approach requires a bulky, complex end-effector and cannot be done outside of an aqueous environment.

MEMS-based microgripping tools with integrated plunging mechanism were demonstrated to impact a microobject to overcome the tool–object adhesion [23]. The technique was experimentally shown to have high release repeatability and accuracy, and amendable to automation in both ambient [24] and SEM environments [25]. However, the performance of this technique has yet to be characterized on soft materials or nonspherical objects.

4.3.3 Depth Detection

Standard microscopy techniques provide two-dimensional (*X–Y*) image feedback with a low depth of field. The depth information (*Z*) is crucial for physically manipulate and efficiently maneuver micro- and nanometer-sized objects. Several methods have been reported for estimating the depth information under optical microscopes. Discussion here focuses on depth estimation techniques used in SEM imaging.

4.3.3.1 Depth from Focus

Depth from focus is a commonly used method under optical microscopy. Applying the same method inside SEM resulted in sacrificed performance because of the significant increase in the depth of field of SEM. The method was used in SEM to determine the relative depth of different objects and can provide coarse depth estimation with a resolution of tens of micrometers [26]. The performance of this approach is highly dependent on SEM imaging conditions.

4.3.3.2 Touch Sensing

Touch sensors based on several working principles have been used for detecting the *Z* positioning of the substrate. For example, microgrippers with integrated capacitive contact sensor with a resolution of tens of nanometers [13]. A vibrating piezo bimorph cantilever changes in vibration amplitude when contacting the substrate [26]. The touch sensing approach offers high sensing resolutions in a wide range of environments, but extra efforts are needed to integrate them with different manipulation tools, such as epoxy gluing.

4.3.3.3 Shadow-Based Detection

As a manipulation tool moves towards a substrate inside an SEM, the shadow of the tool projected onto the substrate converges to a sharper and darker shade resembling the geometry of the tool. This brightness intensity change is due to the fact that the tool tips act as a shield that blocks the secondary electrons (SE) generated by the substrate to reach the SE detector, resulting in a darker region on the substrate. An example of the use of this technique was the alignment between a microgripping tips and CNT, where CNT appeared darker when positioned between the gripping arms [27]. This approach provides a rough estimation of relative depth positioning, but the performance varies greatly in dependence on the position of the SE detector relative to a region of interest.

4.3.3.4 Stereoscopic SEM Imaging

By taking two images of the same region of interest at two different prospectives, three-dimensional stereoscopic images can be constructed to estimate the depth information. This approach was realized by tilting the electron beam using an SEM equipped with special lenses [28], with an added-on beam tilting system [29], or by concentric tilting of the sample stage [30]. These methods require specialized hardware to be installed, and further efforts to ensure the robustness of the method are required.

4.3.3.5 Sliding-Based Detection

As a manipulation tool descends towards the substrate and establishes contact, the sliding movement of the tool on the substrate can be visually detected in the *X–Y* image plane. This approach is commonly done manually by a human operator to estimate tool–substrate contact, and was also automated in both the ambient environment [31] and the SEM environment [32]. The process allows quick and easy estimation of substrate *Z* positioning. The accuracy is dependent upon frictional forces at the tool–substrate interface and the quality of image feedback.

4.4 Conclusion

This chapter discussed a focused topic on methods and challenges in micro-nanomanipulation. The most recent literature on this topic was summarized and critiqued. The chapter discussed the miniaturization of manipulation tools.

Nanoscale physical manipulation requires tools of small sizes, with a few nanometers dimensional tolerance between the gripping tips to ensure properly tip alignment for secured grasping. This has not yet been achieved through existing fabrication processes. Construction of nanogrippers in parallel with a high yield has yet to be realized.

Existing object release techniques reported to date are limited to microscale manipulation. As dimensions of objects to be manipulated are below a few micrometers, the significant increase in surface forces relative to volumetric forces makes object release a major hurdle to overcome.

Depth information extraction is critical for reliable manipulation of micro- and nanometer-sized objects, as well as for automated operation. Most existing methods for estimating the depth information rely on the use of SEM imaging. Other methods involve integrating additional sensors to manipulation tools. The task of depth estimation inside SEM remains to be better solved before reliable nanomanipulation can be realized.

References

1. Kim P, Lieber CM (1999) Nanotube nanotweezers. Science 286:2148–2150
2. Lee J, Kim S (2005) Manufacture of a nanotweezer using a length controlled CNT arm. Sensor Actuator Phys 120:193–198
3. Chang J, Min BK, Kim J, Lee SJ, Lin L (2009) Electrostatically actuated carbon nanowire nanotweezers. Smart Mater Struct 18:065017
4. Boggild P, Hansen TM, Tanasa C, Grey F (2001) Fabrication and actuation of customized nanotweezers with a 25 nm gap. Nanotechnology 12:331–335
5. Yamahata C, Collard D, Legrand B, Takekawa T, Kumemura M, Hashiguchi G, Fujita H (2008) Silicon nanotweezers with subnanometer resolution for the micromanipulation of biomolecules. J Microelectromech Syst 17(3):623–631
6. Chen BK, Zhang Y, Perovic DD, Sun Y (2010) From microgripping to nanogripping. In: IEEE international conference on MEMS, Hong Kong, 24–28 January 2010
7. Miyazaki HT, Tomizawa Y, Saito S, Sato T, Shinya N (2000) Adhesion of micrometer-sized polymer particles under a scanning electron microscope. J Appl Phys 88:3330–3340
8. Saito S, Miyazaki HT, Sato T, Takahashi K (2002) Kinematics of mechanical and adhesional micromanipulation under a scanning electron microscope. J Appl Phys 92:5140–5149
9. Fuchiwaki O, Ito A, Misaki D, Aoyama H (2008) Multi-axial micromanipulation organized by versatile micro robots and micro tweezers. In: IEEE International conference on robotics and automation, Tokyo, 19–23 May 2008
10. SEM-compatible glue (2010) http://www.nanotechnik.com
11. Zhu Y, Espinosa HD (2005) An electromechanical material testing system for in situ electron microscopy and applications. PNAS 102(41):14503–14508
12. Arai F, Andou D, Nonoda Y, Fukuda T, Iwata H, Itoigawa K (1998) Integrated microendeffector for micromanipulation. IEEE/ASME Trans Mechatron 3(1):17–23
13. Kim K, Liu X, Zhang Y, Sun Y (2008) Nanonewton force-controlled manipulation of biological cells using a monolithic mems microgripper with two-axis force feedback. J Micromech Microeng 18:055013
14. Saito S, Himeno H, Takahashi K (2003) Electrostatic detachment of an adhering particle from a micromanipulated probe. J Appl Phys 93:2219–2224

15. Takahashi K, Kajihara H, Urago M, Saito S, Mochimaru Y, Onzawa T (2001) Voltage required to detach an adhered particle by coulomb interaction for micromanipulation. J Appl Phys 90:432–437
16. Saito S, Himeno H, Takahashi K, Urago M (2003) Kinetic control of a particle by voltage sequence for a nonimpact electrostatic micromanipulation. Appl Phys Lett 83(10):2076–2078
17. Saito S, Sonoda M, Ochiai T, Han M, Takahashi K (2007) Electrostatic micromanipulation of a conductive/dielectric particle by a single probe. In: IEEE conference on nanotech, Hong Kong, 2–5 August 2007
18. Haliyo DS, Regnier S, Guinot JC (2003) [mu]mad, the adhesion based dynamic micro-manipulator. Eur J Mech Solid 22:903–916
19. Haliyo DS, Rollot Y, Regnier S (2002) Manipulation of micro objects using adhesion forces and dynamical effects. In: IEEE international conference on robotics and automation, Washington, 11–15 May 2002
20. Zesch W, Brunner M, Weber A (1997) Vacuum tool for handling microobjects with a nanorobot. In: IEEE international conference on robotics and automation, Albuquerque, USA, 20–25 April 1997
21. Arai F, Fukuda T (1997) A new pick up and release method by heating for micromanipulation. In: IEEE international workshop on MEMS, Nagoya, Japan, 26–30 January 1997
22. Lopez-Walle B, Gauthier M, Chaillet N (2008) Principle of a sub-merged freeze gripper for microassembly. IEEE Trans Robot 24:897–902
23. Chen BK, Zhang Y, Sun Y (2009) Active release of micro objects using a MEMS microgripper to overcome adhesion forces. J Microelectromech Syst 18:652–659
24. Zhang Y, Chen BK, Liu X, Sun Y (2010) Autonomous robotic pick-and-place of micro objects. IEEE Trans Robot 26:200–207
25. Jasper D, Fatikow S (2010) Automated high-speed nanopositioning inside scanning electron microscopes. In: IEEE conference on automation science and engineering, Toronto, 21–24 August 2010
26. Eichhorn V, Fatikow S, Wich T, Dahmen C, Sievers T, Andersen KN, Carlson K, Boggild P (2008) Depth-detection methods for microgripper based CNT manipulation in a scanning electron microscope. J Micro-Nano Mech 4:27–36
27. Eichhorn V, Fatikow S, Wortmann T, Stolle C, Edeler C, Jasper D, Boggild P, Boetsch G, Canales C, Clavel R (2009) Nanolab: a nanorobotic system for automated pick-and-place handling and characterization of CNTs. In: IEEE international conference on robotics and automation, Kobe, 12–17 May 2009
28. Abe K, Kimura K, Tsuruga Y, Okada S, Suzuki H, Kochi N, Koike H (2004) Three-dimensional measurement by tilting and moving objective lens in CD-SEM(II). In: Proceedings of SPIE, Santa Clara, USA, 23 February 2004
29. Jahnisch M, Fatikow S (2007) 3-D vision feedback for nanohandling monitoring in a scanning electron microscope. Int J Optomechatron 1(4):26
30. Marinello F, Bariani P, Savio E, Horsewell A, Chiffre LD (2008) Critical factors in SEM 3D stereo microscopy. Meas Sci Tech 19:065705
31. Wang WH, Liu XY, Sun Y (2007) Contact detection in microrobotic manipulation. Int J Robot Res 26:821–828
32. Ru CH, Zhang Y, Sun Y, Zhong Y, Sun XL, Hoyle D, Cotton I (2010) Automated four-point probe measurement of nanowires inside a scanning electron microscope. IEEE Transactions on Nanotechnology 10:674–681

Chapter 5
Digital Microrobotics Using MEMS Technology

Yassine Haddab, Vincent Chalvet, Qiao Chen, and Philippe Lutz

Abstract Microrobotics deals with the design, fabrication, and control of microrobots to perform tasks in the microworld (i.e., the world of submillimetric objects). While end-effectors experienced considerable developments, few works concerned the development of microrobot architectures adapted to the microworld. Most of the current robots are bulky and are based on the miniaturization of traditional architectures and kinematics. In this chapter, we introduce a new approach for the design of microrobot architectures based on elementary mechanical bistable modules. This bottom-up approach called "digital microrobotics" takes advantage of MEMS technology and open-loop (sensorless) digital control to offer a flexible way to experiment various kinematics adapted to the microworld. A microfabricated bistable module is proposed and a complete digital microrobot is designed, modeled and fabricated. Digital microrobotics opens new perspectives in microrobots design and micromanipulation tasks.

5.1 Introduction

During the last decade, significant research activities have been performed in the field of microrobotics which deals with the design, fabrication and control of microrobots. These microrobots are intended to perform various tasks in the so-called Microworld (i.e., the world of submillimetric objects), in particular micromanipulation tasks of single objects (artificial or biological) for positioning, characterizing or sorting as well as for industrial microassembly. When the size of

---

Y. Haddab (✉) • V. Chalvet • Q. Chen • P. Lutz
Automatic Control and Micro-Mechatronic Systems Department (AS2M),
CNRS-UFC-UTBM-ENSMM, FEMTO-ST Institute, 24, rue Alain Savary,
Besançon 25000, France
e-mail: yassine.haddab@femto-st.fr; vincent.chalvet@femto-st.fr; qiaochen555@hotmail.com;
philippe.lutz@femto-st.fr

D. Zhang (ed.), *Advanced Mechatronics and MEMS Devices*, Microsystems,
DOI 10.1007/978-1-4419-9985-6_5,

the handled object is submillimetric, its interaction with the environment is strongly influenced by surface forces and the scale effect influences all microrobotics functions (actuation, perception, handling, and control). Achieving efficient robotic tasks at this scale remains a great challenge and requires some specificities:

- Resolution and accuracy in the submicrometric domain are needed in order to interact with micrometric objects. This is why methods and strategies used to build conventional robots are often not applicable in the microworld.
- New mechatronic approaches, new actuators, and robot kinematics are required. Researches done around the world have shown that the use of active materials to actuate microrobots gives better performances than the use of more traditional actuators. Piezoelectric materials, shape memory alloys (SMA), and active polymers have been successfully used to actuate various types of microrobots. However, despite their intrinsic high resolution, these active materials present some disadvantages, making the design of efficient controllers a hard task. Their behavior is often complex, nonlinear, and sometimes nonstationary. Closed-loop control of the microrobots requires the design and the integration of very small sensors and the use of bulky and expensive instruments for signal processing and real-time operating. Packaging and integration of the sensors and actuators is also a hard problem.

This is why building multiple-degrees-of-freedom microrobots able to perform complex tasks is difficult. Moreover, in many cases, the size of the robot itself has to be very small in order to manipulate microobjects.

The design of a microrobot contains two main parts with specific problems:

- Design of an end-effector and study of its interaction with the manipulated microobject
- Design of a suitable microrobot kinematics

Many research activities have been performed to develop end-effectors compatible with the microworld. Various strategies have been proposed to handle microcomponents. Some of them use microgrippers with shapes adapted to the manipulated objects [1], others use adhesion forces to raise microcomponents [2].

While handling experienced considerable developments, few works concerned the development of microrobot architectures adapted to the microworld. Most of current robots are bulky and based on the miniaturization of traditional architectures and kinematics. Their size is not really compatible with the microworld and limits considerably the use of microrobots to execute complex tasks in confined environments. However, some microrobots based on advanced technology have reached a high miniaturization level and allow performing accurate micromanipulation tasks in confined environments. Figure 5.1 shows two of the most successful microrobots.

In this chapter, we introduce a new approach for the design of microrobots architectures based on elementary mechanical bistable modules. This bottom-up approach called "digital microrobotics" takes advantage of MEMS technology and open-loop (sensorless) digital control to offer a flexible way to experiment various kinematics adapted to the microworld.

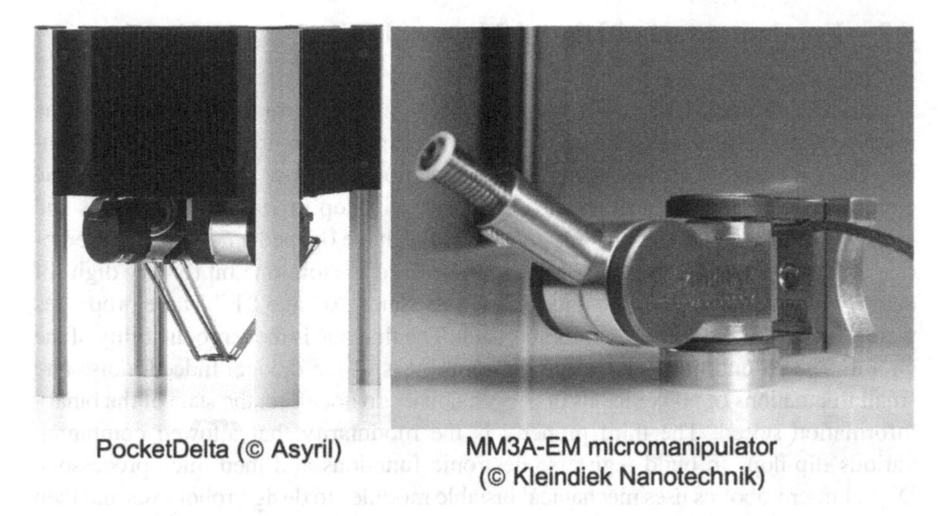

Fig. 5.1 Examples of microrobot architectures

In literature, several conventional robots based on the use of bistable mechanisms are described. However, in the microworld, there is no microrobot based on the bistability concept although bistable structures are widely used. In the macroworld, the most typical discretely actuated robot is the Variable Geometry Truss (VGT) manipulator [3–5]. This planar binary manipulator consists of several modules. Each one includes three binary actuators. As a result, every module has eight states (2^3). G.S. Chirikjian first presented the concept of the binary paradigm for robotic manipulators in [6]. There have been several improvements in this binary hyper-redundant manipulator concept. A lot of work about the calculation of the forward and inverse kinematics is done in [7–10]. Another robot called Binary Robotic Articulated Intelligent Device (BRAID) is described in [11]. It is a 3D configuration of binary actuated manipulators. 2^{15} states are obtained by cascading five modules in the 3D workspace. Nevertheless, the VGT and the BRAID present some drawbacks: trajectory following is costly in term of computation because the number of logical states is huge [7, 8] and the inverse kinematics is hard to obtain as indicated in [9] and [10]. Moreover, the use of joints leads to a repeatability loss due to the joint clearances or gaps. So this kind of structure is not adapted to the microworld. Indeed, assembled manipulators and joints should be avoided and compliant joints are preferred in order to achieve high performances adapted to the microworld [12, 13]. The development of digital microrobotics requires the availability of microfabricated bistable structures that can be connected by flexible joints. MEMS technology is well suited.

5.2 Fundamentals of Digital Microrobotics

Digital microrobotics aims to enable the design of microrobots adapted to the requirements of the work in the microworld. It is partially inspired by digital electronics. Digital electronics has completely revolutionized how to design electronic circuits. The development of the first electronic flip-flop in 1918 by W. Eccles and F.W. Jordan and the numerous improved versions have opened a new way to memorize and process data. A single flip-flop can be used to store one bit (binary digit) of data. The two states are commonly referred as states "0" and "1." Three properties were decisive for the success of this approach. The first one is the reproducibility of the circuit. The second one is the electrical robustness of the device. Indeed, noise and small fluctuations of power levels or trigger signals do not affect the state of the binary information stored. The third property is the modularity that allowed combining various flip-flops to build registers, electronic functions and then microprocessors. Digital microrobotics uses mechanical bistable modules to design robot axes and then complete microrobots and more complex devices.

5.2.1 Mechanical Bistable Module

A digital microrobot is built based on bistable elementary modules. Figure 5.2 shows the principle of a mechanical bistable module. It has two stable states. In state "0," the module has a length of l and in state "1," it has a length of $l + d$. The module can be switched from one stable state to the other by an external signal. The length of the module in the two states must be well known and repeatable.

5.2.2 Robot Axes Based on Bistable Modules

Using several bistable modules, robot axes can be built. The number of logical states is thus increased. The final position of the axis is obtained by the accumulation of the discrete displacements of the modules. The characteristics of the modules and the way they are linked define the characteristics of the reachable

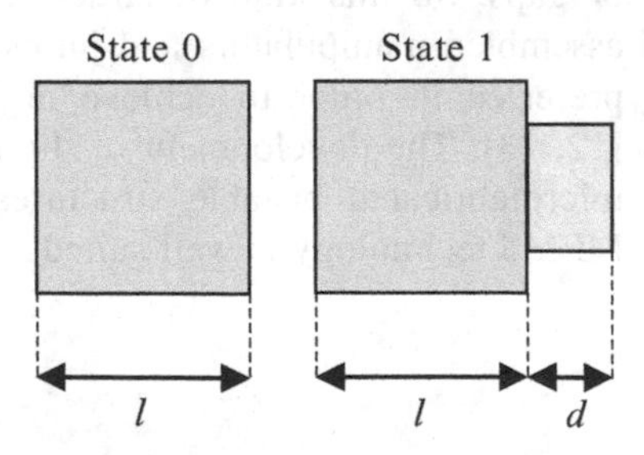

Fig. 5.2 Principle of a mechanical bistable module

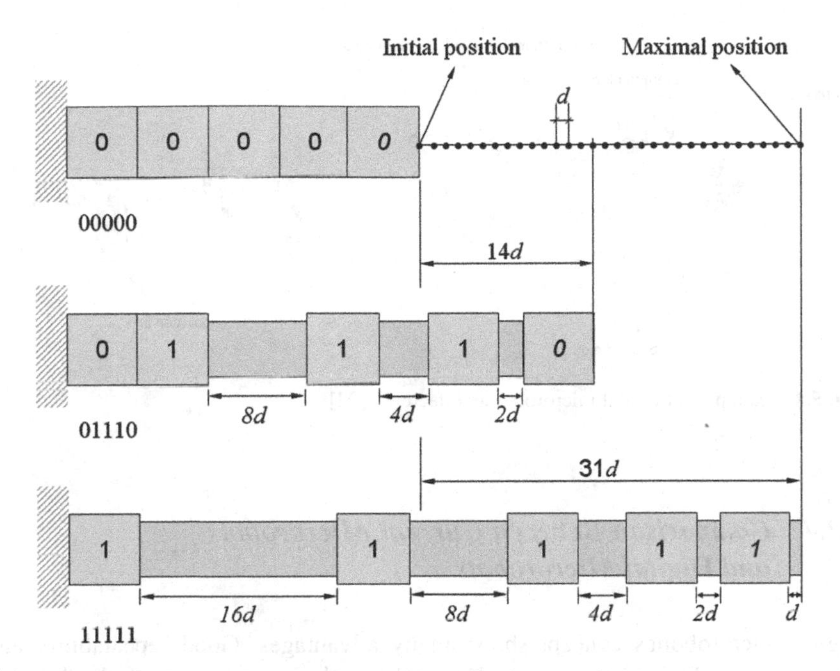

Fig. 5.3 A serial axis with five cascaded bistable modules [23]

workspace of the axes. Although only a discrete space can be reached, the number of logical states is an exponent function of the number of bistable modules. For an axis containing n bistable modules, the number of logical states is 2^n. A robust mechanical design allows high positioning resolution. Figure 5.3 shows a serial axis built using 5 cascaded modules. There are 32 logical states. If the displacements of the modules are defined as: d, $2d$, $4d$, $8d$ and $16d$, where d is the minimum displacement needed, the range of the axis is given by:

$$D = d.(1 \quad 2 \quad 4 \quad 8 \quad 16).\begin{pmatrix} 1 \\ 1 \\ 1 \\ 1 \\ 1 \end{pmatrix} = d.\sum_{i=1}^{5} 2^{i-1} = 31d.$$

The minimum displacement d is the resolution of the axis. The position of the tip of the robot axis is represented by a binary word that can be used as a control instruction.

5.2.3 Digital Microrobots

Combining several axes, various microrobot architectures can be obtained. Figure 5.4 shows examples of robot architectures.

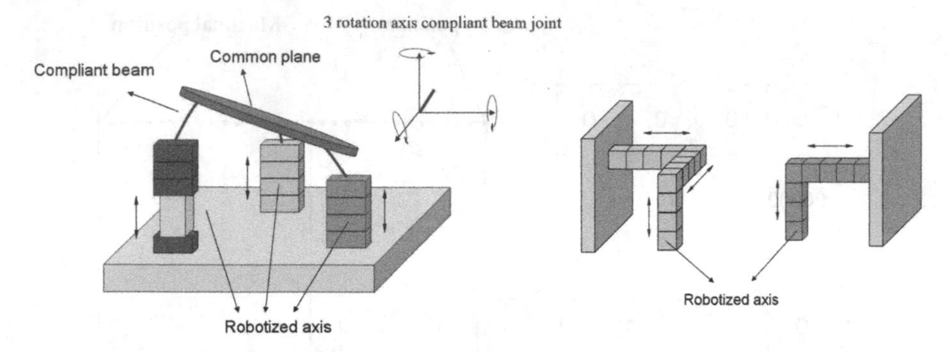

Fig. 5.4 Examples of digital microrobot architectures [23]

5.2.4 Comparison Between Current Microrobots and Digital Microrobots

Digital microrobotics concept shows many advantages. Good repeatability and accuracy are obtained thanks to the mechanical performances of the bistable modules. Neither proprioceptive sensors nor bulky and expensive instruments are needed. Low power consumption can be obtained if external energy is not needed to maintain the modules in a given state but only during the transition phases. Moreover, the immunity to noise and environment changes is improved. Parallel control of the modules allows fast displacements. Using bistable structures give an approach that turns the difficulties of nonlinear control into mechanical structure design. However, some drawbacks exist: error accumulation and discrete reachable area. Table 5.1 gives a comparison between the characteristics of current microrobots and the proposed digital microrobots. This comparison is based on information from [12, 14–17].

5.3 Design and Characterization of a Mechanical Bistable Module

Building bistable modules for microrobotics requires specific features not available in the existing designs. The open loop control approach requires that the two stable positions are well defined. This can be obtained only if the two stable positions are blocked. Control robustness is replaced by mechanical robustness. Generally, in microfabricated bistable modules, only one position is blocked because of the constraints of the monolithic microfabrication process.

Table 5.1 Comparison between current microrobots and digital microrobots

Characteristic	Current microrobots	Digital microrobots
Actuation	Proportional or incremental	Discrete actuation
Size	Medium	Small (microfabrication)
Cost (including control)	High to medium	Low
Control	Closed loop, nonlinear control	Open loop (no sensors needed)
Energy consumption	High to medium	Low
Sensitivity to noise	High to medium	Low
Use of sensors	Yes	No
Displacement	Continuous or discrete	Discrete

5.3.1 Structure of a Module

While the existing modules are adequate for microrelays [18] and microswitches [19–22], new structures with two blocking systems must be developed for microrobotics applications. In addition, the modules must have the following features:

- Easy fabrication using standard microfabrication processes
- Monolithic fabrication
- Easy duplication
- Blocking forces compatible with the range of forces in the microworld
- Possibility to combine several modules to build microrobots and complex functions

Figure 5.5 shows the designed bistable module for microrobotics applications. It is built in a SOI (Silicon On Insulator) wafer and it includes a bistable mechanism, thermal actuators and two stop blocks. The module is designed so that no external energy is required to maintain the mechanism in its stable states. The bistable mechanism is switched using two pairs of actuators. Thermal actuators are among the most used in MEMS. They have been chosen because they are easy to fabricate and allow large displacements and strong forces.

In order to obtain bistability in a monolithically fabricated mechanism and high blocking forces, the module requires an activation procedure (once for all) after fabrication. Figure 5.6 shows an example of fabricated module just after fabrication and after activation (an external force is applied on the shuttle to activate the structure). After that, the module can be switched between the two stable states using the thermal actuators.

In order to define the dimensions of the bistable mechanism, two factors are considered: the fabrication limits that define the minimal width of the beams of the structure and the size of the thermal actuators required to switch the bistable mechanism. According to the pseudo-rigid-body model [13, 19], a calculated

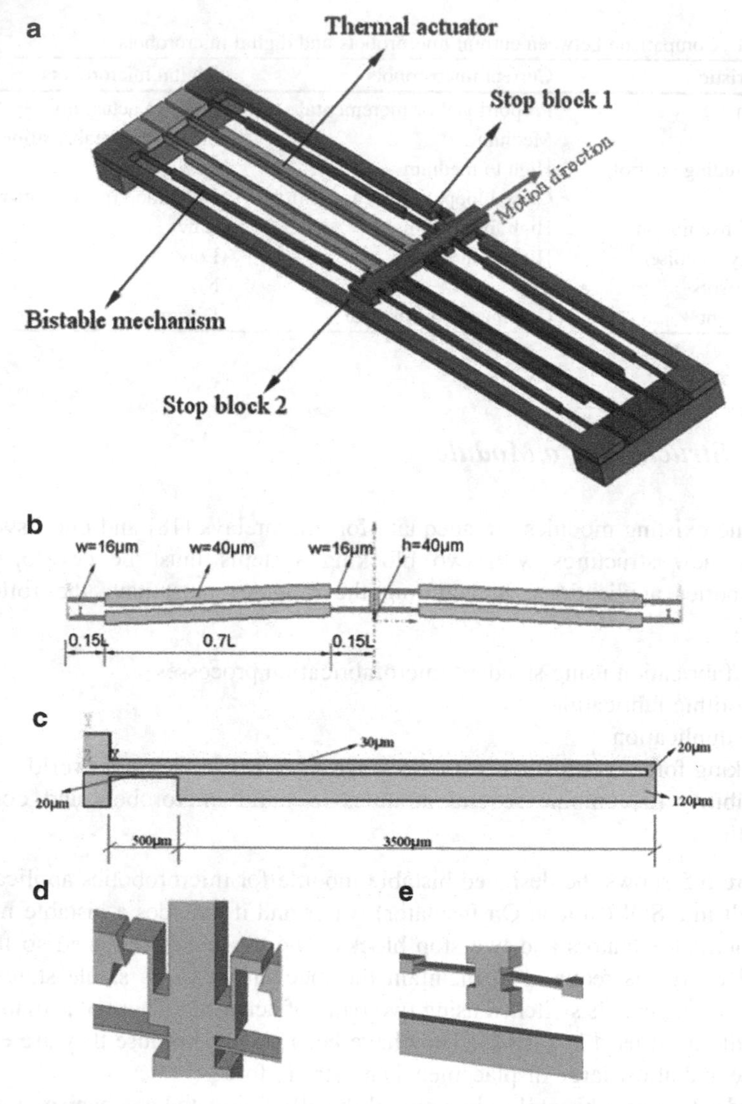

Fig. 5.5 Overall view of a bistable module and details. (**a**) Bistable module, (**b**) bistable mechanism, (**c**) thermal actuators, (**d**) stop block 1, and (**e**) stop block 2

model have been designed [23]. For dimensions close to those shown in Fig. 5.6, displacements from 5 µm to 25 µm and blocking forces[1] from 200 µN to 1.5 mN have been obtained.

[1] The blocking force is the maximum magnitude of the force applied on the shuttle without producing any displacement.

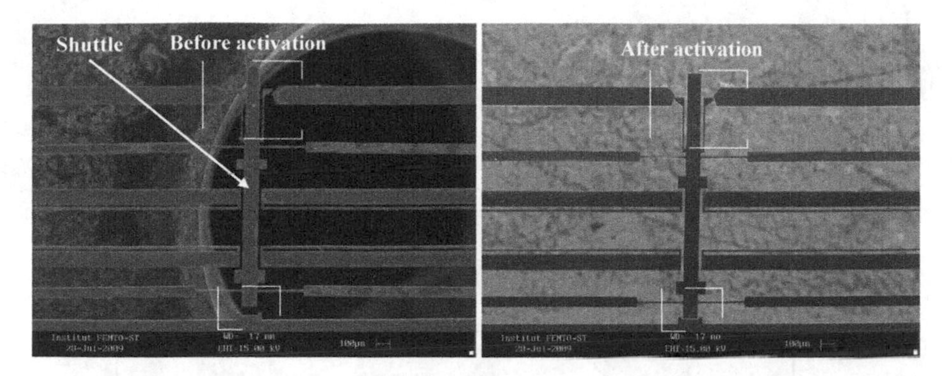

Fig. 5.6 SEM picture of a bistable module before and after activation

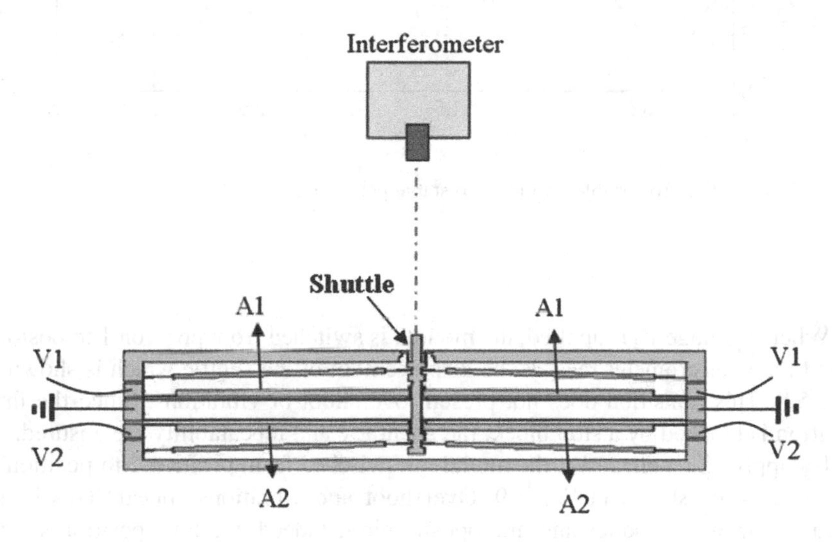

Fig. 5.7 Setup for the analysis of dynamic behavior

5.3.2 Dynamic Characteristics

Static characteristics (displacement and blocking force) and their repeatability and robustness are defined by the mechanical design. However, the use of these modules to perform microrobotic functions requires taking into account the dynamic behavior. Indeed, the operating principle based on mechanical switching may cause vibrations that can disturb the operating of the digital microrobot. In order to analyze the dynamic behavior, the setup shown in Fig. 5.7 have been used. A high resolution (1 nm) laser interferometer from SIOS gives the position of the shuttle.

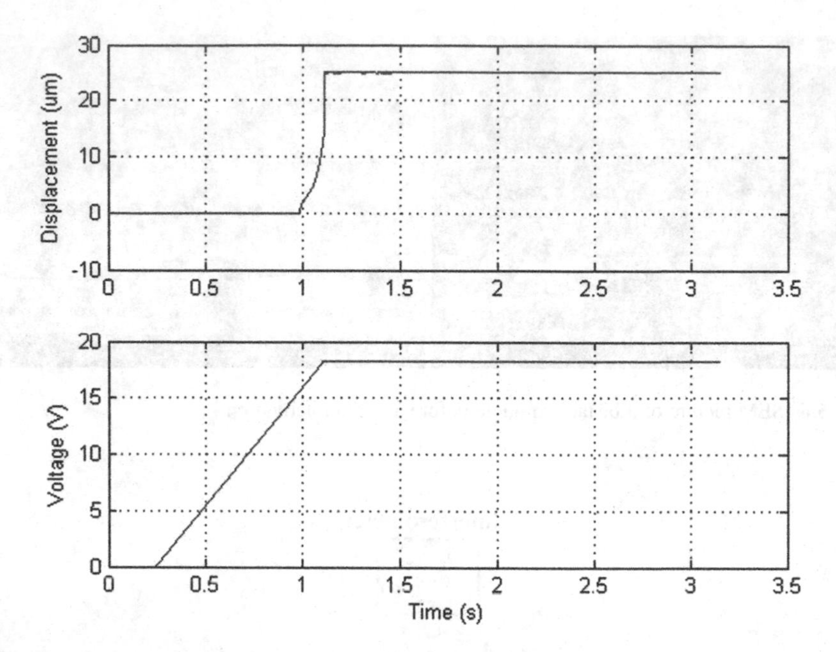

Fig. 5.8 Transition from stable position 1 to stable position 2

When a voltage V_2 is applied, the module is switched from position 1 to position 2 and the interferometer records the displacement of the shuttle which is shown in Fig. 5.8. This transition does not present overshoot or vibrations. Since the final position is blocked by a stop block, the accuracy and repeatability are ensured.

By applying a voltage V_1, the module is switched from position 2 to position 1. The results are shown in Fig. 5.9. Overshoot and vibrations appear. This is not suitable for microrobotics and micropositioning. Indeed, the final position should not be exceeded. An open loop control strategy must be developed for this transition in order to obtain damped response.

5.3.3 *Control Strategy*

In order to control the switching and avoid the overshoot without feedback, we propose an open-loop control strategy to obtain a damped transition. The strategy is based on the use of two pairs of thermal actuators during switching operation. One pair actuates the module and the second one is used to capture the bistable mechanism during the movement. The timing is presented in Fig. 5.10 and the obtained response is shown in Fig. 5.11.

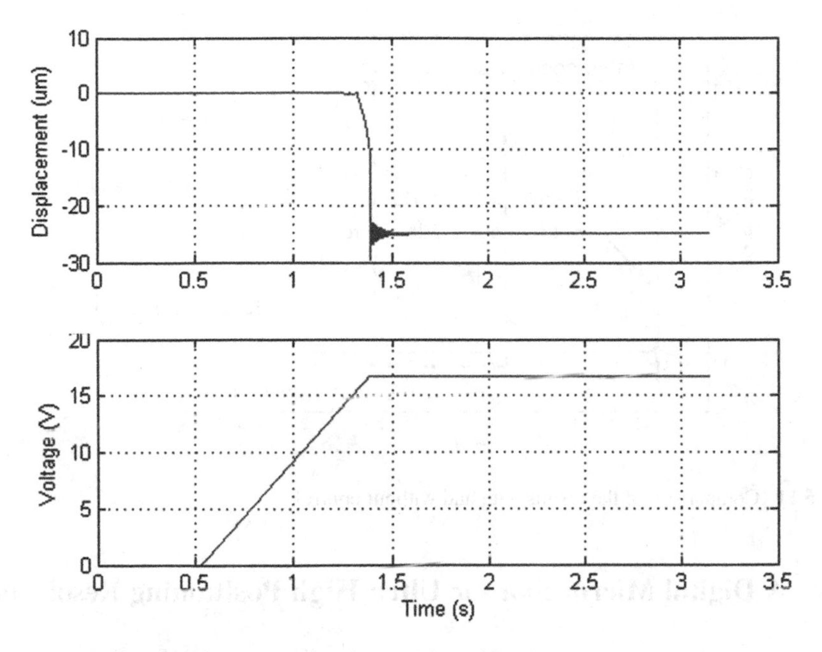

Fig. 5.9 Transition from stable position 2 to stable position 1

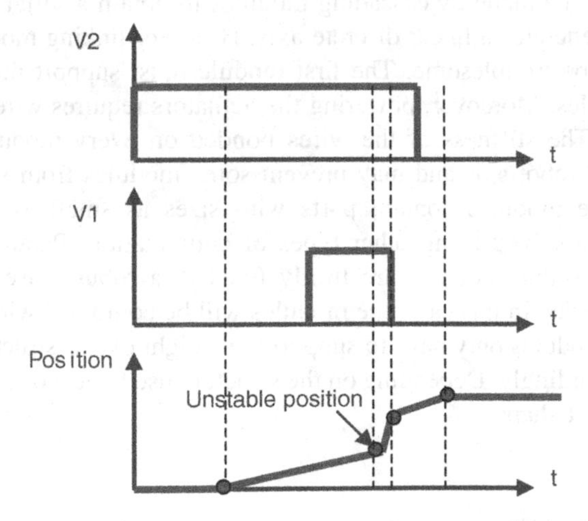

Fig. 5.10 Applied voltages for the control strategy

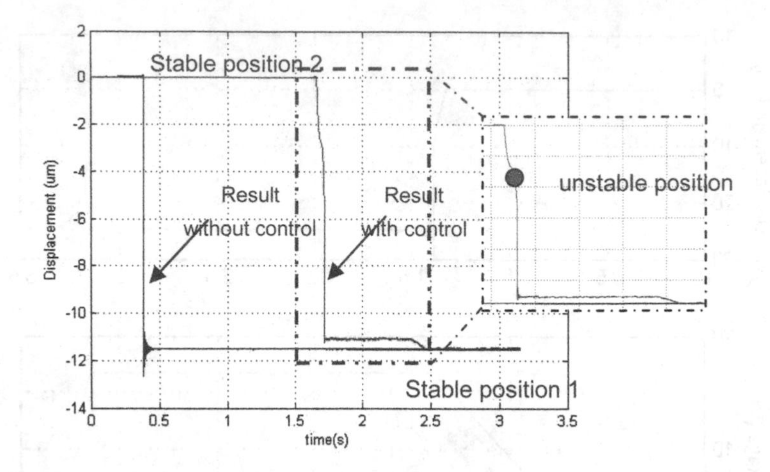

Fig. 5.11 Comparison of the results with and without control

5.4 A Digital Microrobot for Ultra High Positioning Resolution

5.4.1 Module Combination

Building a digital microrobot consists in combining several bistable modules. This combination can be made by cascading modules to obtain a serial axis as seen in Fig. 5.3. This generates a linear discrete axis. However linking modules on top of each other can be troublesome. The first module must support the weight of all following modules. Moreover, powering the actuators requires wires connected to every module. The stiffness of the wires bonded on every module disturbs the operating of the robot axis and may prevent some modules from switching. Note that the bistable modules contain parts with sizes as small as 10 μm. These problems can be solved using other types of combination. Parallel combination where all the bistable modules are firmly fixed to a robust base will solve the previous drawbacks. In this case, the modules will be combined with a mechanical structure. The modules only have to support the weight of this structure which will be designed accordingly. Depending on the structure used, the workspace generated will take different shapes.

5.4.2 The DiMiBot Structure

DiMiBot (Digital MicroRobot) is the name of a particular architecture of a planar microrobot built using bistable modules and a flexible mechanical structure. The design allows monolithic fabrication of the whole robot in a SOI wafer. Figure 5.12 shows the structure of a DiMiBot including six bistable modules.

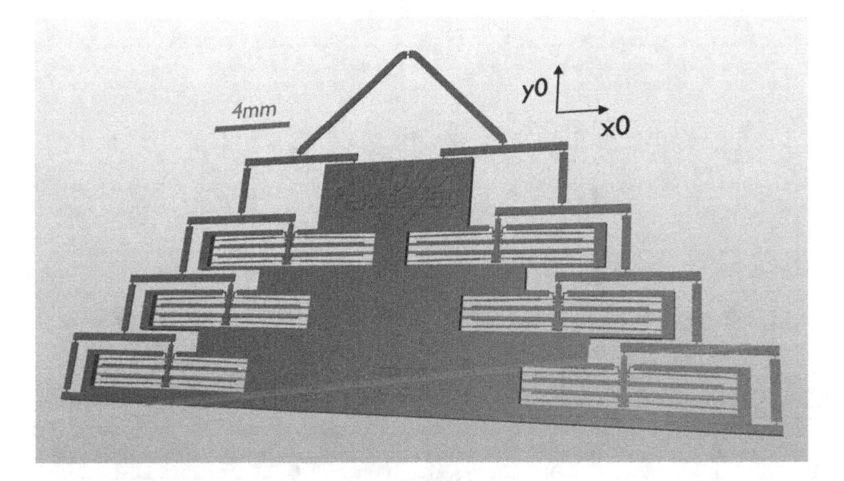

Fig. 5.12 CAD model of a DiMiBot with six modules

All modules have a displacement of 10 μm in the same direction y_0. The robotic structure was designed in order to generate a discrete two dimensional Cartesian workspace. The robot is designed to generate a workspace containing 64 (2^6) distinct reachable positions inside a square of 4 μm side, with a constant resolution of 500 nm. Note that this resolution is smaller than the displacement of the modules. Every reachable position is stable and addressed by a binary word composed of the states of all the modules (see Fig. 5.13).

By adding bistable modules to the microrobot, the resolution of the generated workspace is increased. The resolution is only limited by microfabrication technology. For instance, adding two bistable modules to the structure will give a workspace resolution twice better. Figure 5.14 shows the kinematic scheme of the digital microrobot with six bistable modules, and its extension to eight modules. Including two more modules does not affect the size of the workspace, but only the resolution. The workspace generated is still contained inside a square of 4 μm size, but it includes 256 reachable positions with a resolution of 250 nm (see Fig. 5.15).

5.4.3 Forward and Inverse Kinematics

One of the design criteria of the DiMiBot was to make the calculation of the forward and inverse kinematics easy. The forward kinematics represents the position of the end-effector of the robot (x, y) as a function of the state of each module ("0" or "1").

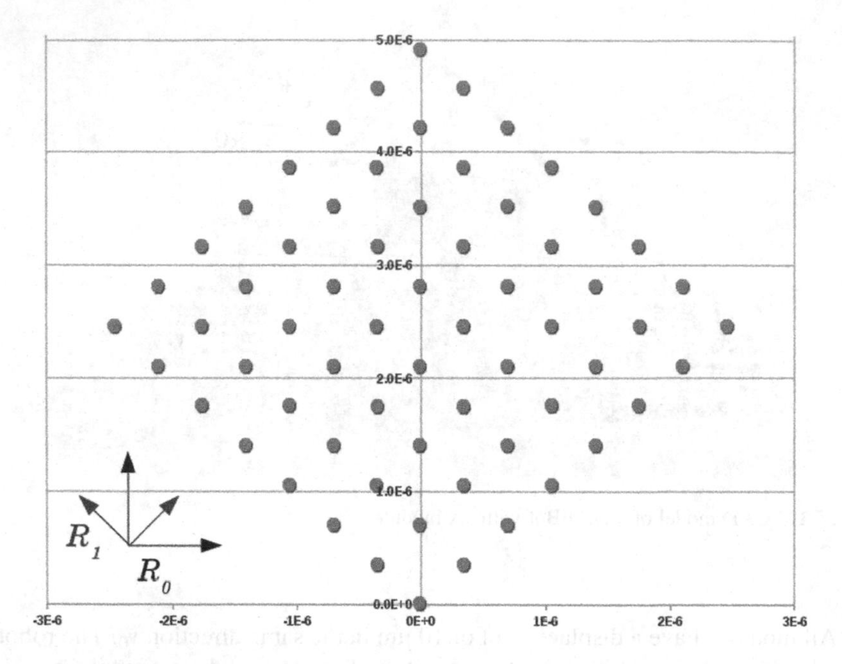

Fig. 5.13 Workspace generated with a DiMiBot with six modules

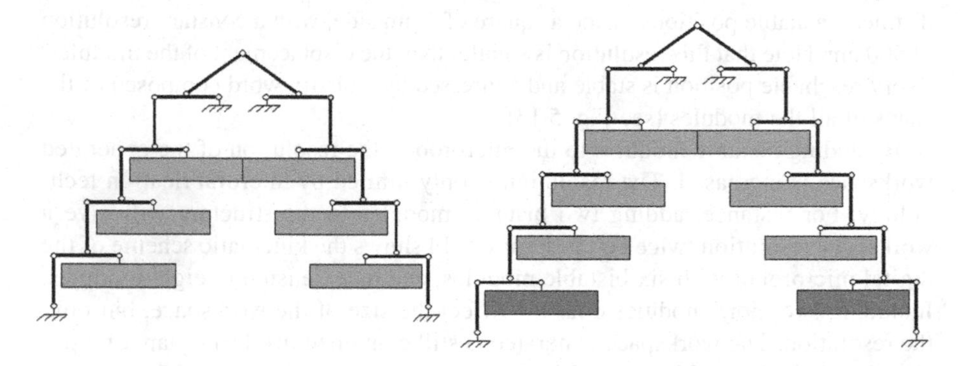

Fig. 5.14 Kinematic scheme of the six modules DiMiBot and its extension to eight modules

For a six modules robot, by naming the bistable modules of the left side of the structure bl_1 to bl_3 from top to bottom, and those of the right side br_1 to br_3 (see Fig. 5.16), the forward kinematic is calculated.

$$\begin{bmatrix} x \\ y \end{bmatrix} = K. \begin{bmatrix} 1 & \frac{1}{2} & \frac{1}{4} & -1 & -\frac{1}{2} & -\frac{1}{4} \\ 1 & \frac{1}{2} & \frac{1}{4} & 1 & \frac{1}{2} & \frac{1}{4} \end{bmatrix} . \begin{bmatrix} \mathrm{bl}_1 \\ \mathrm{bl}_2 \\ \mathrm{bl}_3 \\ \mathrm{br}_1 \\ \mathrm{br}_2 \\ \mathrm{br}_3 \end{bmatrix} .$$

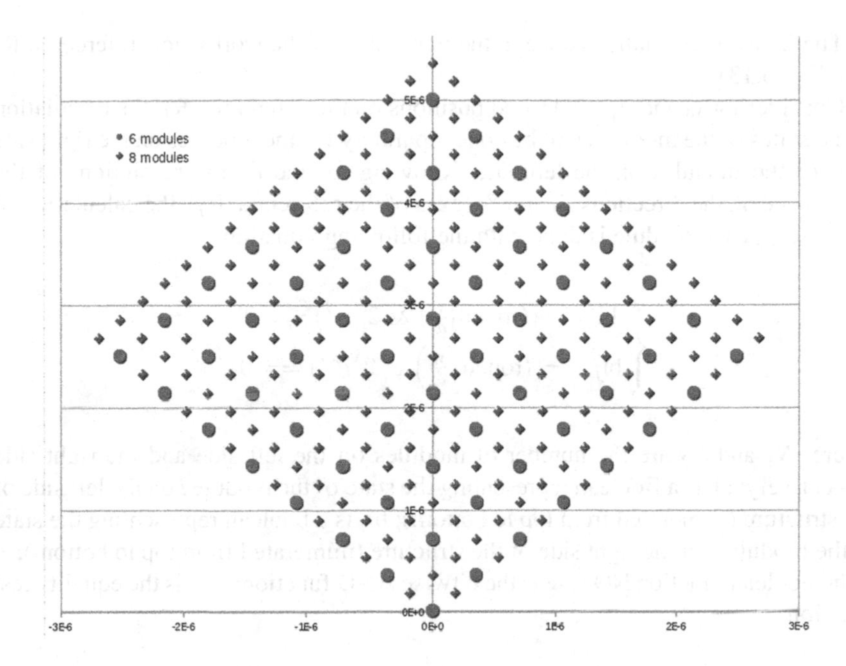

Fig. 5.15 Workspaces generated using six modules and eight modules

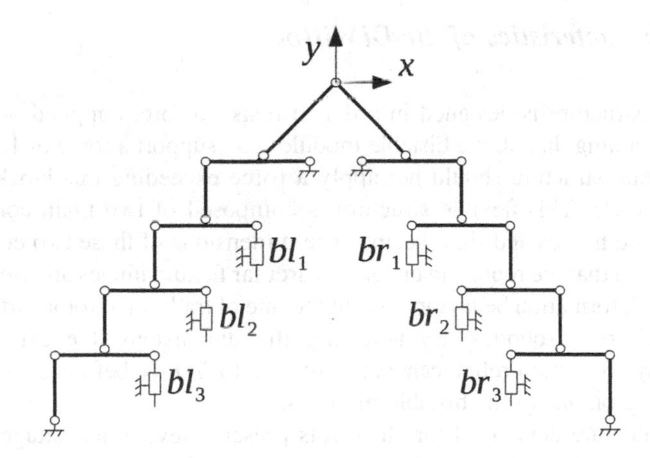

Fig. 5.16 Modules numbering for the models

This equation is reduced to a matrix multiplication in which the constant K depends on the geometric parameters of the structure and on the displacement generated by the bistable modules (which is unique for all the modules). However, it is only valid if the displacement of the modules is small compared to the length of the rigid beams of the robot architecture. This equation is also easily extendable to a higher number of bistable modules.

The inverse kinematic is done in the referential of the workspace, referential R_1 (see Fig. 5.13).

Considering x_d and y_d the desired positions on the referential R_I, the calculation of the states of the modules will be done separately for the modules on the right side and for the modules on the left side. Knowing δx and δy the resolutions of the workspace on the directions X_1 and Y_1 (axes of the referential R_I), the calculation of the state of each module is done with the following equation:

$$\begin{cases} \mathrm{bl}_i = \neg((\mathrm{round}\left(\frac{x_d}{\delta x}\right) \ \& \ 2^{N_1 - i}) == 0), \\ \mathrm{br}_j = \neg((\mathrm{round}\left(\frac{y_d}{\delta y}\right) \ \& \ 2^{N_2 - j}) == 0), \end{cases}$$

where: N_1 and N_2 are the number of modules on the left side and the right side respectively; bl_i is a Boolean representing the state of the module i on the left side of the structure (numerated from top to bottom); br_j is a boolean representing the state of the module j on the right side of the structure (numerated from top to bottom); $\neg$ is the boolean function NOT; & is the bitwise AND function; == is the equality test function.

5.4.4 Characteristics of the DiMiBot

The robotic structure is designed in order to resist to forces applied on the end-effector. Assuming that all the bistable modules can support a force of 1.5 mN, the flexible robotic structure should not apply a force exceeding this blocking force onto any module. This flexible structure is composed of two main components, circular flexure hinges and rigid beams. The dimensions of these two components define the force that the robot can undergo. Circular flexure hinges are used because they present deformation behavior close to the one of traditional rotoid articulations used in traditional robotics. By adjusting the dimensions, the external force supported by this microrobot can vary from 2 to 7 mN before exceeding the blocking force on one of the bistable modules.

The architecture developed for DiMiBots presents several advantages. Its flatness (0.5 mm) allows for working in confined environments such as a TEM (Transmission Electron Microscope). Furthermore, the bistability property allows digital sensorless control. The monolithic property facilitates the microfabrication process in a single wafer avoiding microcomponents assembly. Energy consumption is limited thanks to the use of bistable modules that are powered only for switching. Finally, simplicity of the models and open-loop control allow low cost operating of the system. Figure 5.17 shows a microfabricated DiMiBot with four bistable modules.

Fig. 5.17 Microfabricated DiMiBot with four bistable modules

5.5 Conclusion and Perspectives

In this chapter we presented the basic concepts of digital microrobotics, a new approach for the design and fabrication of microrobots adapted to the microworld. This new approach allows great flexibility in achieving robot kinematics. Performance of today's MEMS technology and open loop control allow monolithic fabrication and control of digital microrobots at low cost. The binary nature of the mechanical bistable module and the control (digital words) breaks the barriers between software and hardware and opens new perspectives for optimal path following and tasks performing in the microworld.

Although the feasibility of such robots has been demonstrated, many perspectives can be considered. Thermal actuators offer enough force but generate heat that may disturb the working environment. The use of different microactuators can be experimented. Study of robot kinematics according to the tasks to perform in the microworld has also a great potential. Moreover, tasks efficiency requires the development of optimal combinatory control algorithms.

References

1. Clévy C (2005) Contribution à la micromanipulation robotisée: un système de changement d'outils automatique pour le micro-assemblage. Thèse de doctorat, Université de Franche-Comté (Besançon, France), décembre 2005
2. Régnier S (2006) La manipulation aux échelles microscopiques. Habilitation à diriger des recherches, Université Pierre et Marie Curie (Paris 6, France), juin 2006

3. Robertshaw H, Reinholtz C (1988) Variable geometry trusses. In: Smart Materials, Structures, and Mathematical Issues. U.S. Army Research Office Workshop, Blacksburg, VA, Selected Papers; 15–16, Sept. 1988, pp 105–120
4. Miura K, Furaya H (1985) Variable geometry truss and its applications to deployable truss and space crane arm. Acta Astro 12(7):599–607
5. Hughes P, Sincarsin P, Carroll K (1991) Truss arm a variable-geometry Trusses. J Intell Mat Syst Struct 1.2:148–160
6. Chirikjian GS (1994) A binary paradigm for robotic manipulators. In: Proceedings of 1994 I.E. conference on robotics and automation, pp 3063–3070
7. Lees D, Chirikjian GS (1996) A combinatorial approach to trajectory planning for binary manipulators. In: Proceedings of 1996 I.E. international conference on robotics and automation, pp 2749–2754
8. Lees D, Chirikjian GS (1996) An efficient method for computing the forward kinematics of binary manipulators. In: Proceedings of 1996 I.E. international conference on robotics and automation, pp 1012–1017
9. Lees D, Chirikjian GS (1996) Inverse kinematics of binary manipulators with applications to service robotics. In: Proceedings of 1996 I.E. international conference on robotics and automation, pp 2749–2754
10. Uphoff IE, Chirikjian GS (1996) Inverse kinematics of discretely actuated hyper-redundant manipulators using workspace densities. In: 1996 I.E. international conference on robotics and automation, 1: 139–145
11. Sujan VA (2004) Design of a lightweight hyper-redundant deployable binary manipulator. J Mech Des 126(1):29–39
12. Gomm T (2001) Development of in-plane compliant bistable microrelays. MS Thesis, Brigham Young University, Provo, UT, 2001
13. Howell LL (2001) Compliant mechanisms. Wiley, New York. ISBN 0-471-38478-X
14. Robert RY (1999) Mechanical digital-to-analog converters In: Proceedings of the international conference on solid-state sensors and actuators. Transducers'99, Sendai, Japan, June 1999, pp 998–1001
15. Qiu J (2004) A curved-beam bistable mechanism. J MicroElectroMech Syst 13(2):137–146
16. Jinni T (2005) Design and experiments of fully compliant bistable micromechanisms. Mech Mach Theor 40:17–31
17. Dong Y (2002) Mechanical design and modeling of MEMS thermal actuators for RF applications. MS Thesis, University of Waterloo
18. Qiu J, Lang JH, Slocum AH (2005) A bulk-micromachined bistable relay with U-shaped thermal actuators. J Microelectromech Syst 14:1099–1109
19. Hwang I-H, Shim Y-S, Lee J-H (2003) Modeling and experimental characterization of the chevron-type bistable microactuator. J Micromech Microeng 13:948–954
20. Freudenreich M, Mescheder U, Somogyi G (2004) Simulation and realization of a novel micromechanical bi-stable switch. Sensor Actuator Phys 114:451–459
21. Kwon HN, Hwang I-H, Lee J-H (2005) A pulse-operating electrostatic microactuator for bi-stable latching. J Micromech Microeng 15:1511–1516
22. Krylov S, Ilic BR, Schreiber D, Seretensky S, Craighead H (2008) The pull-in behavior of electrostatically actuated bistable microstructures. J Micromech Microeng 18:055026
23. Chen Q, Haddab Y, Lutz P (2010) Microfabricated bistable module for digital microrobotics. JMNM 6:1–2. doi:DOI 10.1007/s12213-010-0025-2, Published online: 14 Sept 2010

Chapter 6
Flexure-Based Parallel-Kinematics Stages for Passive Assembly of MEMS Optical Switches

Wenjie Chen, Guilin Yang, and Wei Lin

Abstract A key operation in assembly of MEMS optical switches is to insert fibers into U-grooves on a silicon substrate. Due to the limited positioning accuracy of the handling tool, heavy collision often occurs between fibers and the edges of U-grooves during the insertion operation. Such collisions will not only damage fibers and U-grooves but also sometimes make the fiber skidding from the handling tool. Conventional solutions to the problem involve determining misalignment using machine vision or force sensors, and then positioning fibers accurately by virtue of high precision multiaxis positioning systems (with submicron repeatability). However, such approaches are costly and difficult to implement. In this chapter, we present a cost-effective passive assembly method to solve the problem. It utilizes a specially designed passive flexure-based fixture (stage) to regulate high contact forces and accommodate assembly errors. To determine the design conditions for a successful insertion, the major problems encountered during the fiber insertion are analyzed. A systematic design method is then proposed for a 3-legged Flexure-based Parallel-Kinematics Stage (FPKS) for passive assembly applications. Experimental results show that such a passive assembly approach can effectively and automatically reduce the contact force and accommodate the assembly errors.

6.1 Introduction

Assembly of hybrid MEMS devices often requires a sequential microassembly process, in which MEMS dies are integrated with other microcomponents to form a complete system. Microcomponent insertion is an essential task in such a

---

W. Chen (✉) • G. Yang • W. Lin
Mechatronics Group, Singapore Institute of Manufacturing Technology, 71 Nanyang Drive, Singapore 638075, Singapore
e-mail: wjchen@SIMTech.a-star.edu.sg; glyang@SIMTech.a-star.edu.sg; wlin@SIMTech.a-star.edu.sg

D. Zhang (ed.), *Advanced Mechatronics and MEMS Devices*, Microsystems,
DOI 10.1007/978-1-4419-9985-6_6,

microassembly process. The major challenges in microcomponent insertion include two aspects. First, the required positioning and alignment accuracy is very stringent, typically in a submicrometers range. Thus, a costly ultraprecision positioning systems needs to be employed. Second, the microcomponents in a hybrid MEMS device are normally made from fragile materials. To avoid damage of these components, highly sensitive force/torque sensors are needed if an active force method is used. However, not only such sensors are very expensive, but also the active force control method has problems of poor reliability due to low control bandwidth and instability [1–4].

Optical switches are typical hybrid MEMS devices, in which fibers need to be inserted into U-grooves on MEMS dies for light transmission from the input channel to the output channels. When a fiber is inserted into a U-groove by an automatic means, nasty collision between the fiber and the edge of the U-groove occurs when the positioning errors are larger than assembly tolerances, resulting in either the die damaging or the fiber skidding from the handling tool. Such collision problems can be tackled by utilizing a low-cost passive assembly method, in which a planar compliant stage (fixture) is employed to automatically accommodate the assembly errors in terms of the x, y, and θ offsets so that the excessive contact force can be avoided [5, 6].

The compliance behavior of a stage is generally characterized by its stiffness or compliance matrix expressed in Cartesian space [7, 8]. If the stiffness matrix at a given point is a diagonal matrix, the point is called the Remote Center of Compliance (RCC) [9, 10]. When an external force/torque is applied on the RCC point of a stage, deformation of the stage will follow the direction of the applied force/torque. To achieve passive optical switch assembly, this chapter focuses on the modeling and design of a planar flexure-based parallel stage that can provide the required decoupled compliance in the x, y and θ axes according to the RCC principle [6].

The following sections of this chapter are organized as follows. The problems existed in the optical switch assembly as well as in the required fixture stiffness for passive assembly are introduced in Sect. 2. The stiffness modeling of a general 3-legged flexure-based passive assembly stage is discussed in Sect. 3. In Sect. 4, the geometry and dimensions of the flexural joints in each leg are synthesized and designed based on the stiffness requirements in Cartesian space. Sect. 5 illustrates experimental results. This chapter ends with a conclusion section.

6.2 Fiber Insertion Analysis

6.2.1 Assembly Errors

Figure 6.1 shows a typical optical switch that has one input channel and four output channels. The channels are micromachined to be U-shaped grooves on the die so as to secure fibers firmly [11]. The mating clearance between fibers and grooves is generally less than 1 μm. Collisions will therefore happen between the fibers and the grooves during insertion because of assembly uncertainties.

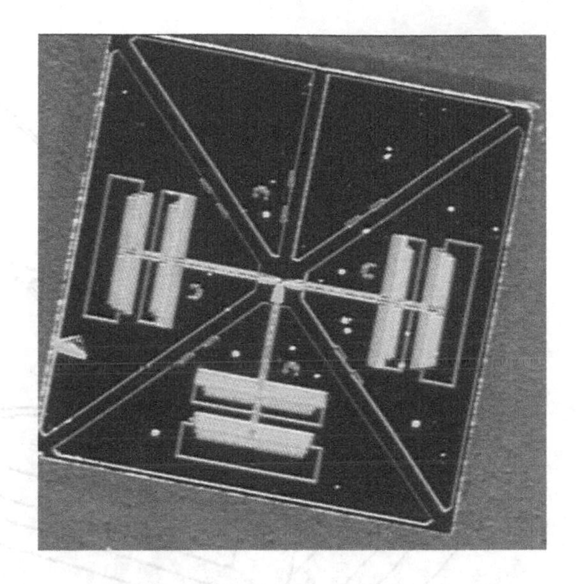

Fig. 6.1 Switch die with one input and four outputs (© IEEE 2008), reprinted with permission

Before a fiber is inserted into a U-groove, it is held along a V-shaped groove on a vacuum head, which is mounted on a pick-and-place system. The overall assembly errors, resulting from positioning errors of the Pick-and-Place (PnP) system and dimensional tolerances of the die and its holder, can be classified as lateral offset, Δx (or e); longitudinal offset, Δz; yaw angle offset, $\Delta\theta_y$; and the tilt angle offset, $\Delta\theta_x$ (see Fig. 6.2). The rotational offset $\Delta\theta_z$ and the vertical offset Δy will not be considered with the assumption that the off-line alignment is to be carried out and the fiber will be inserted all the way down to the bottom of the U-grooves.

Among the offset errors concerned, the lateral offset and the yaw angle offset will cause the most severe collision between the fiber and the edges of the U-groove, while both the longitudinal offset and the tilt-angle offset may just undermine the light coupling. This chapter will, therefore, focus on the studies of accommodating the lateral and yaw angle offsets.

6.2.2 Problems of Fiber Insertion Operation

Two kinds of assembly problems need to be considered during fiber inserting into U-grooves. They are: component damage and fiber skidding out of the grip.

6.2.2.1 Component Damage

Fibers and switch dies are made up of glass and single crystal silica, which have strengths of 228 and 300 MPa, respectively [12]. Due to sharp edge of U-grooves, the die reaches its damage stress earlier than the fiber. To determine the limit of the contact forces, a simulation for insertion process can be conducted using the finite element analysis (FEA). Figure 6.3 illustrates the maximum allowable contact force

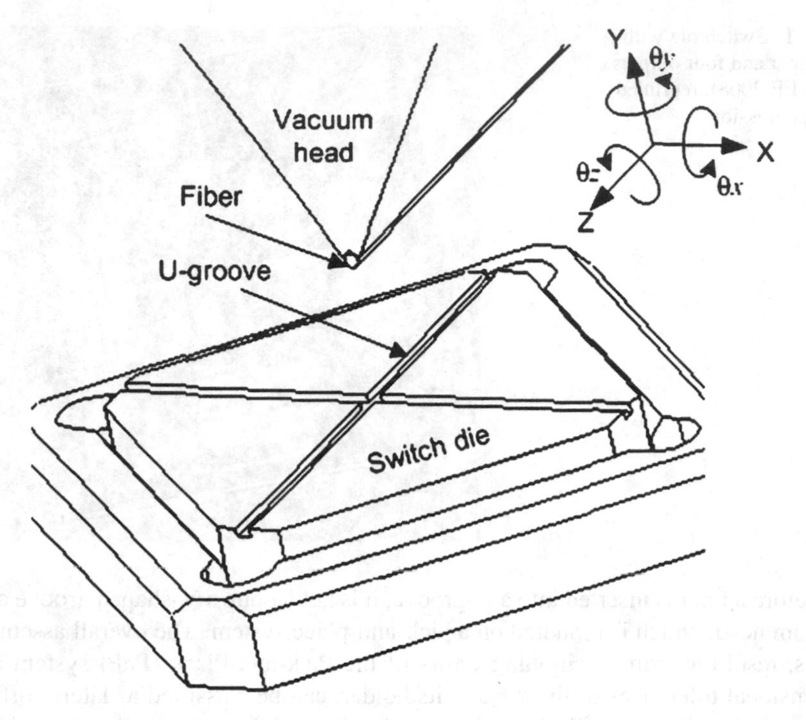

Fig. 6.2 Schematic representation of a switch assembly (© IEEE 2008), reprinted with permission

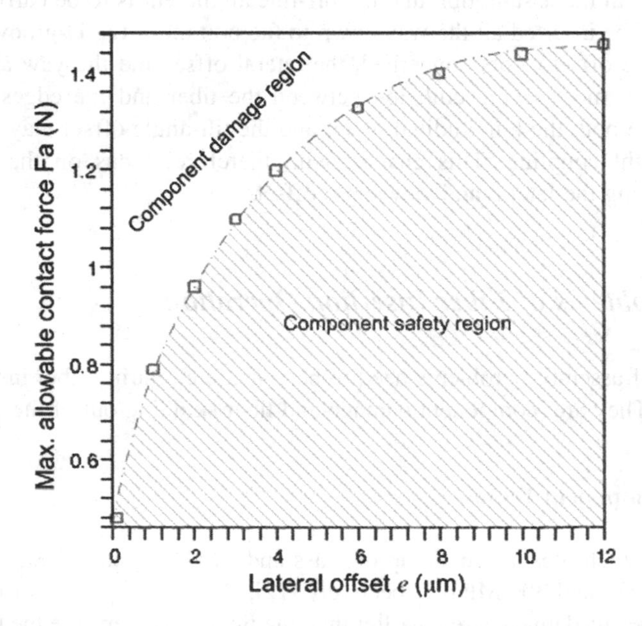

Fig. 6.3 Maximum allowable contact forces on the fiber (© IEEE 2008), reprinted with permission

exerted on the fiber at the normal direction when the die reaches the stress of 300 MPa. It can be seen that the maximum allowable contact force varies with the contact positions. Once the contact force surpasses the maximum allowable contact force, the switch die will be damaged.

6.2.2.2 Fiber Skidding Out of the Grip

During insertion, Fibers may possibly rotate and/or slip relatively to the handling tool, in turn, affecting light coupling efficiency. This phenomenon is termed as fiber skidding [6], which is normally uncontrollable.

Light coupling efficiency of switches depends on alignment accuracy of the input and output fibers. Due to eccentricity of fiber cores, which may be up to ±1 μm, alignment accuracy will vary with the relative rotation of two coupling fibers. In the worst case, alignment error will be the sum of two core offsets. To minimize this alignment error, an off-line alignment method can be employed in which the output fiber is firstly aligned with the input fiber through a proper rotation. After alignment, any motion leading to fiber rotation during insertion is not allowed. In other words, any skidding of the fiber relative to the handling tool is undesirable and has to be prevented in the assembly.

6.2.3 Fiber Skidding Analysis

The contact model and forces exerted on the fiber are shown in Fig. 6.4, where r is the radius of the fiber; e is the lateral offset, i.e., Δx; ϑ is the attack angle which has the relationship of $\cos\vartheta = (r - e)/r$; φ is the half angle of the V-groove of the vacuum head; F_v is the suction force; A, B, and C denote the contact points; F_{a}, F_{b}, and F_{c} are the normal contact forces at points of A, B, and C; and f_{a}, f_{b}, and f_{c} are the friction forces at A, B, and C.

When contacting with the switch die, the fiber is constrained by three points A, B, and C. Under such a constraint, when the contact forces exceed certain values, the fiber will skid out of the vacuum head by the way of rotating about either point A or C. For contact force F_{a}, it should satisfy the following condition:

$$F_{\mathrm{a}} \le F_{\mathrm{a}}^{*} = \min(F_{\mathrm{a}}^{1}, F_{\mathrm{a}}^{2}), \tag{6.1}$$

where F_{a}^{1} and F_{a}^{2} are threshold values due to slipping at point A and C, respectively. They can be calculated by [6]:

$$F_{\mathrm{a}}^{1} = \frac{\cos\varphi}{\sin(\varphi - \vartheta) - \mu_1(1 + \cos(\varphi - \vartheta))} F_v \tag{6.2}$$

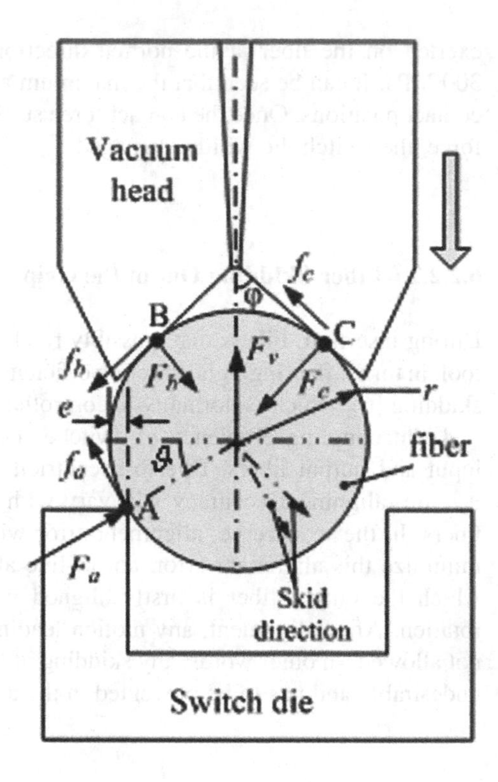

Fig. 6.4 Quasi-static model of contact (© IEEE 2008), reprinted with permission

and

$$F_{\mathrm{a}}^{2} = \frac{\cos\varphi + \mu_2(\sin\varphi + \sin\vartheta)}{\sin(\varphi - \vartheta) - \mu_2(1 + \cos(\varphi - \vartheta))} F_v. \tag{6.3}$$

The value of F_{a}^{*} is a threshold for contact force F_{a}, such that when $F_{\mathrm{a}} \leq F_{\mathrm{a}}^{*}$ the fiber will be held firmly by the vacuum head, and when $F_{\mathrm{a}} > F_{\mathrm{a}}^{*}$ the fiber will rotate relatively to the vacuum head around either point A or point C, resulting in fiber skidding from the vacuum head.

6.2.4 Contact Force vs. Support Fixture Stiffness

In practice, the switch die holder and its support always have some degrees of compliance, which may significantly influence the contact force (Fig. 6.5). Assume that the switch die is rigid and supported by a compliant holder as shown in Fig. 6.5, where the lumped compliance of the holder is modeled by general linear springs. When the fiber is pressed down, the die will be pushed away to the side, leading the

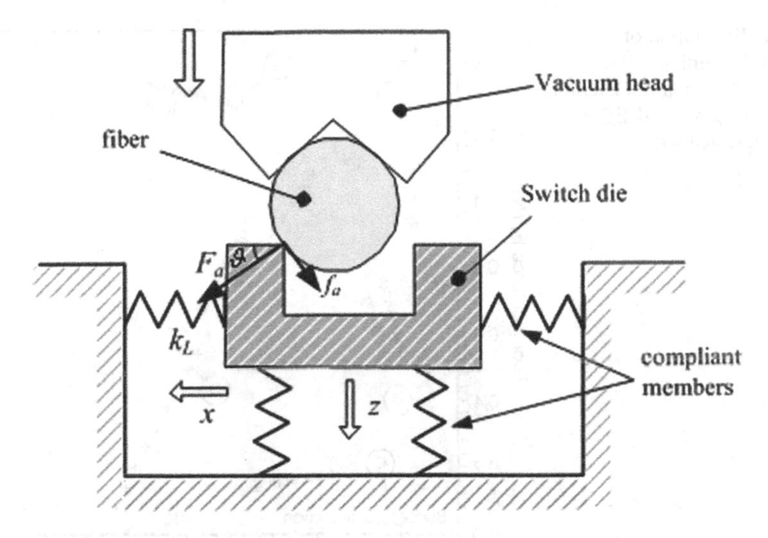

Fig. 6.5 Compliance model of the support fixture (© IEEE 2008), reprinted with permission

fiber to slide into the U-groove. Considering the switch die, a quasi-static force balance in the horizontal direction yields:

$$k_{\mathrm{L}}r(\cos\vartheta - \cos\vartheta_0) = F_{\mathrm{a}}\cos\vartheta - f_{\mathrm{a}}\sin\vartheta,$$
$$f_{\mathrm{a}} = \mu_1 F_{\mathrm{a}}, \tag{6.4}$$

where k_{L} is the stiffness of the general linear spring and ϑ_0 is the initial attack angle. From (6.4), contact force F_{a} can be obtained by

$$F_{\mathrm{a}} = \frac{k_{\mathrm{L}}r(\cos\vartheta_0 - \cos\vartheta)}{\cos\vartheta - \mu_1\sin\vartheta}. \tag{6.5}$$

Equation (6.5) describes how the support fixture stiffness k_{L} influences the contact force F_{a}.

6.2.5 Stiffness Requirements for Passive Fiber Insertion

A successful fiber insertion should avoid both component damage and any fiber skidding from the grip. These requirements can be analyzed by using the plots in Fig. 6.6. In this figure, curve A represents the maximum allowable contact force for die damage, directly coming from Fig. 6.3; curve B represents the contact force that causes the fiber skidding from the grip, plotted according to (6.1); and curves C and D reflect the effect of the supporting (environmental) stiffness on the contact force, plotted based on (6.5).

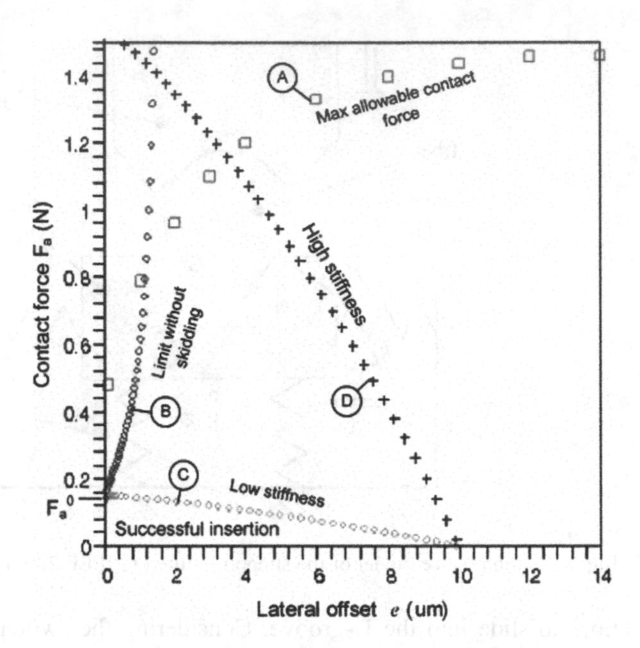

Fig. 6.6 Illustration of successful insertion conditions for an initial lateral offset of 10 μm (© IEEE 2008), reprinted with permission

In the early phase of an insertion, the contact force F_a increases with reduction of the lateral offset e. For low environmental compliance, the force F_a may increase quickly and surpass the maximum allowable contact force of the component before the fiber skidding happens. For high environmental compliance, however, it is possible that damage of components may not happen, but a fiber skidding takes place when the contact force increases to F_a^*. Therefore, for a successful insertion, the environmental compliance must be such a value that the above two situations never occur throughout the entire insertion process. Based on Fig. 6.6, a successful insertion can be achieved if:

$$F_a^0 \geq k_L r(1 - \cos \vartheta_0), \tag{6.6}$$

where F_a^0 is the value of F_a^* when $e = 0$ or $\vartheta=0$

If the maximum allowable lateral offset is denoted as e_{max}, it can be obtained by:

$$k_L \leq k_L^* = \frac{F_a^0}{e_{max}}, \tag{6.7}$$

where k_L^* is the maximum allowable environmental stiffness. Once the environmental lateral stiffness is less than k_L^*, the fiber insertion with the initial lateral error e_{max} will be implemented successfully.

The required environmental torsional stiffness (k_T) can be determined as follows. Assume that the compliance center of the supporting stage is located at

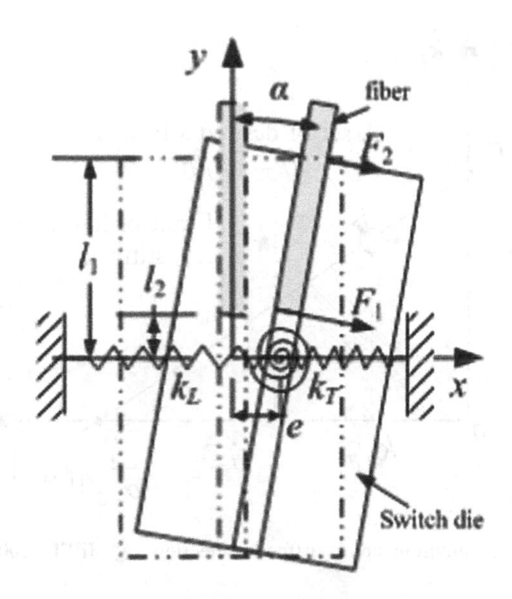

Fig. 6.7 Switch shifting with the angular offset (© IEEE 2008), reprinted with permission

the center of the switch die and the angular compliance is modeled by a general torsional spring. Figure 6.7 illustrates a diagram of contact model including both the lateral offset and the angular offset. Due to existence of the angular offset, the contact forces are concentrated at both ends of the U-groove. The maximum contact force will occur at the inner end of the groove, which can be expressed in the form

$$F_1 = \frac{l_1 \cdot k_\mathrm{L}}{l_1 - l_2} \cdot |e_{\max}| + \frac{k_\mathrm{T}}{l_1 - l_2} \cdot |\alpha_{\max}|, \tag{6.8}$$

where $\alpha_{\max}$ is the maximum allowable angular offset, l_1 and l_2 are respectively the distances from the ends of the groove to the compliance center. Since F_1 must be less than F_a^0, we can obtain

$$k_\mathrm{T} \le \frac{l_1 - l_2}{|\alpha_{\max}|} F_\mathrm{a}^0 - \frac{l_1 k_\mathrm{L}}{|\alpha_{\max}|} |e_{\max}|. \tag{6.9}$$

Once F_a^0, $|e_{\max}|$ and $|\alpha_{\max}|$ are specified, the values of k_L and k_T can be determined based on (6.7) and (6.9). However, the values of k_L and k_T meeting the requirement are not unique. Figure 6.8 shows a region (shadowed area), in which any point may give a set of possible k_T and k_L. A compromise between k_L and k_T is often needed in design. Generally, it is desirable to select the value of k_T as high as possible.

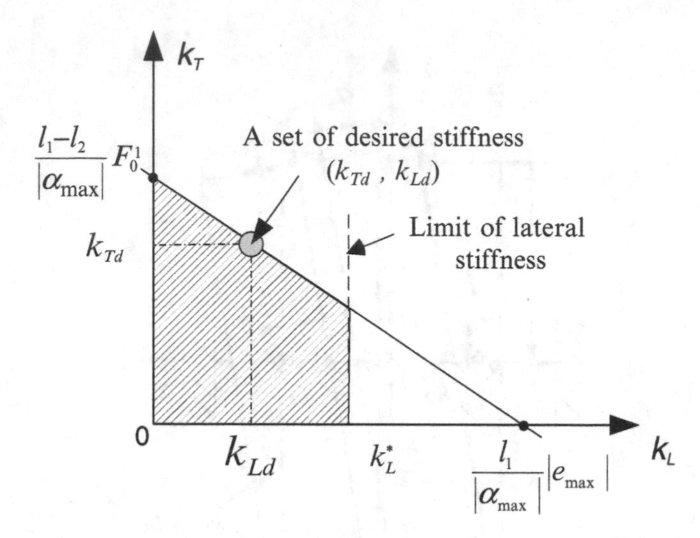

Fig. 6.8 Illustration of environmental stiffness selection (© IEEE 2008), reprinted with permission

6.3 Stiffness of 3-Legged Flexure-Based Parallel Kinematics Stages

As described in the previous section, if the supporting (environmental) compliance satisfies the conditions given by (6.7) and (6.9), fibers can be successfully inserted into U-grooves even though there are lateral and angular offsets. The required environmental compliance can be attained through a flexural mechanism. A flexure obtains its motion through elastic deformation in the material. It is therefore able to offer advantages of zero backlash, free wear, continuous displacement, and embedded return spring effects. A flexure-based mechanism can be designed through the pseudo-rigid-body model (PRBM) approach [13]. This method treats flexural structures as conventional rigid-link mechanisms but attaching torsion or linear springs on corresponding joints (i.e., PRBM). Based on such a model, the performance of flexure-based mechanisms can be analyzed using well-studied rigid linkage kinematics.

6.3.1 Stiffness Modeling

It is straightforward to adopt planar parallel mechanisms with joint compliance to obtain the required 3-DOF planar compliances. Generally, these mechanisms include a platform that is supported by three legs (or chains). Each leg comprises three joints, which are either revolute joints (denoted by *R*) or prismatic joints

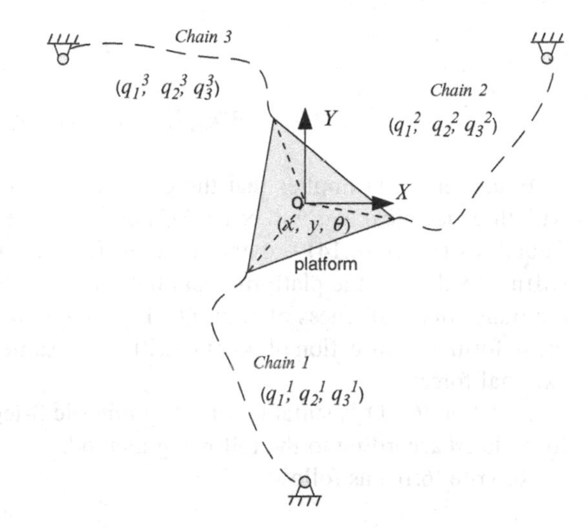

Fig. 6.9 Planar parallel platforms with general leg structure

(denoted by *P*). According to different joint arrangements, there will be at least six types of leg architectures that can provide meaningful 3-DOF motion [14, 15].

Without losing generality, we can consider such a planar parallel mechanism having three generic legs as shown in Fig. 6.9. The platform is at its static equilibrium state, and all legs possess identical configuration but are laid 120° away each other. The three joint variables of the *i*th leg are denoted as $[q_1^i, q_2^i, q_3^i]$ and the corresponding stiffness as k_{q_1}, k_{q_2} and k_{q_3}. The position of the geometric center (point *O*) of the platform in Cartesian space are expressed by x, y and θ with respect to the based frame.

Within the workspace, the velocity vectors of the geometric center in Cartesian space can be correlated with the velocity vector of joint variables of leg *i* by the Jacobian matrix [*J*], i.e.,

$$\begin{bmatrix} \dot{q}_1^i \\ \dot{q}_2^i \\ \dot{q}_3^i \end{bmatrix} = \begin{bmatrix} J_{1x} & J_{1y} & J_{1\theta} \\ J_{2x} & J_{2y} & J_{2\theta} \\ J_{3x} & J_{3y} & J_{3\theta} \end{bmatrix} \begin{bmatrix} \dot{x} \\ \dot{y} \\ \dot{\theta} \end{bmatrix}, \tag{6.10}$$

where J_{ix}, J_{iy}, and $J_{i\theta}$, $i = 1, 2, 3$ are the kinematic influence coefficients of x, y, and θ on the *i*th joint. The elements of Jacobian matrix can be obtained through velocity analysis or differencing the position function.

As shown in the latter end of this section, once (6.10) is obtained, the output stiffness matrix of the platform about point *O* can be calculated in the form:

$$K = \mathrm{diag}(k_x, k_y, k_\theta), \tag{6.11}$$

where

$$k_x = \frac{3}{2}(J_{1x}^2 + J_{1y}^2)k_{q1} + \frac{3}{2}(J_{2x}^2 + J_{2y}^2)k_{q2} + \frac{3}{2}(J_{3x}^2 + J_{3y}^2)k_{q3}, \tag{6.12}$$

$$k_y = k_x, \tag{6.13}$$

$$k_\theta = 3(k_{q1}J_{1\theta}^2 + k_{q2}J_{2\theta}^2 + k_{q3}J_{3\theta}^2). \tag{6.14}$$

Equation (6.11) implies that the geometric center of the planar parallel stages with three identical legs will be a RCC point regardless of types of leg architectures. Equations (6.12)–(6.14) provide a closed-form formulation to predict the output stiffness values of the platform from the joint stiffness. In addition, since $k_x = k_y$, the translational stiffness of point O will be equal in all directions. In other words, the deformation direction of point O will be the same as the direction of the applied external forces.

Equation (6.11) is suitable for all symmetric 3-legged planner stages. It can be formulated according to the following method.

Rewrite (6.10) as follows:

$$\begin{bmatrix} \dot{q}_1^1 \\ \dot{q}_2^1 \\ \dot{q}_3^1 \end{bmatrix} = \begin{bmatrix} J_{1x} & J_{1y} & J_{1\theta} \\ J_{2x} & J_{2y} & J_{2\theta} \\ J_{3x} & J_{3y} & J_{3\theta} \end{bmatrix} \begin{bmatrix} \dot{x} \\ \dot{y} \\ \dot{\theta} \end{bmatrix}$$

or

$$\begin{bmatrix} \dot{q}_1^1 \\ \dot{q}_2^1 \\ \dot{q}_3^1 \end{bmatrix} = \begin{bmatrix} [\mathrm{J}_1] \\ [\mathrm{J}_2] \\ [\mathrm{J}_3] \end{bmatrix} [T],$$

likewise

$$\begin{bmatrix} \dot{q}_1^2 \\ \dot{q}_2^2 \\ \dot{q}_3^2 \end{bmatrix} = \begin{bmatrix} [\mathrm{J}_1] \\ [\mathrm{J}_2] \\ [\mathrm{J}_3] \end{bmatrix} [R_1][T]$$

and

$$\begin{bmatrix} \dot{q}_1^3 \\ \dot{q}_2^3 \\ \dot{q}_3^3 \end{bmatrix} = \begin{bmatrix} [\mathrm{J}_1] \\ [\mathrm{J}_2] \\ [\mathrm{J}_3] \end{bmatrix} [R_2][T],$$

where

$$[T] = \begin{bmatrix} \dot{x} & \dot{y} & \dot{\theta} \end{bmatrix}^{\mathrm{T}},$$

$$[J_i] = [J_{ix} \quad J_{iy} \quad J_{i\theta}], \quad i = 1, 2, 3,$$

$$[R_1] = \begin{bmatrix} \cos(120^\circ) & -\sin(120^\circ) & 0 \\ \sin(120^\circ) & \cos(120^\circ) & 0 \\ 0 & 0 & 1 \end{bmatrix} = \begin{bmatrix} -\sin(30^\circ) & -\cos(30^\circ) & 0 \\ \cos(30^\circ) & -\sin(30^\circ) & 0 \\ 0 & 0 & 1 \end{bmatrix}$$

and

$$[R_2] = \begin{bmatrix} \cos(240^\circ) & -\sin(240^\circ) & 0 \\ \sin(240^\circ) & \cos(240^\circ) & 0 \\ 0 & 0 & 1 \end{bmatrix} = \begin{bmatrix} -\sin(30^\circ) & \cos(30^\circ) & 0 \\ -\cos(30^\circ) & -\sin(30^\circ) & 0 \\ 0 & 0 & 1 \end{bmatrix}.$$

Considering the velocities $\dot{q}_1^i$, $i = 1, 2$ and 3, we have

$$\begin{bmatrix} \dot{q}_1^1 \\ \dot{q}_1^2 \\ \dot{q}_1^3 \end{bmatrix} = \begin{bmatrix} [\mathrm{J}_1] \\ [\mathrm{J}_1][R_1] \\ [\mathrm{J}_1][R_2] \end{bmatrix} [T] = [\Omega_1]\,[T].$$

The stiffness contribution of variables $\dot{q}_1^i$, $i = 1, 2$ and 3, can be obtained by [8]

$$\begin{aligned} K_1 &= k_{q1}[\Omega_1]^{\mathrm{T}}[\Omega_1], \\ &= k_{q1}([\mathrm{J}_1]^{\mathrm{T}}[\mathrm{J}_1] + [R_1]^{\mathrm{T}}[\mathrm{J}_1]^{\mathrm{T}}[\mathrm{J}_1][R_1] + [R_2]^{\mathrm{T}}[\mathrm{J}_1]^{\mathrm{T}}[\mathrm{J}_1][R_2]). \end{aligned}$$

Arranging K_1 in matrix form gives

$$K_1 = \begin{bmatrix} \frac{3}{2}k_{q1}(J_{1x}^2 + J_{1y}^2) & 0 & 0 \\ 0 & \frac{3}{2}k_{q1}(J_{1x}^2 + J_{1y}^2) & 0 \\ 0 & 0 & 3k_{q1}J_{1\theta}^2 \end{bmatrix}.$$

Likewise, K_2 and K_3, the stiffness contributed by variables $\dot{q}_2^i$ and $\dot{q}_3^i$, $i = 1$, 2 and 3, can be calculated. The system total stiffness matrix can be combined as

$$K = K_1 + K_2 + K_3.$$

Finally, the above stiffness matrix can be written in the form

$$K = \mathrm{diag}(k_x, k_y, k_\theta),$$

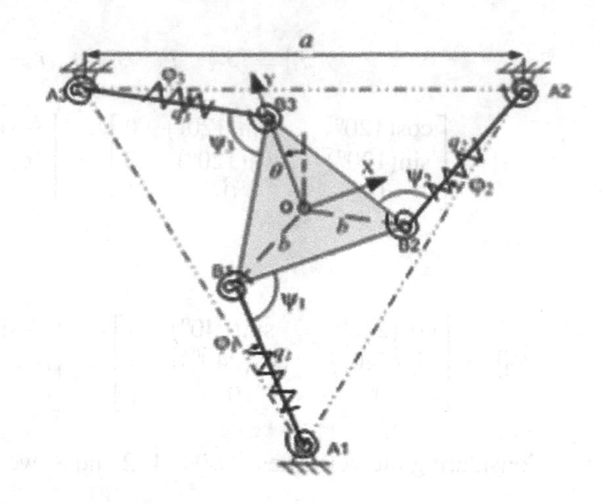

Fig. 6.10 A planar isotropic parallel platform with three RPR legs

where

$$k_x = \frac{3}{2}(J_{1x}^2 + J_{1y}^2)k_{q1} + \frac{3}{2}(J_{2x}^2 + J_{2y}^2)k_{q2} + \frac{3}{2}(J_{3x}^2 + J_{3y}^2)k_{q3},$$
$$k_x = k_y,$$
$$k_\theta = 3(k_{q1}J_{1\theta}^2 + k_{q2}J_{2\theta}^2 + k_{q3}J_{3\theta}^2).$$

6.3.2 *Case Studies*

A leg configuration is generally characterized by its Jacobian matrix. For platforms with identical leg structure, only the Jacobian matrix of one of the three legs is needed when using (6.11)–(6.14). As examples, two 3-legged platforms, having the leg configurations of RPR and RRR, respectively, are used to demonstrate how to apply the above equations to obtain the output stiffness matrix.

6.3.2.1 RPR Leg Configuration

The leg with RPR configuration includes a prismatic joint (P) and two revolution joints (R), arranged in R–P–R sequence. A compliant planar parallel mechanism with three RPR legs is shown in Fig. 6.10. Since only one leg need to be considered, the vector of joint variables of each leg can be denoted as $[q, \varphi, \psi]^{\mathrm{T}}$ and the corresponding joint stiffness as k_q, k_a, and k_b. Considering leg 1, its Jacobian matrix is

$$[J] = \begin{bmatrix} -S(30^{\circ} - \varphi - \theta) & C(30^{\circ} - \varphi - \theta) & -bC(\varphi + \theta) \\ C(30^{\circ} - \varphi - \theta)/q & S(30^{\circ} - \varphi - \theta)/q & bS(\varphi + \theta)/q \\ C(30^{\circ} - \varphi - \theta)/q & S(30^{\circ} - \varphi - \theta)/q & (bS(\varphi + \theta) + q)/q \end{bmatrix},$$

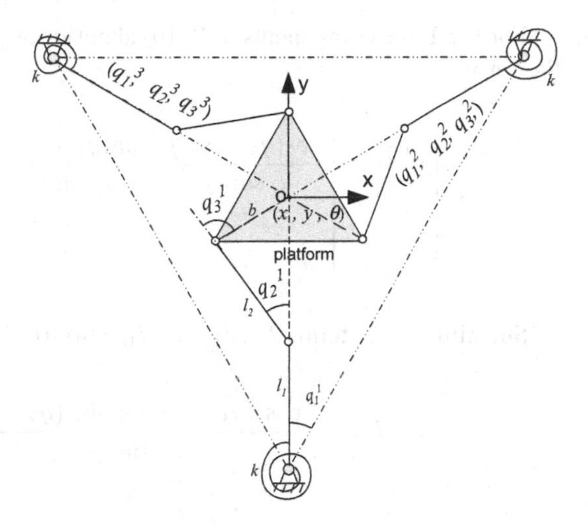

Fig. 6.11 Planar symmetric parallel platform with RRR leg

where S and C stand for *sine* and *cosine* functions respectively, θ is the rotational angle of the platform and b is the distance from geometric center O to the end-joints of the legs. From this matrix, the output stiffness of the platform can be directly obtained by using (6.11)–(6.14), i.e.,

$$K = \operatorname{diag}\left(k_x, k_y, k_\theta\right),$$

where

$$k_x = \frac{3}{2}\left(k_q + \frac{1}{q}k_a + \frac{1}{q}k_b\right), \tag{6.15}$$

$$k_y = \frac{3}{2}\left(k_q + \frac{1}{q}k_a + \frac{1}{q}k_b\right), \tag{6.16}$$

$$k_\theta = 3k_q b^2 \cos^2(\varphi + \theta) + \frac{3k_a}{q^2} b^2 \sin^2(\varphi + \theta) + \frac{3k_b}{q^2}\left(b \sin(\varphi + \theta) + q\right)^2. \tag{6.17}$$

6.3.2.2 RRR Leg Configuration

In an RRR leg configuration, all three joints are revolute joints. Figure 6.11 is an example of a platform supported by three RRR legs. This example was also used in [16]. Please note that only joint 1 has compliance in this mechanism. Since only one joint contributes compliance, it is not necessary to consider all elements in the Jacobian matrix.

For leg 1, the components in (6.10) about q_1 or joint 1 can be easily obtained in the form

$$\begin{bmatrix} \dot{q}_1^1 \\ \dot{q}_1^2 \\ \dot{q}_1^3 \end{bmatrix} = \begin{bmatrix} \dfrac{\cos(q_3+30^o)}{l_1 \sin q_2} & \dfrac{\sin(q_3+30^o)}{l_1 \sin q_2} & -\dfrac{b \sin q_3}{l_1 \sin q_2} \\ ? & ? & ? \\ ? & ? & ? \end{bmatrix} \begin{bmatrix} \dot{x} \\ \dot{y} \\ \dot{\theta} \end{bmatrix}.$$

Substituting the terms J_{1x}, J_{1y} and $J_{1\theta}$ into (6.11)–(6.14) gives:

$$k_x = k_y = \frac{3}{2}\frac{\cos^2(q_3+30^o)+\sin^2(q_3+30^o)}{l_1^2 \sin^2 q_2} k = \frac{3k}{2l_1^2 \sin^2 q_2},$$

$$k_\theta = \frac{3kb^2 \sin^2(q_3)}{l_1^2 \sin^2(q_2)}.$$

The results are the same as those presented in [16], but the development is more straightforward.

6.4 Design of a 3-Legged Flexure-Based Parallel Kinematics Stage

The major purpose of the flexure-based stage design here is to determine its configuration and key dimensions so that its output stiffness can meet the specified stiffness requirements. According to the process analysis of fiber insertion in the MEMS switch assembly, to ensure successful assembly, the required x or y stiffness, k_L, is about 20 mN/μm and the required torsion stiffness, k_T, is about 12 mN m per degree [6]. Further analysis of these specifications shows that the torsion stiffness requirement is relatively stringent comparing to the translational stiffness, requiring prior considerations in the design.

6.4.1 Supporting Leg Configuration

Due to relatively simple kinematics, the leg configuration with RPR architecture is adopted in our fixture mechanism design. From (6.17), it can be found that when $\varphi + \theta = 90°$ the prismatic joint will have no contribution to torsion stiffness k_θ as

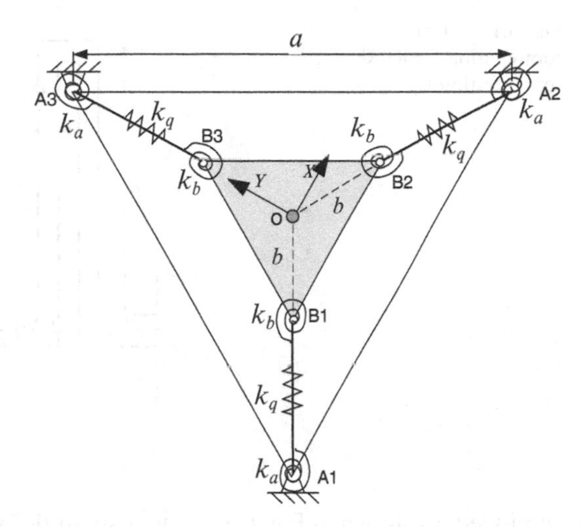

Fig. 6.12 Minimum torsion stiffness configuration

shown in Fig. 6.12. Actually, this is a singularity configuration, having minimum overall torsion stiffness. Thus, (6.15)–(6.17) become

$$k_x = k_y = \frac{3}{2}(k_q + \frac{1}{q}k_a + \frac{1}{q}k_b),$$
$$k_\theta = \frac{3k_a}{q^2}b^2 + \frac{3k_b}{q^2}(b+q)^2. \tag{6.18}$$

To simplify the design, two revolute joints are assumed having the same compliance. Accordingly, the joint stiffness can be determined from (6.18) in the form

$$k_a = \frac{q^2 k_\mathrm{T}}{3[b^2 + (b+q)^2]},$$
$$k_q = \frac{2}{3}k_\mathrm{L} - \frac{q k_\mathrm{T}}{b^2 + (b+q)^2}. \tag{6.19}$$

Equation (6.19) provides a formulation to calculate the joint stiffness from the given output stiffness. The joint stiffness depends on not only the output stiffness in Cartesian space, but also the structure dimensions. Thus even if the value of k_T is small, a high value of k_a is still achievable by selecting the suitable structure dimension. It will increase manufacturability of the flexure-based fixture using a relatively high value of torsion stiffness k_a for joints.

6.4.2 Flexure Design

In general, flexural elements can be made in the forms of notch hinges and leaf-springs types as well as the combination of them. To form a RPR-configuration leg, a combined flexural structure including two notch hinges and one dual-leaf spring is

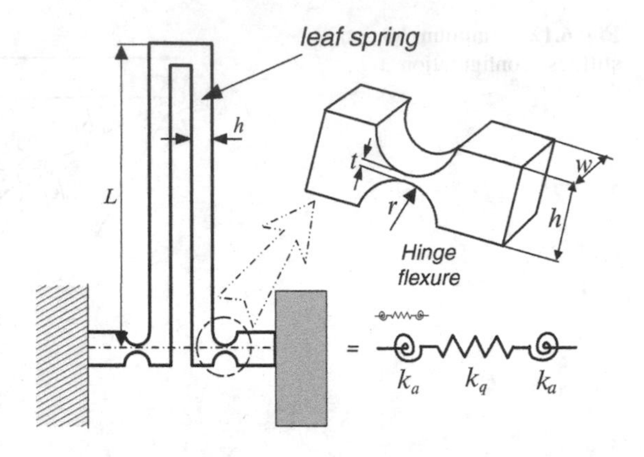

Fig. 6.13 Flexure combination for RPR configuration leg

employed, as shown in Fig. 6.13. Under small deformation, we can treat the notch hinges as revolute joints and the dual-leaf springs as a prismatic joint. Their joint stiffness values are obtained by [17]

$$k_a = \frac{2Ewt^{5/2}}{9\pi r^{1/2}}, \tag{6.20}$$

$$k_q = \frac{Ew}{8}\left(\frac{h}{L}\right)^3, \tag{6.21}$$

where E is the Young's module of the material, L is the length of leaf spring, w, h, t and r are the flexure geometric dimensions (see Fig. 6.13). Substituting (6.20) and (6.21) into (6.19), the flexure dimensions can be determined according to the values of k_{T} and k_{L}. Figure 6.14 shows the curves of k_{T} and k_{L} against the flexure dimensions.

6.4.3 FEA Simulation

FEA simulation aims to check the platform's compliance performance and to fine tune flexure dimensions so that the specified requirements can be satisfied at the design phase. It can be implemented through the software ANSYS. In general, the flexure sizes obtained from the simplified model in the previous sections are treated as the initial dimensions (Fig. 6.15). Based on these dimensions, a 3D model of the stage is created using CAD tools such as SoildWork or Pro/E. The model is then simulated by using ANASYS in terms of the output stiffness of the platform and the maximum stress of the flexure joints (Fig. 6.16). An iterative design approach can be utilized to modified the flexure sizes and determine the final flexure dimensions.

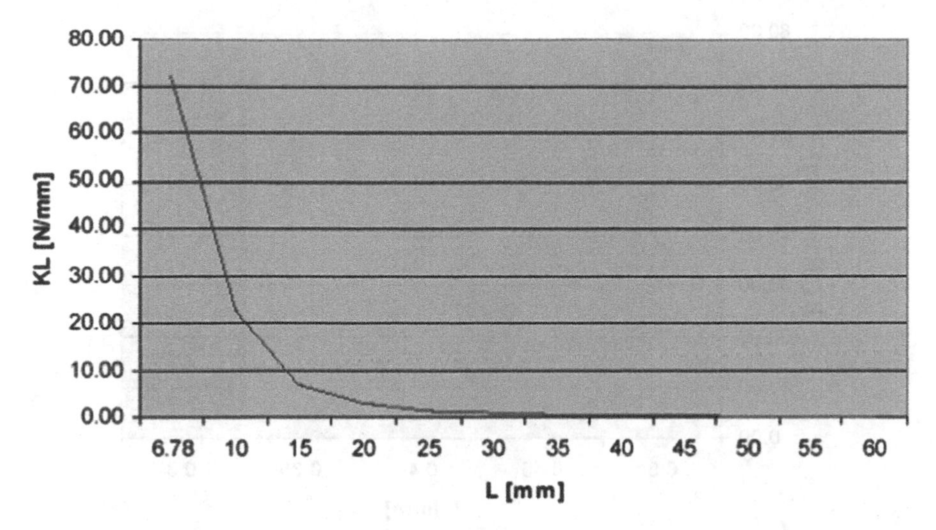

Fig. 6.14 Output stiffness k_L, against flexure length L

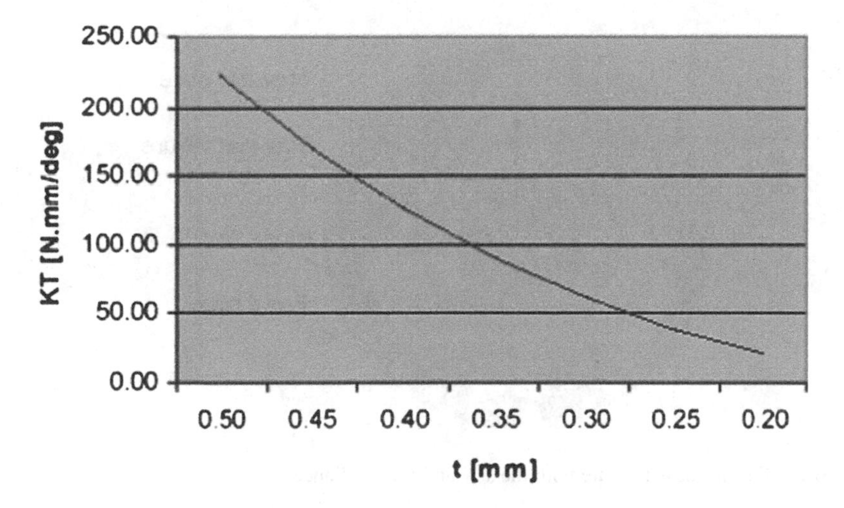

Fig. 6.15 Output stiffness k_T, against flexure thickness t

Figure 6.17 shows a final design of the flexure-based stage that is used for the optical MEMS assembly.

In simulation, the maximum stress of flexure joints is requested less than the elastic limit of the material so as to ensure the flexure deformation within the elastic range. In addition, the ratio of the maximum stress to the yield strength should be as small as possible. According to the theory of fatigue failure, when this ratio is less than 10–15% [13], flexure performance can remain stable for long period of time (without fatigue failure).

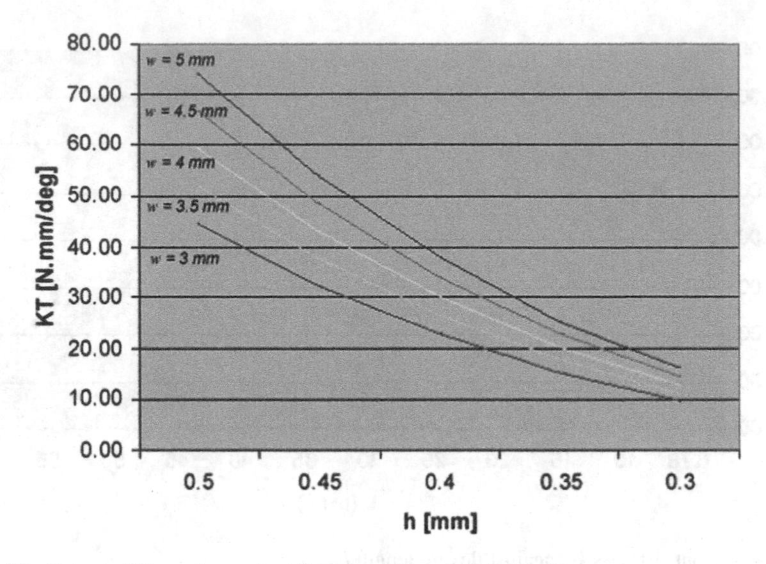

Fig. 6.16 Output stiffness k_T, against flexure width w and height h

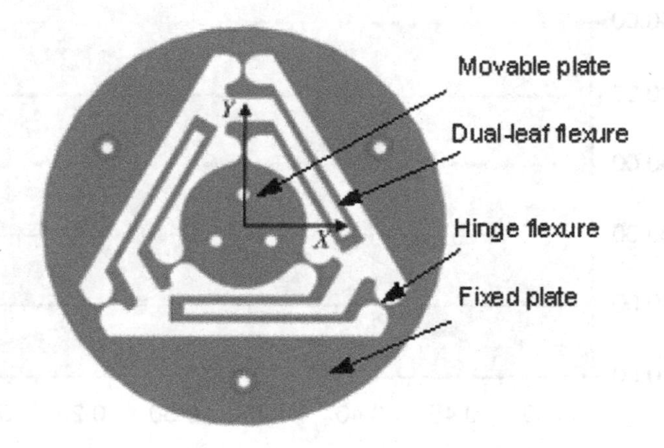

Fig. 6.17 Flexure-based fixture with the x, y and θ compliance

For the stage as shown in Fig. 6.17, when the translation distance reaches 20 μm and the twist angle is 0.1°, i.e., the required motion range of the platform, the maximum stress on the flexural hinges is 0.035 GPa. It is far less than the elastic limit (0.48 GPa) of the material. The ratio of the maximum stress to the yield strength is 9.3%.

The stage prototype is fabricated with a piece AL plate through the wire-cut machining. The compliance behaviors obtained from the simulation and the actual measurement are shown in Fig. 6.18. The translational stiffness of the platform of the fixture is 19.38 N/mm and the torsion stiffness is 11.62 mN m per degree.

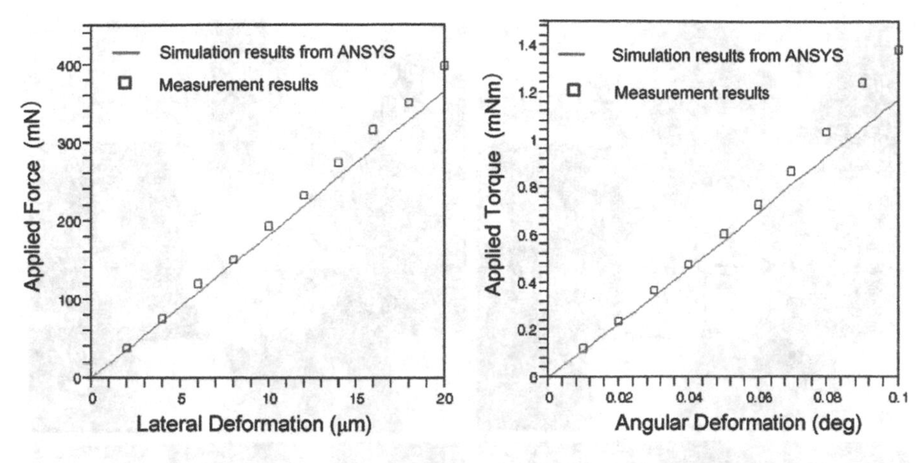

Fig. 6.18 Compliance behavior of the flexural fixture (© IEEE 2008), reprinted with permission

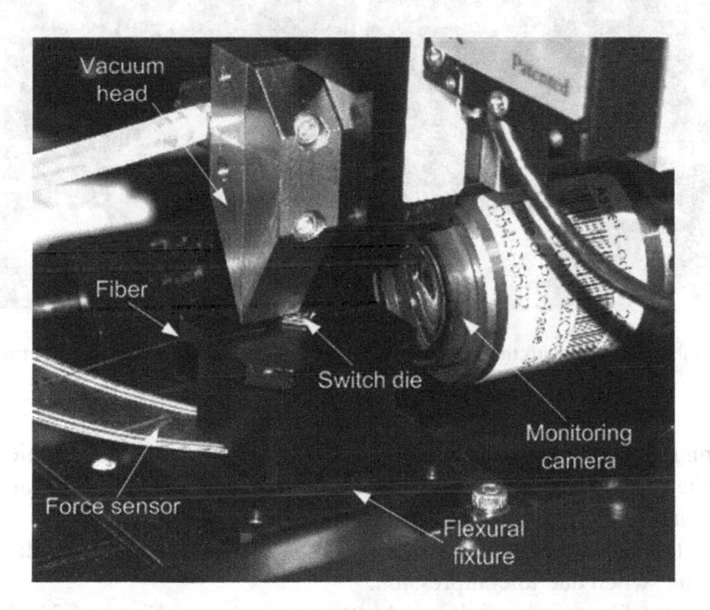

Fig. 6.19 Passive microassembly system with the flexural fixture (© IEEE 2008), reprinted with permission

6.5 Experimental Studies

To evaluate the performance of the flexure-based fixture, an experiment is setup as shown in Fig. 6.19. The process of fiber insertion with the help of the flexural fixture under an initial lateral offset of 12 μm is exhibited in Fig. 6.20. Figure 6.20 (a) is

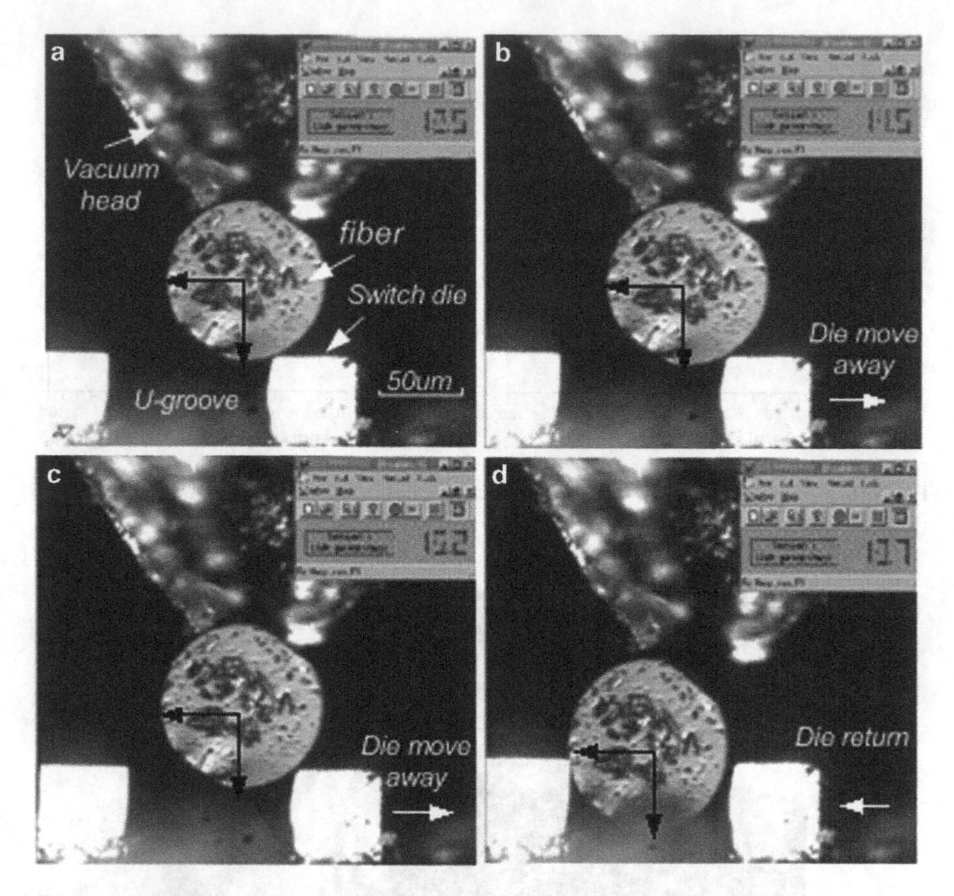

Fig. 6.20 Fiber insertion with the flexural fixture (© IEEE 2008), reprinted with permission

the beginning of the insertion; Fig. 6.20 (b) and (c) show the switch die moving towards the right side to accommodate the initial offset; Fig. 6.20 (d) is at the state of insertion completion. In the entire insertion, no fiber rotation (or skidding) relative to the vacuum head was observed. In addition, there were no damages in the fiber and switch due to compression.

The contact forces are monitored by the force sensor. Figure 6.21 shows the curves of the maximum contact forces for the cases with and without the use of the flexural fixture. The maximum contact force without the use of the fixture is much higher than that with the fixture and increases sharply when the lateral offset is bigger than 6 μm. With the designed passive compliance in the fixture mechanism, the maximum contact force is regulated to a low level. The experimental result shows that even if the lateral offset is up to 12 μm or yaw angle offset is up to 0.07° the maximum contact force is still less than 100 mN, which is well below the fiber skidding threshold of 160 mN and the damage limit of 480 mN in the switch die (Fig. 6.21).

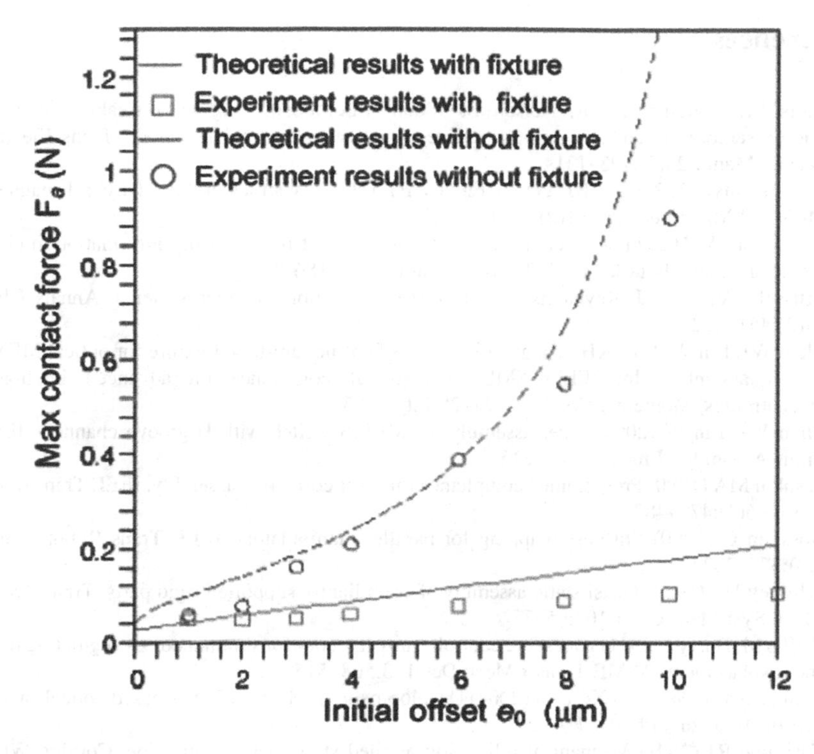

Fig. 6.21 Comparison of the Max contact force with and without fixture (© IEEE 2008), reprinted with permission

6.6 Conclusions

This Chapter describes a low-cost passive method to eliminate the problems of fiber skidding and component damage during assembly of MEMS optical switches with U-groove channels. In order to understand the problems affecting the fiber assembly, fiber insertion process was analyzed. It was found that high contact forces caused by misalignment may result in component damage or fiber skidding from the holder. The controlled environmental compliance can regulate the high contact force and accommodate the assembly errors, therefore, ensuring the fiber assembly successfully implemented. A 3-legged flexure-based parallel kinematics stage (or fixture) was developed to compensate the assembly errors in terms of the lateral offsets and the angular offsets. The fixture was designed based on the conditions of successful fiber insertion and the principle of flexures. It was shown that with the help of the fixture, the fiber can be successfully inserted into the U-groove even with the presentence of lager lateral and the angular misalignments. Hence, it is concluded that the high-end MEMS optical switch assembly can be realized through relatively inexpensive tooling.

References

1. Ralis SJ, Vikramaditya B, Nelson BJ (2000) Micropositioning of a weakly calibrated microassembly system using coarse-to-fine visual serving strategies. IEEE Trans Electron Packag Manuf 23(2):123–131
2. Vikramaditya B, Nelson BJ, Ge Y, Enikov ET (2001) Microassembly of hybrid magnetic MEMS. J Micromechatron 1(2):99–116
3. Yamamoto Y, Hashimoto T et al (2002) Measurement of force sensory information in ultra-precision assembly tasks. IEEE Trans Mechatron 7(2):186–189
4. Brussel HV, Peirs J, Reynaerts D et al (2000) Assembly of microsystems. Annals CIRP 49(2):451–472
5. Chen WJ, Lin W, Low KH, Yang G (2005) A 3-D of flexure-based fixture for optical MEMS switch assembly. In: IEEE/ASME international conference on advanced intelligent mechatronics, Monterey, CA, USA, 24–28 July 2005
6. Chen WJ, Lin W (2008) Fiber assembly for MEMS switch with U-groove channels. IEEE Trans Autom Sci Eng 5(2):207–215
7. Peshkin MA (1990) Programmed compliance for error corrective assembly. IEEE Trans Robot Autom 6(4):473–482
8. Gosselin C (1990) Stiffness mapping for parallel manipulators. IEEE Trans Robot Autom 6(3):377–382
9. Whitney DE (1982) Quasi-static assembly of compliantly supported rigid parts. Trans ASME J Dyn Syst Meas Contr 104:65–77
10. Griffis M, Duffy J (1991) Kinestatic control: a novel theory for simultaneously regulating force and displacement. ASME Trans J Mech Des 113:508–515
11. Wang ZF, Cao W, Shan XC et al (2004) Development of a 1 × 4 MEMS-based optical switch. Sensor Actuator 114:80–87
12. Castilone RJ (2001) Mechanical reliability: applied stress design guideline, Corning White Paper, WP5053
13. Howell LL (2001) Compliant mechanisms. Wiley, New York
14. Lipkin H, Petterson T (1992) Generalized center of compliance and stiffness. In: Proceedings of IEEE international conference on robotics and automations, Nice, France, vol 2. pp 1251–1256, May 1992
15. Bonev LA, Zlatanov D, Gosselin CM (2003) Singularity analysis of 3-D of planar parallel mechanisms via screw theory. ASME Trans J Mech Des 125:573–580
16. Kim WK, Yi BJ, Cho W (2000) RCC characteristics of planar/spherical three degree-of freedom parallel mechanism with joint compliances. Trans ASME J Mech Des 122:10–16
17. Alexander HS (1992) Precision machine design. Prentice Hall, NJ. ISBN 0-13-690918-3

Chapter 7
Micro-Tactile Sensors for In Vivo Measurements of Elasticity

Peng Peng and Rajesh Rajamani

Abstract In this chapter, a sensing approach for the measurement of both contact force and elasticity is introduced and discussed. By using the developed method, the elasticity of various objects (e.g., tissue) can be measured by simply touching the targeted object with the sensor. Each developed sensor consists of a pair of contact elements that have different values of stiffness. During contact, the relative deformation of the two sensing components can be used to calculate the Young's modulus of elasticity. Several prototypes of tactile sensors have been fabricated through various MEMS processes. One of the prototypes developed through a polymer MEMS process has a favorable flexible structure, which enables the sensor to be integrated on end-effectors for robotic or biomedical applications. Finally, the tactile sensor has been attached on a touch probe and tested in a handheld mode. An estimation algorithm for this handheld device, which employs a recursive least squares method with adaptive forgetting factors, has also been developed. Experimental results show that this sensor can differentiate between a variety of rubber specimens and has the potential to provide reliable in vivo measurement of tissue elasticity.

7.1 Introduction

Elasticity measurement is important in biomedical sensing, robotics, and various industrial applications. For instance, in robotic manipulation, knowledge of elasticity of the targeted object would enable better control of the contact force in a precision grasp [1]. In biomedical applications, in vivo measurement of tissue elasticity can facilitate doctors to reach a reliable palpation diagnosis [2], because

P. Peng • R. Rajamani (✉)
Department of Mechanical Engineering, University of Minnesota, Minnesota, USA
e-mail: pengpeng@me.umn.edu; rajamani@me.umn.edu

D. Zhang (ed.), *Advanced Mechatronics and MEMS Devices*, Microsystems,
DOI 10.1007/978-1-4419-9985-6_7, © Springer Science+Business Media New York 2013

many diseases change the physical properties, especially elasticity, of natural tissues or organs [3, 4].

The knowledge of tissue elasticity is also a valuable tool in minimally invasive surgery (MIS) and telerobotic surgery. MIS offers many advantages over open surgery including fewer complications, less discomfort after the operation, faster recovery times and lower health care costs. However, one significant existing difficulty is the loss of the intraoperative tactile feedback [5] which is readily obtained in open surgery by touching the tissue. In order to recover tactile sensing during MIS, measurements of contact force and elasticity are vital. In addition to MIS, elasticity measurements could also be used in other biomedical applications such as ligament tension measurement during knee implant surgery, early detection of compartment syndrome and cartilage hardness measurements.

One major technology for elasticity or stiffness measurement of tissues is based on measurement of force-deformation response. In this method, a controlled load or displacement is first applied on the tissue site and the corresponding deformation or force is then recorded [6–10]. Indentation type tissue stiffness tests have been performed on cancerous breast tissue by estimating elastic moduli from measured force displacement curves [10]. In telerobotics research, a laparoscopic grasper attached to a robot arm [6] has been designed to provide force and vision feedback. The stiffness of the testing object was tested by measuring the applied force and the angular displacement of the jaw. Another work on tissue based measurements employs piezoelectric cantilevers to investigate the force-deformation response of soft tissue [9]. As an active sensing alternative, an air-driven oscillating indenter was developed to detect bulk tissue compliance such as soft tissue compliance measurements in stumps of amputated lower limbs [11]. The pressure was controlled by a flexible rubber hose and the displacement was recorded by an electromagnetic sensor.

Another approach for elasticity measurement is based on magnetic resonance elastography [12] or ultrasonic shear wave elasticity imaging [13, 14]. In these methods, a harmonic mechanical excitation (up to a 1,000 Hz) is applied on the target material or tissue and then magnetic resonance imaging (MRI) or ultrasonic imaging can be used to probe local tissue deformation. The most significant benefit of this method is to enable noninvasive in vivo measurement of mechanical properties of tissues. Furthermore, a focused ultrasound beam can be used to generate a localized strain of tissue in the vicinity of its focal point. This local strain could be measured by optical method, MRI, or a separate ultrasound detection beam. Recent advances [14, 15] with this technique also enable the measurement of viscoelasticity properties of the tissue, which is believed to be another useful index of tissue health.

Another technical approach of measuring stiffness/softness of the targeted object is based on the measurement of contact impedance [16]. A ceramic piezoelectric transducer operating at its resonance frequency has been developed to detect this contact impedance [17]. This type of sensor is composed of two pieces of piezoelectric components connected through a feedback circuit, which make the sensor vibrate at its resonance frequency. The resonance frequency shifts when the sensor contacts an object due to the change of contact impedance. This frequency change, at a constant load or a constant contact area, is related to the spring stiffness of the object [18]. In order to measure microscale local elasticity, a glass needle with a pin head of 10 μm can be attached on the sensor to form a contact interface [19]. By utilizing this type of sensor along with a step-motor controlled platform, a tactile mapping system has been developed to obtain a contour image and topographical Young's modulus information. Characterization of tissue slices such as porcine heart [20], human prostate [21], and vascular segments [22] has been performed by this system.

The tactile sensor that will be discussed extensively in this chapter is based on a method for in vivo tissue elasticity measurement which does not involve applying a controlled load or displacement. This technology differs from the rest since no displacement information of the end-effectors needs to be obtained. Dargahi et al. [23] proposed a softness sensor which consists of two coaxial cylinders with different stiffnesses. Piezoelectric elements (PVDF) are placed beneath the cylinders for force measurements. A dynamic load driven by a vibrating unit is applied during the contact in order to generate sinusoidal shape signals to excite the piezoelectric units. There are challenges associated with miniaturization of this sensor for endoscopic applications.

Due to a size constraint for surgical tools, a tactile sensor developed through a MEMS process would be desirable for practical usage. Early tactile sensors have been developed through conventional MEMS processes, either by bulk-micromachining [24, 25] or by surface-micromachining techniques [26, 27]. These tactile sensors are designed for contact force measurement and cannot measure elasticity. The sensing elements of these microsensors usually consist of sensing diaphragms constructed from silicon-based materials. The deflection of these sensing diaphragms can be interpreted precisely by either capacitive sensing techniques or piezo-effects of certain structural materials. The most significant features of these tactile sensors include good sensitivity, low mechanical cross-talk due to isolated sensing cells, and considerably high sensor density stemming from the fine precision of the semiconductor fabrication process. However, pilot research in this team shows some drawbacks for these types of sensors, which includes the fragile nature of the sensing diaphragms and wire-bonding interfaces.

Micro-tactile sensors built by flexible materials (e.g., polymers) would address the aforementioned bottlenecks. Moreover, a flexible sensor can adhere well onto the curved surfaces of end-effectors of various geometries. A number of flexible tactile sensors have been developed by utilizing various polymer materials such as a broad piece of fabric [28], polyimide (PI) [29, 30], or polydimethylsiloxane (PDMS) [31]. Again, these sensors are capable purely of measuring contact force, not tissue elasticity. This chapter will cover the development of micro-tactile sensors for measurement of both force and elasticity. These sensors are developed through both surface-micromachining and polymer MEMS processes. Each type of sensor is elaborated and evaluated individually throughout this chapter.

7.2 Sensing Principles

7.2.1 A Spring-Pair Model

In its simplest embodiment, the proposed tactile sensor consists of two sensing diaphragms with different stiffness values. A simple spring model is illustrated in Fig. 7.1. In this model, the tissue under investigation can be viewed as an elastic spring with a stiffness value of k_t. The sensor is modeled by two springs of different spring constants, k_h and k_s. By pushing the sensor towards the targeted tissue, this spring pair will have different amounts of deflections. As shown in Fig. 7.1b, the soft spring (k_s) will undergo a larger deflection compared to the hard counterpart (k_h). A compatibility condition can be derived from the fact that the sensor has a solid base, thus resulting in the same displacements at the bottom ends of the sensing springs. This condition is shown in (7.1).

$$\begin{aligned} &\delta x_1 = \delta x_2, \\ &\Rightarrow \frac{F_h(k_h + k_t)}{k_h k_t} = \frac{F_s(k_s + k_t)}{k_s k_t}. \end{aligned} \tag{7.1}$$

The tissue stiffness can then be calculated from (7.2), which is obtained by rearranging (7.1):

$$k_t = \frac{(F_h/F_s - 1)k_h k_s}{k_h - F_h/F_s k_s}, \tag{7.2}$$

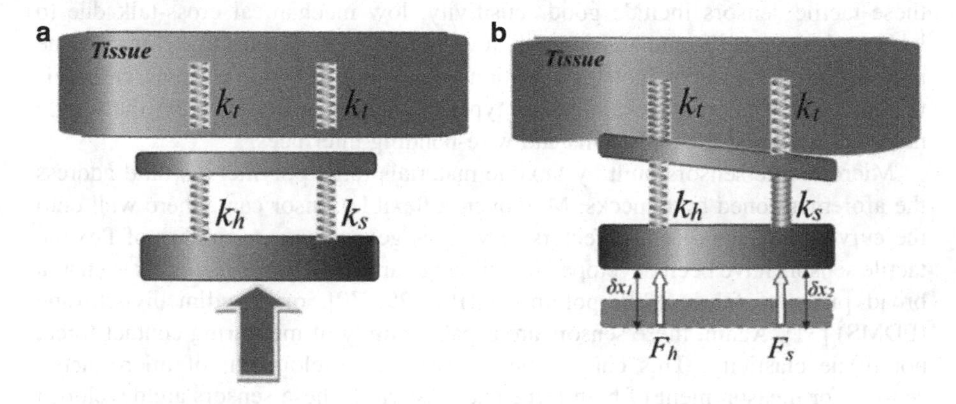

Fig. 7.1 (**a**) Schematic diagram of the tactile sensor and the tissue under investigation. (**b**) Schematic diagram of the contact condition between the tactile senor and targeted tissue (© IEEE 2009), reprinted with permission

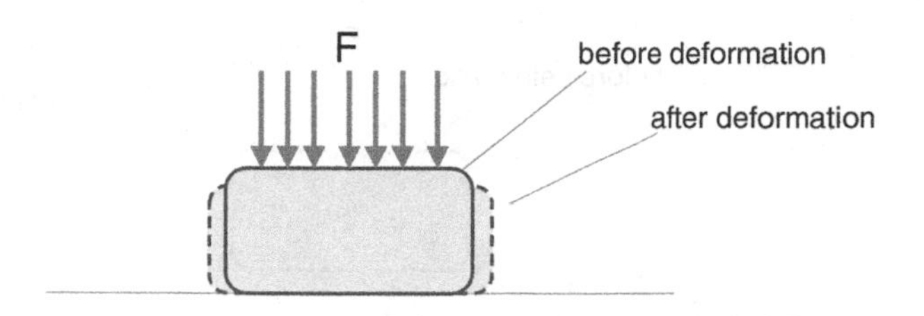

Fig. 7.2 Tissue modeled as a one-degree-of-freedom system under uniform load

where F_{h} and F_{s} are the forces applied on the hard spring and soft spring, respectively. If the defections of the springs can be measured, then with $F_h = k_h \delta x_h$ and $F_{\mathrm{s}} = k_{\mathrm{s}} \delta x_{\mathrm{s}}$ the equation can be further derived as shown in (7.3):

$$k_{\mathrm{t}} = \frac{k_{\mathrm{s}} - k_{\mathrm{h}}(\delta x_{\mathrm{h}}/\delta x_{\mathrm{s}})}{(\delta x_{\mathrm{h}}/\delta x_{\mathrm{s}}) - 1}, \tag{7.3}$$

where δx_{h} and δx_{s} are deflections of the hard spring and soft spring, respectively.

7.2.2 A More Precise Contact Model

Since this tactile sensor is designed to recover haptic sensory feedback, the operating load is comparable to the contact force generated by human hand. Therefore it is reasonable to assume that the tissue under contact by this sensor deforms within its linear range, which is similar to the deformation provided during a hand touch. If the size of the contact area of the sensor is close to the dimension of the tissue site, the tissue can be modeled as a one-degree-of-freedom system under uniform load, as can be seen in Fig. 7.2. Then the magnitude of deformation depends on both the tissue elasticity and the tissue thickness.

In this case, the tissue deformation magnitude can be represented as (7.4):

$$y_{\mathrm{t}} = \frac{F}{E_{\mathrm{t}}A/L}, \tag{7.4}$$

where E_{t} is the Young's modulus of the tissue, A is the cross-section area and L is the thickness of the tissue sample. On the other hand, if the deforming force does not occur uniformly over the entire cross section area of the tissue, but instead occurs over a very small area compared to the size of the tissue cross-section, then the tissue will undergo local flexible body deformation, as shown in Fig. 7.3.

In this case the deformation of the tissue at the center of the loading area is given by

$$w_{\mathrm{t}}|_{z=0} = \frac{2(1 - v^2)qa}{E_{\mathrm{t}}}, \tag{7.5}$$

where E_{t} is the tissue Young's modulus, υ is the Poisson's ratio, q is the force per unit area, a is the radius of the loading zone, and w_{t} is the tissue deformation as illustrated in Fig. 7.4 [32].

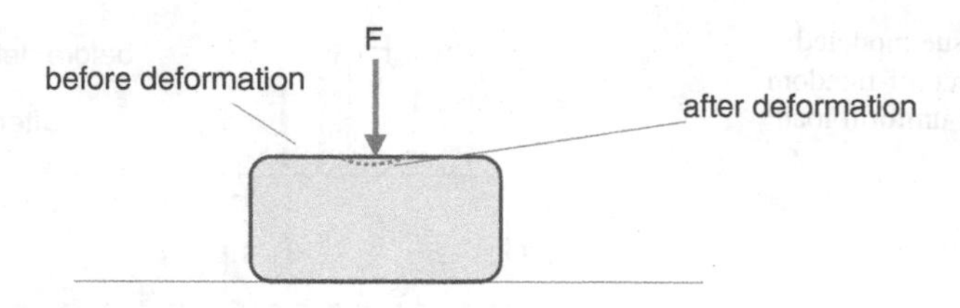

Fig. 7.3 Local deformation on a tissue site

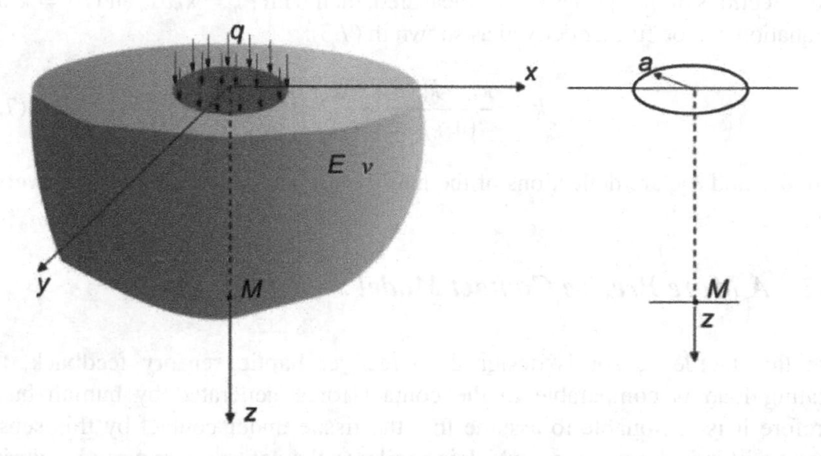

Fig. 7.4 Contact model of tissue site with a local deformation (© Elsevier 2010), reprinted with permission

Assume that both sensing membranes in the tactile sensor have the same cross section area and thickness. Let the force per unit area on the soft sensor be q_s and on the hard sensor be q_h. Then, using $\delta x_1 = \delta x_2$, it can be seen that

$$\frac{2(1-v^2)q_h a}{E_t} + y_h = \frac{2(1-v^2)q_s a}{E_t} + y_s \tag{7.6}$$

or

$$\frac{2(1-v^2)q_h a}{E_t} + \frac{q_h A}{E_h A/L} = \frac{2(1-v^2)q_s a}{E_t} + \frac{q_s A}{E_s A/L}, \tag{7.7}$$

where E_h and E_s represent the values of Young's modulus for the hard and soft sensing nodes, respectively. The tissue elasticity can then be calculated as in (7.8):

$$E_t = \frac{2(1-v^2)a}{L}\left(\frac{q_h/q_s - 1}{(1/E_s) - (1/E_h)(q_h/q_s)}\right). \tag{7.8}$$

As can be seen, the tissue elasticity can be obtained by measuring the ratio of the force distributions q_s and q_h. It is worthwhile to mention that the elasticity can be measured instead of stiffness if the assumption that the dimension of the targeted tissue is much larger than the contact area of the sensor is satisfied. This may not always be the contact condition for tissue characterizations. However, a MEMS sensor would definitely bring the advantage of small size, and thus results in a contact model close to the above assumption.

7.3 Prototype Micro-Tactile Sensors

As presented in (7.3), the deflections of the sensing elements need to be measured to calculate the tissue elasticity. These deflections can be measured by a membrane based capacitive sensor as shown in Fig. 7.5 in the first generation prototype sensors developed by this team [33, 34]. In this simple configuration, a sensing membrane (diaphragm) is suspended over the substrate. There are two electrodes with one on top of the membrane and the other on the substrate. The deflection of the membrane causes a change of the capacitance between the two electrodes which can be measured and be converted back into a deflection readout. The stiffness of this membrane will be defined by the geometry (i.e., diameter) of the membrane. Several membranes with identical membrane thickness and gap height but different diameters will have different compliances under the same load.

The prototype MEMS sensors have been fabricated using a surface micromachining process. The fabrication process is shown schematically in Fig. 7.6 [35]. The sensing membrane is made of silicon nitride while the top and bottom electrodes are made of gold.

The fabrication process starts with a silicon wafer which is first covered with 6,000Å of silicon nitride (SiNx) layer for passivation. To deposit this layer, plasma enhanced chemical vapor deposition (PECVD) is employed (Fig. 7.6a). Then a Cr–Au metal layer is E-beam evaporated to form the bottom electrodes with a thickness of 2,600Å (Fig. 7.6b). This is followed by a sacrificial aluminum

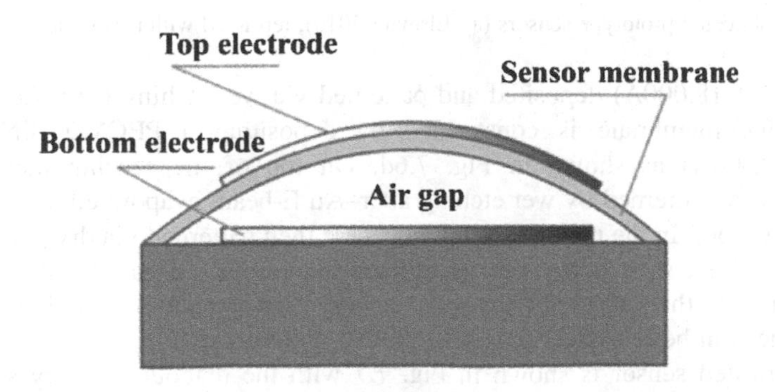

Fig. 7.5 Conceptual diagram of a capacitive sensing membrane

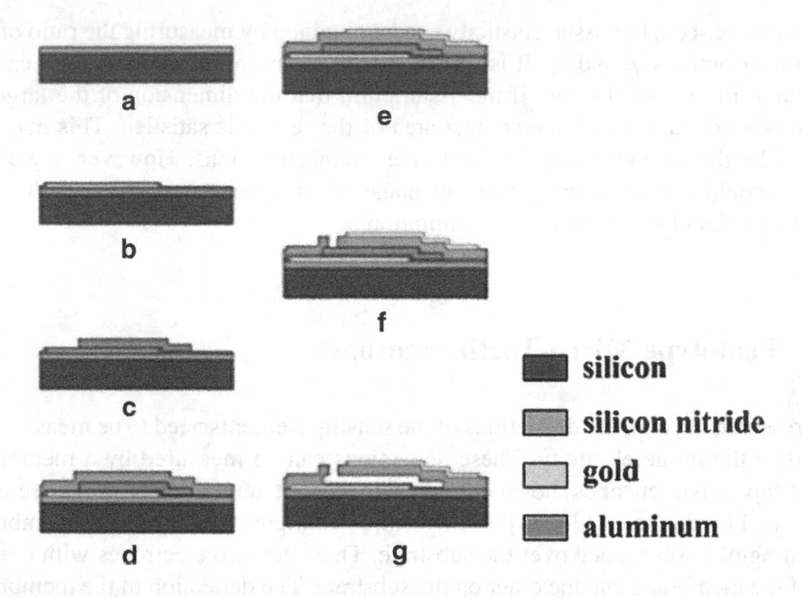

Fig. 7.6 Fabrication process for the first prototype MEMS sensors (© Elsevier 2010), reprinted with permission

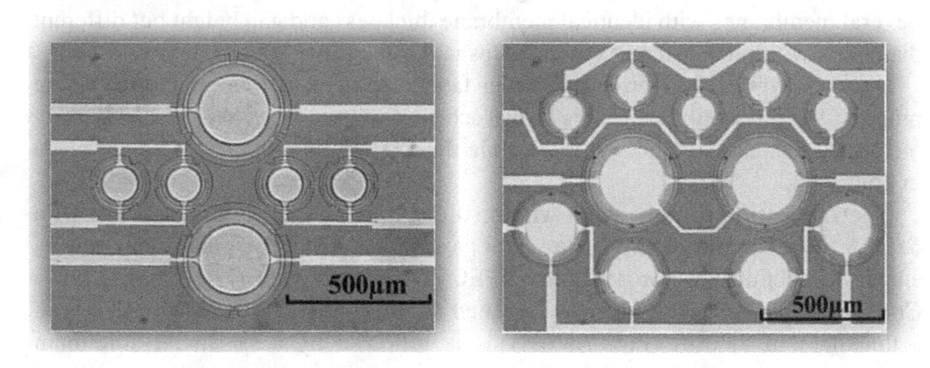

Fig. 7.7 Fabricated prototype sensors (© Elsevier 2010), reprinted with permission

layer (8,000–18,000Å) deposited and patterned via wet etching (Fig. 7.6c). Then the sensing membrane is constructed by depositing a PECVD SiNx layer (5,000–10,000Å) as shown in Fig. 7.6d. On top of the sensing membranes, electrodes are patterned by wet etching a Cr–Au E-beam evaporated metallization layer (Fig. 7.6e). In the next step, etch holes are then patterned via dry plasma etch (Fig.7.6f). These etch holes will be utilized to remove the sacrificial aluminum layer. At last, the sacrificial layer is etched in wet etching solution and the membranes can be released by using a critical point dryer (Fig. 7.6g).

A fabricated sensor is shown in Fig. 7.7 with the readout circuitry shown in Fig. 7.8. As can be seen, the sensor is composed of a set of circular membranes of

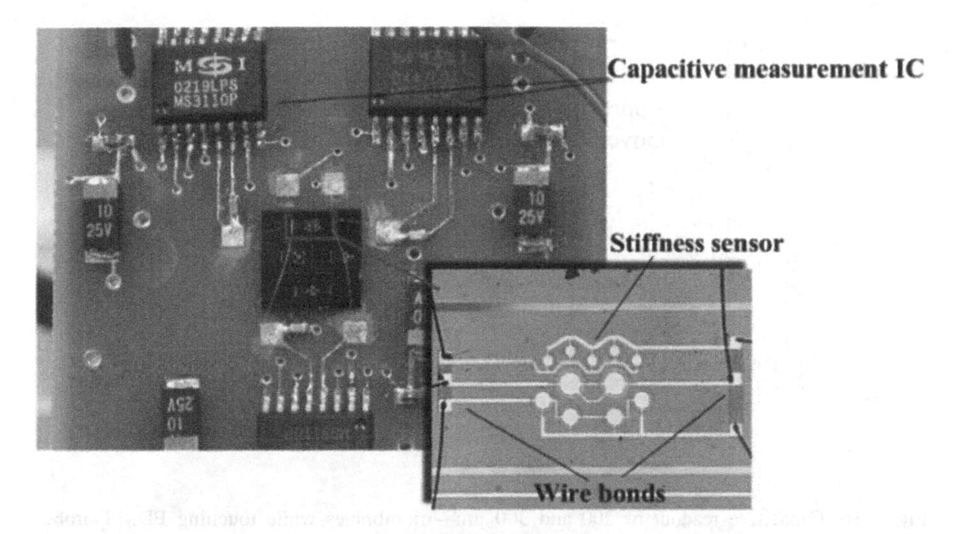

Fig. 7.8 Readout circuitry for prototype sensor (© Elsevier 2010), reprinted with permission

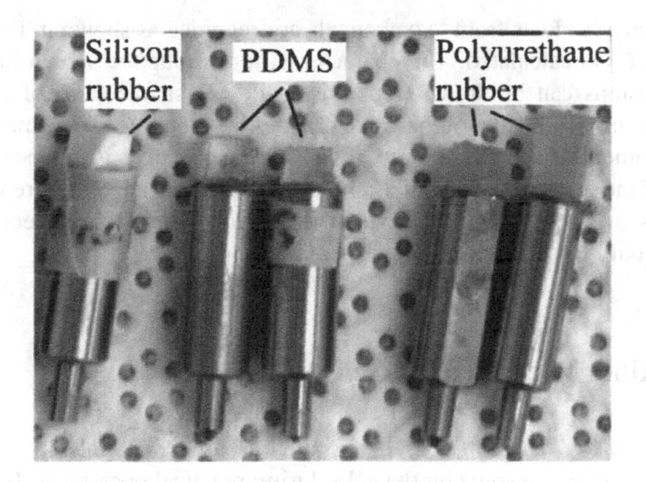

Fig. 7.9 Probes of different polymer materials and different thicknesses (© Elsevier 2010), reprinted with permission

various sizes. The diameters of these membranes are designed to be 200, 300 and 400 μm, respectively. Under the same amount of stress, the larger membrane will experience larger deflection compared to its softer counterpart, which enables the hard-soft spring pair design. As can be seen, more numbers of smaller membranes can be utilized to make the capacitance readout comparable (equally sensitive) to those from larger membranes.

To characterize the fabricated sensor, experiments were conducted by touching material samples of different elasticity and thickness to evaluate the ability of the sensors for elasticity measurement. Fig. 7.9 shows the different polymer materials that were used with the probes.

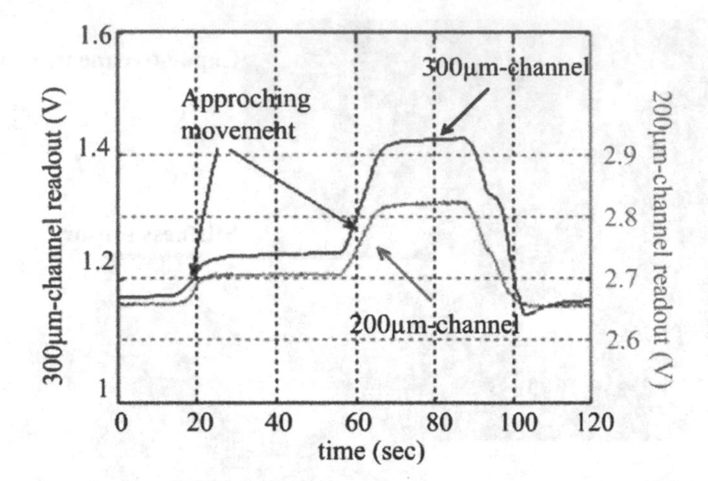

Fig. 7.10 Capacitive readout of 200 and 300 μm—membranes while touching PDMS probe (© Elsevier 2010), reprinted with permission

Capacitance readouts from two channels are recorded as shown in Fig. 7.10 and the ratio C_1/C_2 is calculated. The characterization results are shown in Fig. 7.11. Two conclusions can be drawn from the results. First, the ratio of capacitance changes is a unique function of the value of Young's modulus for the materials used in the experiments. Second, the ratio of capacitance changes is independent of the thickness of the material used in the sample. These results are therefore in line with the analysis in Sect. 7.2.2, (7.8), where the ratio of forces is expected to be a function of only the elasticity of the target material.

7.4 Flexible Micro-Tactile Sensors

The tactile sensors fabricated by using silicon-based materials in the previous section may experience some bottlenecks during practical operations. For instance, wire bonding is employed in such silicon-based devices to establish electrical connection between the MEMS sensors and the readout circuitry. Since the wire diameter in such wire bonds is only around 25 μm, the wires break easily under contact loads. In addition, the sensing membranes constructed of silicon-based materials (e.g., silicon nitride) have a thickness of around 1 μm or up to a few microns. While the membrane can handle normal loads, the application of shear loads causes the membrane to fail. This could be attributed to the ultrathin thickness of the sensing membranes and the brittleness of the structural material.

To develop a more robust and reliable tactile sensor, a polymer-based sensor has been prototyped by this team [36], as shown in Fig. 7.12. PDMS is chosen as the structural material due to its favorable properties such as flexibility, ductility and biocompatibility. This sensor is designed to utilize the same capacitive sensing

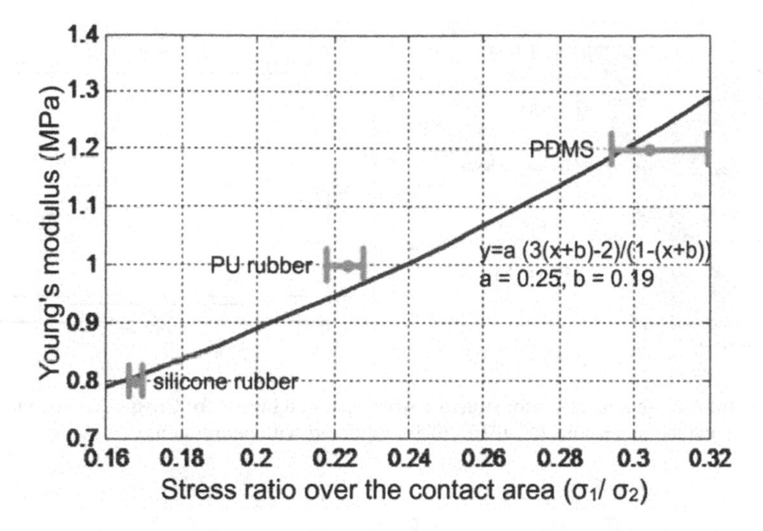

Fig. 7.11 Estimated σ_1/σ_2 values vs. Young's modulus of various materials (© Elsevier 2010), reprinted with permission

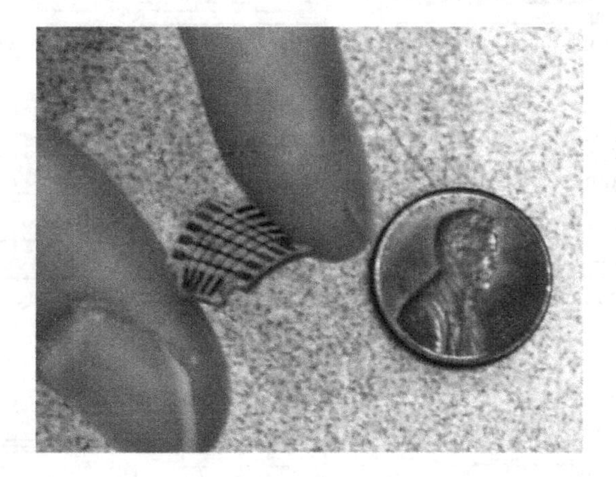

Fig. 7.12 Fabricated PDMS tactile sensor (© IEEE 2009), reprinted with permission

technique to measure membrane deflections as the previous prototype. To integrate capacitors into the polymer structure, a five-layer design is employed as shown in Fig. 7.13. To form a capacitor, embedded electrodes are built on the top and bottom PDMS layers. A spacer layer is sandwiched between the electrodes to define the membrane size. An insulation layer is also highly suggested to prevent the shorting of electrodes which could be a consequence when large deflection of sensing diaphragms occurs. Finally a bump layer is utilized for contact points.

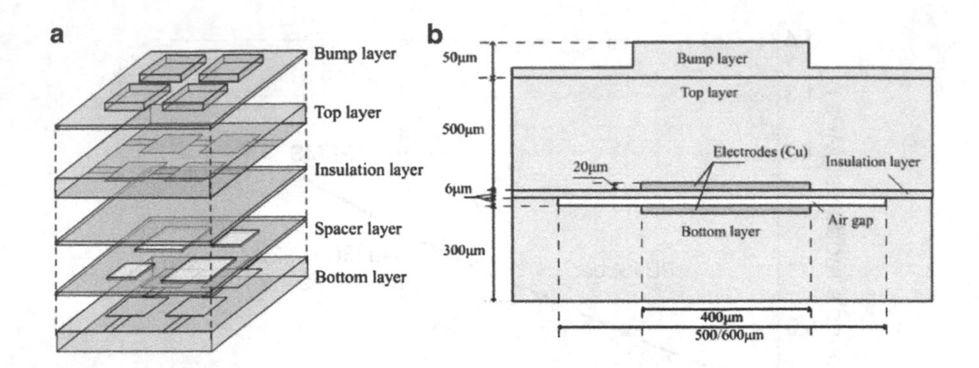

Fig. 7.13 (**a**) A close look at sensor structure with separated layers. (**b**) Cross-sectional view of a sensing cell and its dimensions (© IEEE 2009), reprinted with permission

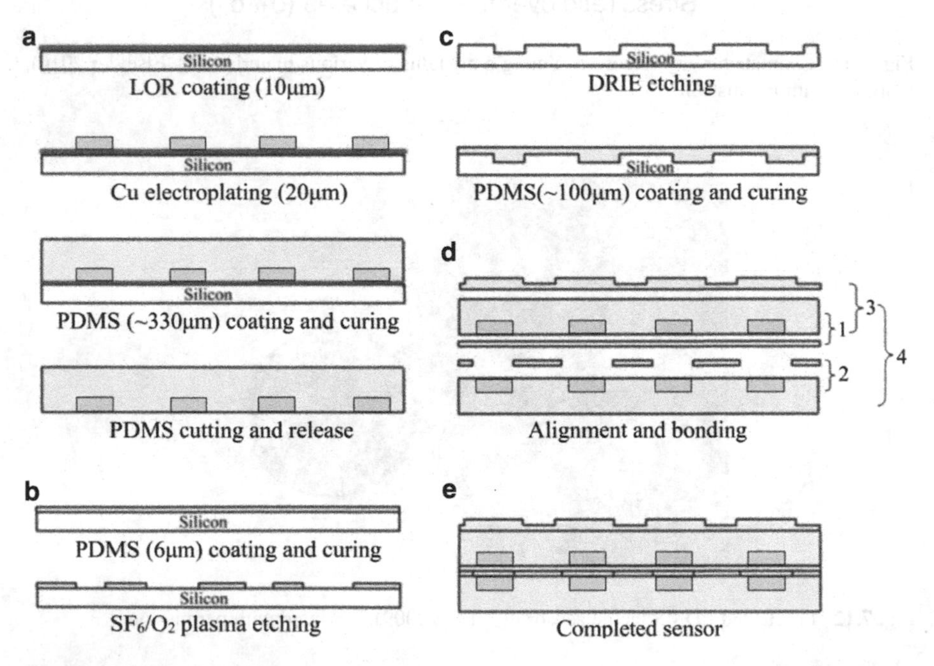

Fig. 7.14 Fabrication process. (**a**) electrode layer, (**b**) insulation and spacer layer, (**c**) bump layer, (**d**) alignment and bonding, and (**e**) completed sensor (© IEEE 2009), reprinted with permission

A schematic of the sensor fabrication process is shown in Fig. 7.14. Each PDMS layer is processed separately and finally the layers are bonded together with the aid of oxygen plasma treatment [37]. The fabrication starts from coating one sacrificial layer of lift-off resist (LOR) on the silicon wafer surface. Then copper wires are electroplated on LOR utilizing a through-mask electroplating process. After building the electroplated copper electrodes, an adhesion layer of titanium is sputtered

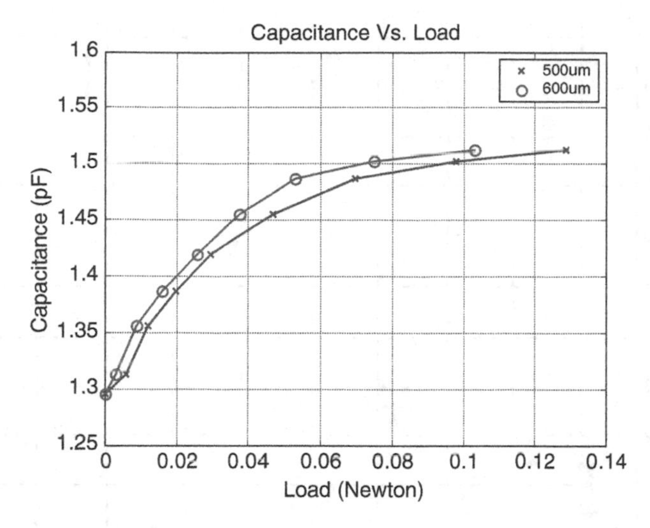

Fig. 7.15 Capacitance (pF) vs. load (Newton) for a single membrane (500 or 600 μm) (© IEEE 2009), reprinted with permission

before PDMS coating. Liquid PDMS (Sylgard 184, Dow Corning) is then spin-coated and cured at room temperature. These electrode layers can be finally peeled off the substrates when curing is completed (Fig. 7.14a).

The spacer and insulation layers are fabricated by coating and curing an ultrathin PDMS layer (~6 μm) on silicon substrates. Before spin-coating the PDMS layers, trichlorosilane vapor serving as a self-assembly monolayer (SAM) treatment is used on the bare silicon substrate. After curing the liquid PDMS, SF_6/O_2 plasma etching is utilized to pattern the spacer layer which will be used to define the dimensions of air gaps (Fig. 7.14b). The bump layer starts with a clean wafer patterned with desired bump molds (Fig. 7.14c). These concaves are etched using a DRIE process. The same SAM treatment is then conducted on the bump mold surface followed by spin-coating PDMS with a thickness of approximately 100 μm. At last, alignment of the fabricated layers is completed on a conventional contact aligner. The sequence of alignment steps is specified in Fig. 7.14d. It is worthwhile mentioning herein that all the PDMS layers should be treated in advance through oxygen plasma to form hydroxy keys. This chemical connection is crucial to achieve an interlayer bonding with desirable mechanical strength.

As shown in Fig. 7.13, the openings on the spacer layer can be used to define the size of membranes. As aforementioned, a larger membrane will perform as a soft element while a smaller membrane will perform as a hard one. The membranes in this sensor are designed to have square shapes with a side length of 500 and 600 μm. Each sensing membrane has been characterized through a force gauge as shown in Fig. 7.15. As can be seen, the capacitor cell with a membrane size of 600 μm × 600 μm has larger capacitive change than the one of 500 μm × 500 μm, which indicates a larger deflection occurring under the same magnitude of loads. This result validates our capacitor–pair design with different stiffnesses.

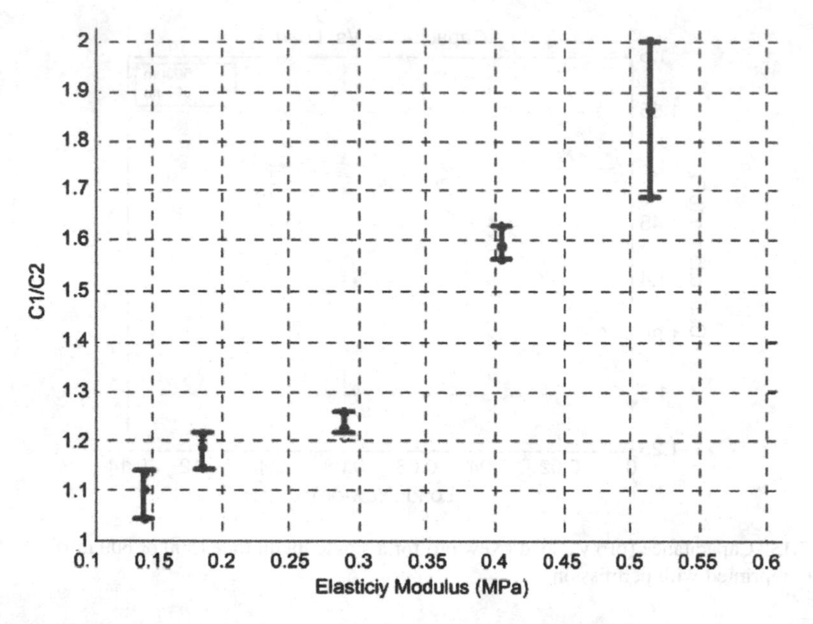

Fig. 7.16 Ratio of capacitive change vs. Young's modulus of polymer specimens (© IEEE 2009), reprinted with permission

Further characterization is aimed at examining the sensing capability for elasticity measurement. Sorbothane rubber specimens with the value of Young's modulus ranging from 0.1 to 0.5 MPa are used for calibrating the sensor. Experimental results are shown in Fig. 7.16. As can been seen from the results, the sensor readout ($\Delta C_1/\Delta C_2$) has a small deviation for the same rubber specimen under different magnitudes of load. This deviation is likely due to the oblique contact between the rubber specimen and the sensor. The resolution of measurement for this sensor is around 0.1 MPa in the range of 0.1 –0.5 MPa.

7.5 Ultralow-Cost Sensors for Handheld Operation

In this section, an ultralow-cost prototype of the micro-tactile sensor is presented [38]. The relatively more straightforward fabrication process of this sensor enables a shorter development cycle. This sensor has also been attached on a touch probe and operated as a handheld device. In order to provide a reliable estimation of the capacitance ratio, a recursive least square (RLS) algorithm has been developed to process the two capacitance signals. Further, an adaptive algorithm for choosing forgetting factors is also employed to achieve fast detection of elasticity change as well as high immunity to noise, which are important properties for in vivo tissue characterization.

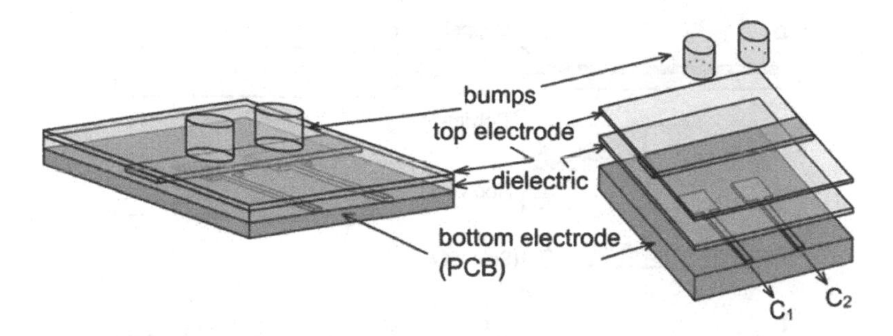

Fig. 7.17 Schematic diagram of sensor structure (© IEEE 2011), reprinted with permission

7.5.1 Sensor Structure and Fabrication Process

In order to understand how the capacitance readouts are generated, a schematic diagram of the sensor structure is shown in Fig. 7.17. The capacitors are constituted by a top electrode layer as a common electrode and two separated electrodes on the bottom. An insulator layer is sandwiched between the two electrodes as the dielectric. Two bumps are mounted on top of the capacitors serving as the contact interface. The two bumps are designed to have different stiffness values.

A brief description of the fabrication process is shown in Fig. 7.18. As can be seen, the fabrication of the top electrode starts by patterning the copper layer on a polyimide substrate (DuPont™ Pyralux® AC 182500R) through photolithography and etching. To fabricate the bumps, an acrylic mold with concaves for bumps is first made by a computer-controlled driller. Urethane rubber compound (PMC-724, Smooth-On Inc.) is then dripped on the substrate to fill the concaves. The hardness values of this rubber compound are carefully adjusted to Shore 40A and Shore 6A, respectively, thus creating a hard and a soft bump. A blade is used to squeegee the substrate surface to remove the extra rubber compound. These bumps should then be aligned and bonded with the polyimide substrate within half an hour before the rubber starts curing. After the bonding, the bumps are left for curing overnight and then peeled off from the mold. The final step is to bond the top and bottom electrodes. The bottom electrodes are designed and fabricated through a printed circuit board (PCB) manufacturer. Rubber compound with a hardness value of Shore 40A is then poured onto the bottom electrode to form the dielectric. Finally, the top electrode and the bottom electrode are aligned and bonded together to complete the sensor. The fabricated sensor with the readout circuitry is shown in Fig. 7.19.

The fabricated tactile sensor is then tested by pushing against a variety of sorbothane rubber specimens (Part No. 8450K3, McMaster-Carr) as shown in Fig. 7.20. As aforementioned, the sensor readout is composed of two channels of capacitance values, and the ratio (r_c) can be used to represent the elasticity of the targeted rubber sample. However, due to motions of the hand, a considerable

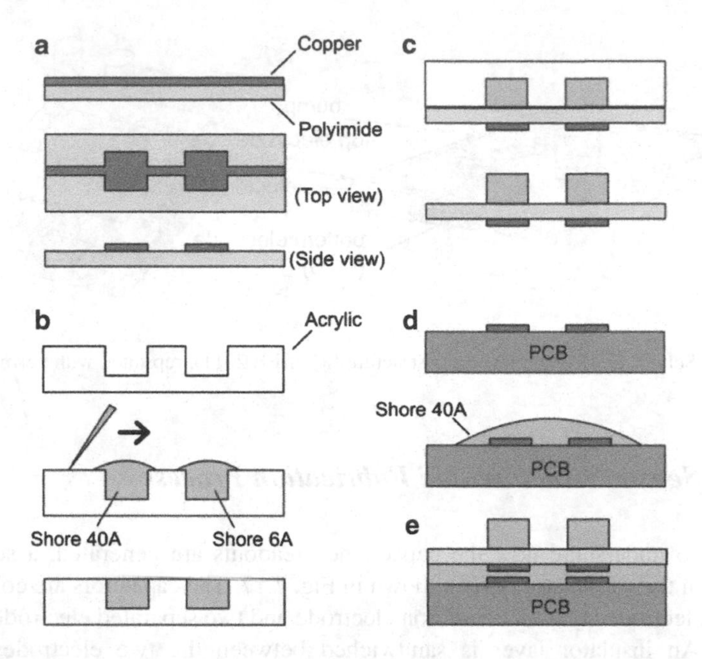

Fig. 7.18 Fabrication of the tactile sensor: (**a**) top electrode, (**b**) bumps, (**c**) bonding of top electrode and bumps, (**d**) bonding of top electrode and bottom electrode, (**e**) completed sensor (© IEEE 2011), reprinted with permission

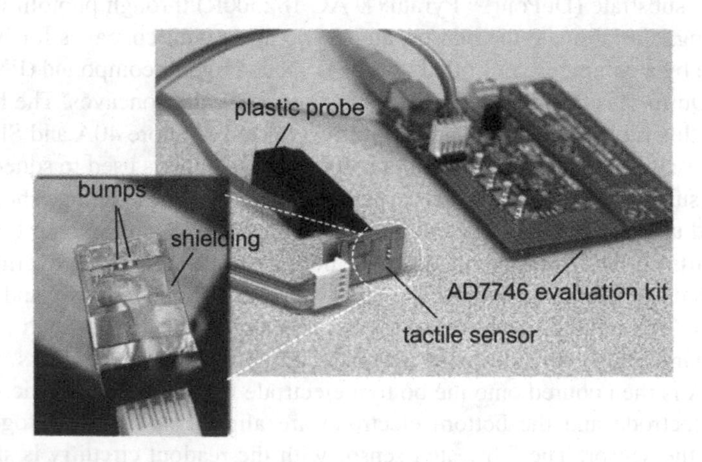

Fig. 7.19 Fabricated tactile sensor attached on a plastic probe

amount of interference can be observed at the capacitance signals, and therefore results in a varying value of capacitance ratio. To alleviate this problem, an algorithm using RLS method enhanced by adaptive forgetting factors has been developed.

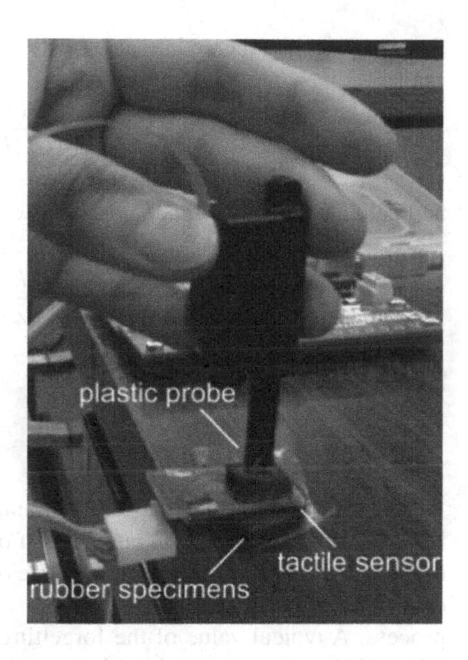

Fig. 7.20 Tactile sensor pushing against rubber specimens in a handheld mode (© IEEE 2011), reprinted with permission

7.5.2 Estimation Algorithm for Measurement of Capacitance Ratio

The capacitance relations can be formulated in an identification form shown in (7.9):

$$C_1(k) = C_2^T(k) r_c(k) + e(k), \tag{7.9}$$

where $C_1(k)$ and $C_2(k)$ represent the capacitance values for the two sensing membranes which can be viewed as the output and input data for an dynamic identification model. It can also be seen that $r_c(k)$ serves as the estimated parameter, while $e(k)$ is the identification error.

By implementing a RLS algorithm[39–41], the unknown parameter $r_c(k)$ can be iteratively updated at each sampling time. Through this process, the sum of estimation errors can be minimized. The procedure of identifying the capacitance ratio can be described in the following steps.

Step 1: Read the sensor readouts, $C_1(k)$ and $C_2(k)$.

Step 2: Calculate the identification error, $e(k)$, which is the difference between $C_1(k)$ at this sample and the estimated $C_1(k)$, which is the product of $C_2(k)$ and the estimated ratio in previous sample $r_c(k-1)$, i.e.,

$$e(k) = C_1(k) - C_2^T(k) r_c(k-1). \tag{7.10}$$

Step 3: Calculate the update gain vector, $K(k)$, as

$$K(k) = \frac{P(k-1)C_2(k)}{\lambda + C_2^T(k)P(k-1)C_2(k)} \tag{7.11}$$

and calculate the covariance matrix, $P(k)$, using

$$P(k) = \frac{1}{\lambda}\left[P(k-1) - \frac{P(k-1)C_2(k)C_2^T(k)P(k-1)}{\lambda + C_2^T(k)P(k-1)C_2(k)}\right]. \tag{7.12}$$

Step 4: Update the estimated parameter, $r_c(k)$, as

$$r_c(k) = r_c(k-1) - K(k)e(k). \tag{7.13}$$

The parameter, λ, in the above equations is known as the forgetting factor. By properly adjusting λ, the influence of old data, which may no longer be relevant to the model, can be suppressed. The use of the forgetting factor not only prevents a covariance wind-up problem, but also allows a fast tracking of the changes in process. A typical value of the forgetting factor is suggested to be in the interval 0.9–1 [39]. It can also be intuitively understood that the RLS algorithm utilizes a batch of $N = 2/(1-\lambda)$ samples to update the current estimation. When $\lambda = 1$, all the previous data collected will be used. A smaller λ value usually results in a faster convergence of estimates. However, a reduced value of λ increases the sensitivity of estimation to measurement noise, which may cause oscillatory estimation. Therefore a tradeoff between the fast-tracking capability and high immunity to noise should be considered in the system design, which will be addressed next.

An adaptive algorithm for the forgetting factor [42, 43] has been used with the estimation algorithm to achieve both the favorable properties of fast-tracking and immunity to measurement noise. The identification error $e(k)$ is monitored throughout the period of contact. An alarm is signaled if the identification error has been larger than a threshold for a certain amount of time. The recursive formula of this method is shown below.

$$\begin{aligned} a_k &= \max(a_{k-1} + |e_k| - d, 0), \quad k = 1, 2, \ldots, \\ a_0 &= 0. \end{aligned} \tag{7.14}$$

As can be seen, given the identification error e_k calculated in an ordinary RLS as the input, an output alarm signal can be generated. If the alarm value $a_k > h$, a smaller forgetting factor will be chosen in the RLS. Here, the threshold value h is used to determine when the forgetting factor should be adjusted in the condition that an alarm signal has been on for a sufficiently long time. The other threshold value d in the above equation is used to judge when to turn on the alarm. This makes the process ignore errors smaller than d. If the estimation system can swiftly track any abrupt change in capacitance ratio, the identification error will drop below a certain level, thus resulting in a zero value of the alarm signal. At this stage, the alarm is turned back off and a larger forgetting factor is chosen for its high immunity to noise at a steady-state.

To demonstrate this algorithm, an experimental test on touching a rubber sample is conducted and the results are illustrated in Fig. 7.21. As shown in Fig. 7.21a, capacitance values of the hard and soft elements are plotted and the estimation is performed by using RLS with adaptive forgetting factors. It can be seen that contact occurs at around the 150th sample, where the estimation starts. At the beginning the estimation error is larger than 0.01, which triggers the alarm signal (Fig. 7.21b). When the alarm signal is on for a while, a relatively small forgetting factor ($\lambda = 0.9$) is chosen. After the identification error drops to a certain level, the alarm signal is turned off, and a larger forgetting factor ($\lambda = 0.995$) is then applied throughout the rest of the estimation process. To illustrate the benefit of this adaptive algorithm, estimation results of ordinary RLS with $\lambda = 0.995$, $\lambda = 0.9$, and adaptive forgetting factors are shown in Fig. 7.21c. As shown in the figure, the estimation results generated by $\lambda = 0.9$ show the favorable property of fast-tracking of changes. However, it is susceptible to measurement noise, which makes it difficult to identify the true value of the capacitance ratio. The algorithm of using an adaptive forgetting factor inherits the desirable property of fast convergence at the beginning stage, and then trends to the curve of $\lambda = 0.995$, which holds a relatively constant value at the steady state.

With the developed estimation algorithm, the tactile sensor is characterized by touching a variety of rubber specimens. Each rubber sample has been touched four times and the estimated capacitance ratio for each test has been recorded. As shown in Fig. 7.22, the capacitance ratio shows an overall increasing value as the rubber sample becomes harder. It can also be observed that some measurements have a considerable amount of standard deviation as depicted by error bars on the plot. This variation is likely due to oblique contact between the sensor and rubber samples as well as the nonlinearity of the capacitance response to the value of applied load on the bump, which cannot be solved solely by using the developed estimation algorithm. Experiments also show that these variations can be significantly reduced by mounting the sensor on a test stage and applying constant loads throughout all the tests.

7.6 Summary

This chapter presented a sensing technology for in vivo measurement of tissue elasticity. By utilizing this method, the value of Young's modulus can be measured without applying a controlled load or a controlled displacement, which has been required for conventional approaches. First, a prototype sensor fabricated through a surface micromachining process has been demonstrated. Then flexible sensors are developed through a polymer MEMS process, which includes ultrathin polymer layer fabrication, bonding, alignment, and metallization technique. Finally, an

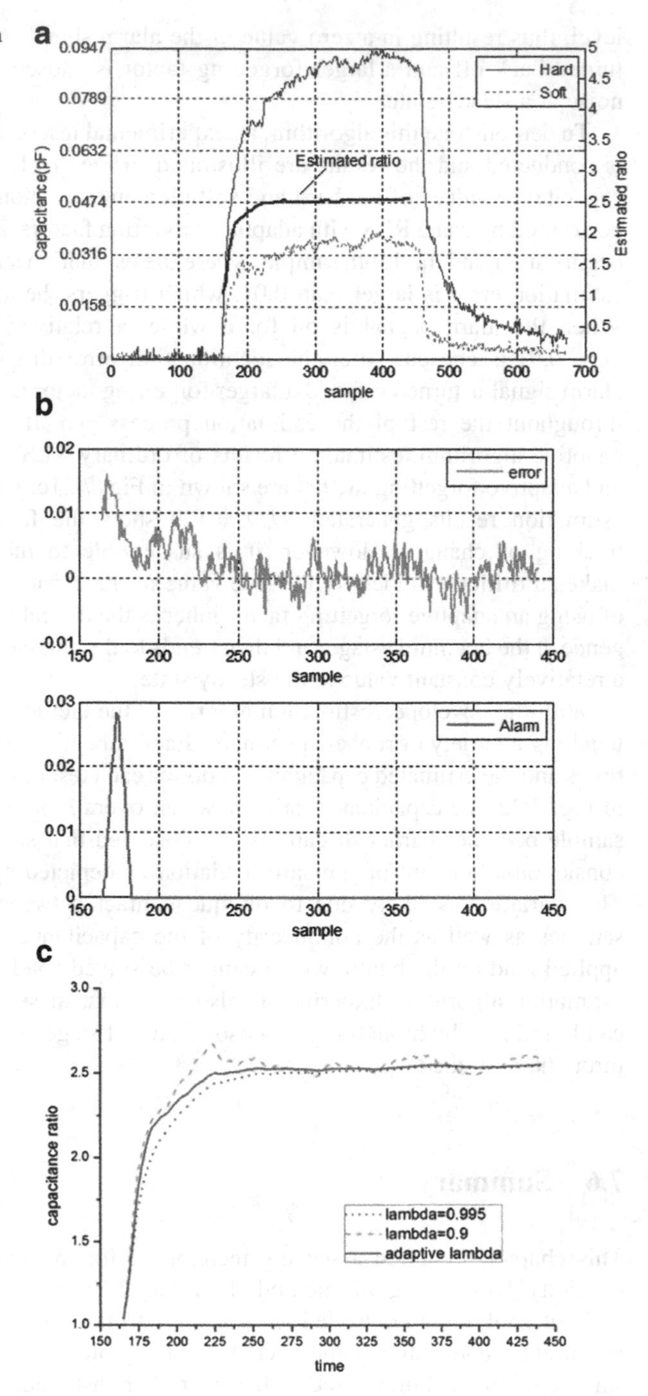

Fig. 7.21 Sensor readouts on sample 70OO: (**a**) estimation of capacitance ratio using RLS with adaptive forgetting factors; (**b**) identification error and alarm signal; (**c**) comparison of estimation results of ordinary RLS with $\lambda = 0.995$, $\lambda = 0.9$, and adaptive forgetting factors (© IEEE 2011), reprinted with permission

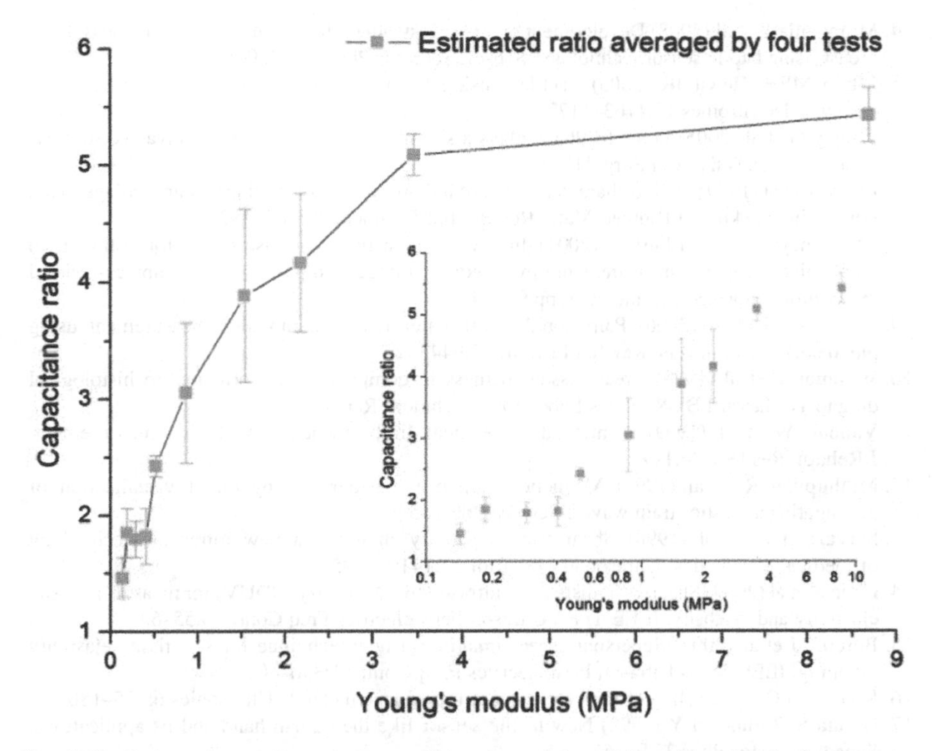

Fig. 7.22 Estimated capacitance ratio vs. Young's modulus of the rubber sample. Insets: data plotted in ln-log scale (© IEEE 2011), reprinted with permission

ultralow-cost prototype sensor has been made by employing a PCB manufacturer and commercially available polyimide products. This type of sensor is especially useful for research or development without access to high-end microfabrication facilities. The developed sensors have been characterized by touching a variety of rubber specimens in a handheld operating mode. To facilitate this operation, an estimation algorithm has been presented, which can achieve both fast-detection of changes in elasticity and alleviation of the noise due to hand motions. To the best knowledge of the authors, this is the first micro-tactile sensor for elasticity measurement that operates as a handheld device.

References

1. Tegin J, Wikander J (2005) Tactile sensing in intelligent robotic manipulation—a review. Ind Robot 32:64–70
2. Aoyagi R, Yoshida T (2004) Frequency equations of an ultrasonic vibrator for the elastic sensor using a contact impedance method. Jpn J Appl Phys 43:3204–3209
3. Bab A et al (2008) Design and simulation of a tactile sensor for soft-tissue compliance detection. IEEJ Trans Sensor Micromachine 128:186–192

4. Murayama Y et al (2008) Development of a new instrument for examination of stiffness in the breast using haptic sensor technology. Sensor Actuator Phys 143:430–438
5. Eltaib MEH, Hewit JR (2003) Tactile sensing technology for minimal access surgery—a review. Mechatronics 13:1163–1177
6. Tholey G et al (2005) Force feedback plays a significant role in minimally invasive surgery: results and analysis. Ann Surg 241:102
7. Lyyra T et al (1999) In vivo characterization of indentation stiffness of articular cartilage in the normal human knee. J Biomed Mater Res B Appl Biomater 48:482–487
8. Ottensmeyer M, Salisbury J (2001) In vivo data acquisition instrument for solid organ mechanical property measurement. In: Medical image computing and computer-assisted intervention, Springer, Heidelberg, pp 975–982
9. Szewczyk S et al (2006) Palpationlike soft-material elastic modulus measurement using piezoelectric cantilevers. Rev Sci Instrum 77:044302
10. Wellman P et al (1999) Breast tissue stiffness in compression is correlated to histological diagnosis. Harvard BioRobotics Laboratory Technical Report
11. Vannah W et al (1999) A method of residual limb stiffness distribution measurement. J Rehabil Res Dev 36:1–7
12. Muthupillai R et al (1995) Magnetic resonance elastography by direct visualization of propagating acoustic strain waves. Science 269:1854
13. Sarvazyan A et al (1998) Shear wave elasticity imaging: a new ultrasonic technology of medical diagnostics. Ultrasound Med Biol 24:1419–1435
14. Chen S et al (2009) Shearwave dispersion ultrasound vibrometry (SDUV) for measuring tissue elasticity and viscosity. IEEE Trans Ultrason Ferroelectrics Freq Contr 56:55–62
15. Bercoff J et al (2004) Supersonic shear imaging: a new technique for soft tissue elasticity mapping. IEEE Trans Ultrason Ferroelectrics Freq Contr 51:396–409
16. Kleesattel C, Gladwell G (1968) The contact-impedance meter-1. Ultrasonics 6:175–180
17. Omata S, Terunuma Y (1992) New tactile sensor like the human hand and its applications. Sensor Actuator Phys 35:9–15
18. Jalkanen V (2010) Hand-held resonance sensor for tissue stiffness measurements—a theoretical and experimental analysis. Meas Sci Tech 21:055801
19. Murayama Y, Omata S (2004) Fabrication of micro tactile sensor for the measurement of micro-scale local elasticity. Sensor Actuator Phys 109:202–207
20. Murayama Y et al (2005) Development of tactile mapping system for the stiffness characterization of tissue slice using novel tactile sensing technology. Sensor Actuator Phys 120:543–549
21. Murayama Y et al (2007) High resolution regional elasticity mapping of the human prostate. Conf Proc IEEE Eng Med Biol Soc 2007:5803–5806
22. Oie T et al (2009) Local elasticity imaging of vascular tissues using a tactile mapping system. J Artif Organs 12:40–46
23. Dargahi J et al (2007) Modelling and testing of a sensor capable of determining the stiffness of biological tissues. Can J Electr Comput Eng 32:45–51
24. Beebe D et al (1995) A silicon force sensor for robotics and medicine. Sensor Actuator Phys 50:55–65
25. Kim K, Lee K, Kim Y, Lee D, Cho N, Kim W, Park K, Park H, Park Y, Kim J (2006) In: 19th IEEE International Conference on Micro Electro Mechanical Systems, IEEE, Istanbul, pp. 678–681.
26. Gray BL, Fearing RS (1996) In: IEEE International Conference on Robotics and Automation, IEEE, Minneapolis, MN, USA pp. 1–6.
27. Leineweber M et al (2000) New tactile sensor chip with silicone rubber cover. Sensor Actuator Phys 84:236–245
28. Sergio M, Manaresi N, Tartagni M, Guerrieri R, Canegallo R (2002) In: IEEE Sensors, IEEE, Kissimmee, Florida, USA pp. 1625–1630.

29. Engel J et al (2003) Development of polyimide flexible tactile sensor skin. J Micromech Microeng 13:359
30. Engel J et al (2005) Polymer micromachined multimodal tactile sensors. Sensor Actuator Phys 117:50–61
31. Lee H et al (2006) A flexible polymer tactile sensor: fabrication and modular expandability for large area deployment. J Microelectromech Syst 15:1681–1686
32. Johnson K (1987) Contact mechanics. Cambridge University Press, Cambridge
33. Peng P et al (2009) Novel MEMS stiffness sensor for in-vivo tissue characterization measurement. Conf Proc IEEE Eng Med Biol Soc 1:6640–6643
34. Peng P et al (2010) Novel MEMS stiffness sensor for force and elasticity measurements. Sensor Actuator Phys 158:10–17
35. Sezen A et al (2005) Passive wireless MEMS microphones for biomedical applications. J Biomech Eng 127:1030
36. Peng P et al (2009) Flexible tactile sensor for tissue elasticity measurements. J Microelectromech Syst 18:1226–1233
37. Jo B et al (2002) Three-dimensional micro-channel fabrication in polydimethylsiloxane (PDMS) elastomer. J Microelectromech Syst 9:76–81
38. Peng P, Rajamani R (2011) Handheld micro tactile sensor for elasticity measurement. IEEE Sensor, Vol. 11, no. 9, pp 1935–1942, Sept 2011
39. Wang J et al (2004) Friction estimation on highway vehicles using longitudinal measurements. J Dyn Syst Meas Contr 126:265
40. Sastry S, Bodson M (1989) Adaptive control: stability, convergence, and robustness. Prentice-Hall, NJ
41. Gustafsson F (2001) Adaptive filtering and change detection. Wiley, Chichester
42. Page E (1954) Continuous inspection schemes. Biometrika 41:100
43. Rajamani R (2002) Radar health monitoring for highway vehicle applications. Veh Syst Dyn 38:23–54

Chapter 8
Devices and Techniques for Contact Microgripping

Claudia Pagano and Irene Fassi

Abstract The gripping and manipulation of microparts significantly differs from the handling and assembly of macroscopic components. In the macroworld gravity dominates, whereas in the microdomain, it becomes negligible, and superficial forces dominate pick and place operations. Releasing a part from the grasp of a microgripper is not a simple task as the part may stick to the gripper due to the presence of these adhesive forces. For this reason, beside the numerous attempts of downscaling traditional grippers also innovative actuation strategies have been proposed. The chapter critically reviews some of the most widely used micromanipulation techniques with contact, highlighting their advantages and disadvantages and describing some innovative solutions based on capillary forces.

8.1 Introduction

Although some MEMS are usually fabricated via massively parallel photolithographic techniques, sequential assembly might be required in some instances, such as if different materials or high-aspect-ratio structures are needed and the silicon-based fabrication is not suitable or not possible. For these microdevices, assembly and packaging of is still time consuming and costly, often contributing for the largest part of the total cost of the microproduct. In order to increase the manufacturing throughput and reduce costs, a flexible assembly scheme is requires, to allow a quick adaptation to various part geometries and configurations. This need has been recognized by private industries and government agencies, leading to considerable progresses in the development of visually served robotic systems. At present, the main issue to be addressed is the development of end-effector and the corresponding position-sensing technology. Several solutions have been proposed,

C. Pagano • I. Fassi (✉)
National Research Council of Italy, Piazzale Aldo Moro 7, Roma 00185, Italy
e-mail: claudia.pagano@itia.cnr.it; irene.fassi@itia.cnr.it

D. Zhang (ed.), *Advanced Mechatronics and MEMS Devices*, Microsystems,
DOI 10.1007/978-1-4419-9985-6_8,

including downscaled tweezers and vacuum microgrippers, together with innovative handling systems based on superficial forces.

Indeed, the gripping and manipulation of microparts significantly differs from the assembly macroscopic devices. The main difference stems from the increased role of superficial forces, electrostatic, van der Waals and surface tension forces, and the reduced influence of volume forces such as gravity. Due to charging effects, the parts often "jump" into their energetically most favorable configuration, and this results in an uncontrolled grasping or release. Moreover, due the adhesion forces object sticks to the gripper, preventing the release or the correct positioning. These issues can affect one or more of the three main manipulation phases (grasping, handling, release) and have to be taken into account during the design of microgripping systems. Several studies have been carried on so far, mainly focused on the grasping phase, which have led to a variety of prototypes.

8.2 Gripping Techniques with Physical Contact

Manipulation of microparticles can be done using several physical principles and methods. For manipulation of a microstructure under specific ambient conditions or in liquid; suction, cryogenic, electrostatic, and friction are the most often considered methods. Hereafter, a brief critical review is presented.

8.2.1 Friction Microgrippers

In analogy with the mechanical grippers traditionally used for the manipulation of macroscopic objects, several types of microtweezers have been fabricated to handle microscopic objects on the base of the friction between them and the fingers of the gripper. Several fabrication methods have been presented for such a gripper, including lithographic technologies [1–7], LIGA, [8, 9], laser micromachining [10, 11], Electro-Discharge Machining (EDM) [12–14], and many kinds of actuation have been proposed such as electrostatic [15], pneumatic [16], piezoelectric [13], thermal [17, 20], and SMA [20] based controls. However, although downsizing the dimensions of traditional mechanical grippers is not straightforward and their actuation can be an issue, the main disadvantage of microtweezers is the presence of the superficial and uncontrolled forces, due to which the parts stick to the gripper, preventing their release or accurate placement. The main advantages is the ability to center the part between the gripping jaws and to align it parallel to the jaws for a precise handling, but they are not suitable for very fragile parts due to the likely of damaging components during grasping and holding. In order to control the exerted force and avoid squeezing delicate microparts, several mechanical grippers provided with an embedded sensor system, have been proposed. They are mainly based on piezoelectric devices, but also optical detectors have been implemented.

For instance, a capacitive force sensor, able also to measure adhesive forces between the fingers of the gripper was integrated on an electrostatic gripper [15]; a microfabricated and electrothermally actuated gripper has been implemented with a piezoresistive readout which measures the gripper deflection to estimate the applied forces [17]; a sensing microgripper was presented by Arai et al. [18], who enhanced the capability of their silicon microgripper, manufacturing its surface with several micropyramids coated with gold, in order to reduce the superficial force, and adding a semiconductor strain gauge to measure the force. Park and Moon [19] have developed an hybrid sensorized microgripper, made of several parts manufactured with different technologies, in order to improve their functionality; the microgripper consists of two silicon cantilevers fingertips, fabricated by micromachining technology to integrate a very sensitive force sensor and piezoelectric actuators, produced by a conventional manufacturing process to achieve large gripping forces and moving strokes. A superelastic alloy (NiTi) microgripper, fabricated by electro-discharge machining, was integrated with a piezoelectric force sensing capability [21]. Moreover, an on-chip optical detection was proposed on an electrothermally actuated microgripper [20]: the microgripper was fabricated by CMOS process, using the dielectric layers of the process to protect the conductive parts and make the gripper suitable for biomanipulation in aqueous solutions, and, beneath the gripping sites, a n-well/p-substrate diode is placed to provide optical feedback during the manipulation. Petrovic et al. [14] glued the optical sensors on both arms of a steel mechanical microgripper, in order to estimate the force exerted by each arm from the voltage output of the sensors. Furthermore, a polymeric sensor was bonded on a superelastic NiTi alloy microgripper using nonconductive glue [12]: the sensor is made of a film of Polyvinylidene Fluoride (PVDF), which exhibits strong piezoelectric proprieties, high flexibility and biocompatibility.

8.2.2 Pneumatic Grippers

Suction grippers are commonly used for the manipulation of fragile object and can easily be miniaturized, even though, like all the contact grippers at the microscale, the adhesion forces significantly affect the release so that the precise positioning is difficult to achieve, moreover, both the gripper and the part need to have smooth surfaces to prevent air leakage. On the other hand, they can be very cheap as consist mainly of a micropipette connected to a vacuum pump. An estimation of the optimal tip diameter was empirically obtained by Zesch et al. [22]; they proposed a vacuum tool made of a glass pipette and a computer-controlled vacuum supply to manipulate 80–150 μm sized diamond crystals using pipettes with varying tip diameters and concluded that the tip diameter should be 25–50% of the object size. They also observed the effect of the adhesive forces, since the success rate of the operations was affected by humidity, and, in some cases, the release of an object had to be assisted applying a short pressure pulse, which affects the accuracy of the positioning.

Petrovic et al. [14] proposed to reduce the adhesive effect due to the electrostatic mutual attraction between the gripper and the micropart, coating the glass pipette for the suction with a conductive layer of gold in order to connect them both to the ground. Vikramaditya and Nelson [23], used a vacuum gripper to assemble magnetic particles and, therefore, took into account the magnetic interaction together with the force of the negative pressure. A sensorized vacuum gripper has been developed by Wejinya et al. [24], who integrated it with an in situ PVDF beam force sensor on the gripper to study the suction force in order to automates the manipulation operations.

In the previous works the pressure controlled for the manipulation is generated by a pump, however, a connection between the vacuum pump and the very small tube, which can easily be obstructed by small particles has to be done. In order to overcome these problems, Arai et al. [25], proposed to generate the pressure changing the temperature inside several microholes made on the end-effector surface of the gripper. As for the thermal actuation of friction grippers, this method takes advantage of the reduced dimensions to obtain a quick response, due to the fast heat transmission. The end-effector consists of a silicon cantilever, micromachined to obtain several microholes at its bottom; in order to grasp the micropart the temperature of the end-effector is increased up to 85°C, and, after the contact between the gripper and the part, it is decreased and a negative pressure is generated inside the microholes, achieving the grasp of the part. Then, the end-effector is heated again for the release. This method is, evidently, not suitable for heat sensitive parts.

8.2.3 Adhesive Gripper

Taking advantage of the small dimensions of the parts and the predominance of the superficial interactions, the capillary force can be controlled and exploited to manipulate components. Indeed, a micropart can be grasped simply touching it in a humid environment or a dispenser can be added to the gripper in order to form a small drop at the surface of the gripper and increase the superficial tension between the gripper and the object. Adhesive grippers as well as the vacuum grippers are suitable for low aspect ratio and fragile components, for which tweezers are not adequate, and, due to their compliancy in the horizontal plane and stiffness in the vertical one, and are self-centering, but they are generally not suitable for electrical components and leave traces of liquid on the component surface and contamination problems can occur.

Several works have investigated the capillary force as mean to grasp components, considering different geometries of the gripper and the parts [26–34], so that a variety of geometries and sizes of the gripper have been proposed [27, 31], in order to make the grasp of parts more efficient according to their geometry. However, as for all the contact grippers, the main issue is not the grasp, but the release, which cannot be based on the gravity force.

A lot of solutions have been presented for the release of the components, including dynamical effects [26, 30], gluing [28], mechanical release [29] and gas injection [29]. Moreover, a promising solution for the release is the variation of the properties of the gripper tip: Vasudev et al. [35] fabricated a complex capillary microgripper able to change the contact angle between the liquid bridge and the gripper surface via electrowetting, whereas Biganzoli et al. [36] conceived a capillary gripper able to release parts changing its shape in order to reduce the contact surface and therefore the gripping force. These solutions require advanced systems for the grasp of components but can be more precise in positioning and do not need external apparatus.

8.2.4 Phase Changing

Cryogenic gripper are based on freezing a drop of liquid between the tool and the object in order to increase the adhesion force, passing from a liquid–solid contact to solid–solid contact; the release occurs heating or breaking the solid intermediate material. They are also referred to as ice gripper, since the first idea was to use water as the intermediate liquid [37], but other types of materials have been, more recently, proposed [38]. These grippers, like the capillary ones are self-centering, but can be not suitable for parts sensitive to temperature changes or where liquid contact has to be avoided. To freeze the liquid the methods proposed use the Peltier effect [39] or a low-temperature gas to convectively cooling the material [40]. Beside ice grippers working in air, a more recent paper shows the possibility of applying the phase changing principle also in a fluid, using the liquid environment to generate an ice microsurface [41].

8.2.5 Electric Grippers

Electrostatic microgrippers grasp microcomponents using the electrostatic forces exerted by electrodes; applying a voltage to the electrodes, an electric field, homogeneous or inhomogeneous according to the materials of the components, is generated. A variety of grippers based on the electrostatic forces have been studied so far, with a unipolar or a bipolar design and able to handle a variety of objects. They are generally made of a conductive part, connected to the supply power and an insulator part avoiding the direct contact of the electrode with the component.

Fantoni and Biganzoli [42] manufactured a unipolar microgripper, using photolithgrafic processes. It consists of a thin cross shaped electrode deposited on a glass substrate, and, due to the inhomogeneous field, offers centering capability.

A similar shaped gripper was proposed by Hesselbach et al. [43], depositing a gold electrode on a Pyrex substrate. They proposed five shapes of concentric ring electrodes, in order to manipulate parts with different geometries and dimensions.

Enikov and Lazarov [44] measured the force exerted by their optically transparent electrostatic microgripper, the residual charge on the gripper and time to release the components due to the residual charge. Since it was about 47 min, the gripper had to be lightly tapped in order to release the part. On the other hand, electrostatic grippers can be useful for handling in vacuum equipment, such as in SEM [45] and TEM, where suction grippers are not suitable, but one of their main disadvantages, consists on the residual charge, so that in many cases switching off the voltage is not enough to achieve the release and other solutions were applied, which often leads to a non precise positioning of the component. Furthermore, the presence of the other superficial forces (van der Waals and capillary) can affect their performances as well; indeed, the success of the release, strongly depended on the humidity, due to the contribute of the superficial tension to the adhesion of the part to the gripper [44].

Possible solutions for the release were proposed, including the implementation of a piezoelectric element directly on the gripper to reduce the friction using ultrasonic excitation [46] and a gripper with a third movable electrode [47]; during the grasp of the part, the voltage is applied to the two outer electrodes whereas the inner one is connected to the ground, in order to release the component, the voltage is turned off and simultaneously the inner electrode is moved backwards breaking the electric field and rapidly reducing the residual charges on the object.

8.3 A Case Study: Development of a Variable Curvature Microgripper

In this paragraph, as an application example, a capillary gripper, whose lifting force is controlled changing its shape, is presented. The gripper switches from a flat configuration, exerting its maximum lifting force, to an hemispherical shape that reduces the capillary force and releases the object (Fig. 8.1).

This configuration accomplishes the release avoiding the use of external auxiliary tools, complex external actuation apparatus or electric field, preventing from eventual spare charges.

Figure 8.2 shows the capillary force due to an axisymmetric fluid capillary bridge formed between a sphere (the gripper) and a flat surface (the object) (inset

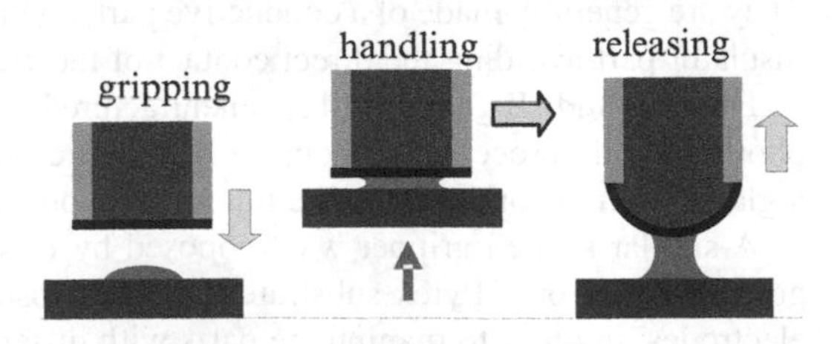

Fig. 8.1 Sketch of the working principle of the gripper

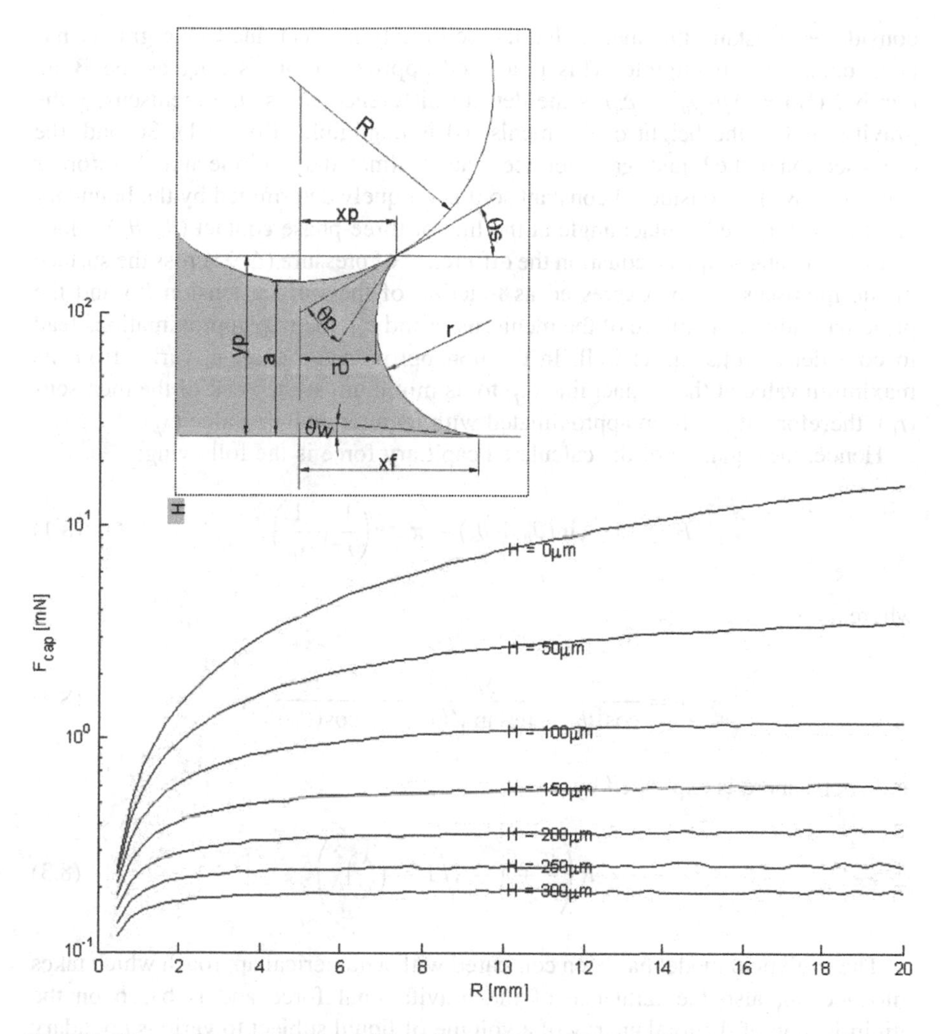

Fig. 8.2 Plot of the ratio of the capillary forces (F_{Cap}) and the surface tension (γ) versus the curvature radius (R) of the sphere for different value of the distance (h) between the wet surfaces. Contact angles (θ_p and θ_w) are equal to 30° and the volume of liquid (V) is 0.1 mm^3; *inset plot*: draw of the geometry taken into consideration

Fig. 8.2). The plot highlights that the capillary force significantly increases when the radius of the sphere increases and it reaches its highest value when the curvature radius approaches infinity (the sphere degenerates into a plane).

The force has a contribute due to the axial component of the surface tension acting at the liquid–gas interface ($F(\gamma)$) and one to the pressure difference across the curved surface of the meniscus ($F(p)$).

In order to solve the problem analytically some approximations have been made. First, the difference of pressure Δp along the meniscus interface has been considered constant, this means that the deformation effect due to the gravity has been considered negligible. This is a good approximation as long as the Bond number (Bo $= \Delta\rho g {y_p}^2/\gamma$, $\Delta\rho$ is the density difference across the meniscus, g the gravity, and y_p the height of the meniscus) is negligible (Bo $<<1$). Second, the cross section of the liquid–gas interface is approximated by a circle and, therefore, r (inset Fig. 8.2) is considered constant so it is uniquely determined by the boundary conditions for fixed contact angle at the line of three-phase contact (θ_p, θ_w). Since from the Young–Laplace equation the difference of pressure (Δp) across the surface of the meniscus can be expressed as function of the surface tension (γ) and the principal radii of curvature of the meniscus (r and r_a), the two approximations lead to consider r_a constant as well. In the nonapproximated case, r_a varies from its maximum value at the contact line (x_f) to its minimum at the neck of the meniscus (r_0), therefore, it has been approximated with its intermediate value (x_p).

Hence, the equation of the calculated capillary force is the following:

$$F = 2\pi x_p^2 \sin(\theta_p + \theta_s) - \pi x_p^2 \gamma \left(\frac{1}{r} + \frac{1}{x_p}\right), \tag{8.1}$$

where

$$r = \frac{y_p(x_p)}{\cos\left[\theta_p + \arctan y'(x_p)\right] + \cos(\theta_w)}, \tag{8.2}$$

and each sphere is expressed by:

$$y = R\left(\frac{a}{R} + 1 - \sqrt{1 - \left(\frac{x}{R}\right)^2}\right). \tag{8.3}$$

The analytical model has been compared with a numerical approach which takes into account also the influence of the gravitational force and is based on the minimization of the total energy of a volume of liquid subject to various boundary conditions and constraints. A public domain software package, Surface Evolver^{TM}, has been used to numerically compute the capillary force. In Fig. 8.3 the results obtained from the analytical (lines) and numerical (dots) models are shown. The results are in good agreement and confirm the possibility of using the capillary force as a mean for manipulation through a gripper with variable curvature.

The gripper can be made of a circular membrane of radius R_g able to decrease its curvature radius from a flat configuration ($R = \infty$) to an hemispherical configuration ($R = R_\mathrm{g}$) as the result of stresses induced by an external force. In Table 8.1 the maximum and minimum handleable weights, obtained from the numerical analyses, and the corresponding R_g for a certain amount of liquid deposited (V) are reported.

A prototype has been actualized in order to experimentally test the working principle of such a gripper. A gripper able to handle millimetric objects has been

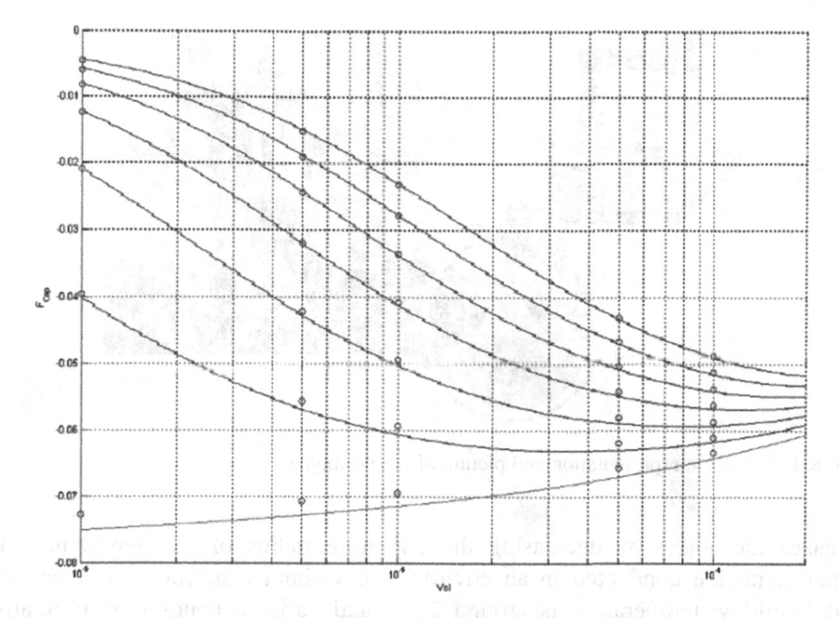

Fig. 8.3 Capillary force exerted by a sphere of radius R ($R = 0.1$ mm) versus the drop volume, for different values of the distance between the sphere and the plane ($h = 0$ (red line)/0.006 mm); the surface tension of the distilled water (γ) is equal to 0.073 mN/mm and θ_p, θ_w are equal to 30°

Table 8.1 The maximum and the minimum force and handleable weights with the corresponding R_g and V with $h = 0$ mm, $\gamma = 0.073$ mN/mm, $\theta_p = \theta_w = 30°$

R_g (mm)	V (mm^3)	F_{min} (mN)	Weight$_{min}$ (mg)	F_{max} (mN)	Weight$_{max}$ (mg)
0.05	1.00E − 04	0.03	2.67	0.37	37.7
0.1	8.00E − 04	0.05	5.34	0.74	75.4
0.2	6.40E − 03	0.10	10.7	1.48	150
0.4	5.12E − 02	0.21	21.4	2.96	301
0.8	4.10E − 01	0.42	42.8	5.92	603

fabricated to carry on the experimental tests. The prototype is provided of two DOF (vertical and horizontal translation) and an elastic membrane, with radius (R_g) equal to 0.8 mm, is fixed to its bottom tip (Fig. 8.4). An hydraulic actuation system controls the shape of the membrane and, in particular, realizes the transition between the flat and the hemispherical configuration. The elastic membrane is mechanically fixed to the tip of a microsyringe filled with an uncompressible liquid (water); in this way the curvature of the membrane is controlled by the displacement of the plunger, actuated by a microcomparator.

The procedure carried out for the experiments is the following: first, a drop is deposited on the object; second, the membrane in planar configuration approaches the object and grips it; finally the gripper moves towards the target position and

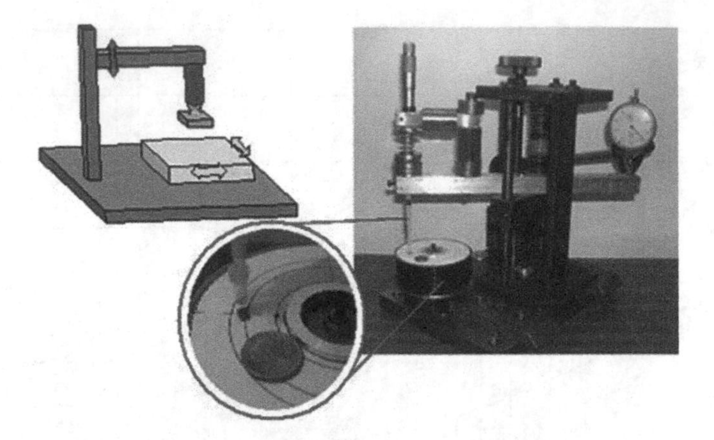

Fig. 8.4 Schematic representation and picture of the prototype

releases the object by decreasing the curvature radius of its membrane. The experiments are conducted in an environment without restrictions on cleanness and humidity: temperature is around 25°C and variable conditions of relative humidity (40 up to 70%) are encountered during the experiments. The manipulation sequence was considered to be successfully accomplished when both lifting and releasing operations succeeded. The experimental handleable weight range was between 48 and 198mg. Indeed, objects lighter than 48mg were successfully grasped but not released, and the other way round for components heavier than 198mg.

The experimental minimum weight is very close to the predicted one (42.8mg) whereas the maximum value is considerably lighter since 603mg is the theoretical value. The experimental setup, indeed, guarantees a good control of the geometrical configuration for the minimum curvature radius, thus the experimental conditions for the release are very close to theoretical model; while, in the lifting configuration the finite size of the membrane, the interaction with the mechanical fixture and the roughness of the surfaces can interfere with the capillary bridge and thus reduce the experimental capillary force.

Moreover, it has been observed that: the drop should be placed as close as possible to the center of gravity of the object in order to avoid tilting during the grasp and the release of the objects, although an autocentering capability has been showed; for the grasp of the object the gripper tip has to be in contact with the surface of the object ($h = 0$) to achieve the configuration of maximum capillary force; since positioning and releasing phases are decoupled, an accurate placement of objects has been obtained.

Moreover, liquids with different surface energy (water, oil, soapy water, alcohol) have been examined: the gripping principle is valid for any fluid; the magnitude of the capillary force changes according to the surface tension of the liquid and the handleable range of weights can be changed choosing the suitable liquid: the lower the surface tension of the liquid is, the lighter the maximum and minimum liftable weights are. Furthermore, very volatile fluids can be used to leave no traces of liquid on the components, but attention has to be paid in order to avoid unstable handling. On the other hand, appropriate oil can be used for the manipulation and left as lubricant (e.g., for the assembly of watch gears). Furthermore the influence of the humidity has to be taken into account, and the choice of the liquid has to be done according to the wettability of both the surfaces.

The values of the contact angles can be controlled using appropriate coatings on the surfaces in contact with the liquid. Finally, for the manipulation of nonflat objects their shape has to be taken into account in the evaluation of the capillary force.

As a conclusion, a variable curvature gripper based on capillary force has been demonstrated for the handling of microobjects. Theoretical and numerical models, together with experiments on millimetric objects have proved that the capillary forces can be controlled varying the shape of the gripper.

Future works will involve the study of smart materials for the actuation of a prototype of the gripper with smaller dimensions ($R = 50$ μm), which, according with the theoretical studies, will be suitable for the manipulation of micrometric objects. Electroactive polymers (EAP) are under study for this application, and two configurations of the gripper actuated by EAP are under investigation. These are based on a thin layer of dielectric elastomer, coated with compliant electrodes (first configuration) or surrounded by stiff electrodes (second configuration), which control its shape. In order to test these configurations, though, first, suitable EAPs have to be characterized, since these are quite innovative materials and no exhaustive information can be already found in the literature. The actual research is, thus, focused on the characterization of these smart materials aiming to design a gripper actuated by them.

8.4 Conclusion

Manipulation and assembly of microparts is still an issue, since no versatile and completely automatic solutions have been so far proposed. This chapter presented a brief overview of the main strategies for the manipulation of microcomponents using tools in physical contact with the parts. The methods above described have the main challenge of accurately releasing components in a precise position controlling the influence of all the superficial, and therefore dominant at the microscale, forces. Other methods based on a contactless approach can be found in literature, but were beyond the purpose of this chapter.

References

1. Van Brussels H et al (2000) Assembly of micro-systems. Annals CIRP 49(2):451–472
2. Kim C-J, Pisano AP, Muller RS (1992) Silicon-processed overhanging microgripper. Trans ASME J Microelectromech Syst 1(1):31–36

3. Chu PB, Pister KSJ (1994) Analysis of closed-loop control of parallel-plate electrostatic microgrippers. In: Proceedings of the IEEE international conference on robotics and automation, San Diego, California, 8–13 May 1994, pp 820–825
4. Jericho SK, Jericho MH, Hubard T, Kujath M (2004) Micro-electromechanical systems microtweezers for the manipulation of bacteria and small particles. Rev Sci Instrum 75 (5):1280–1282
5. Salim R, Wurmus H, Harnisch A, Hulsenberg D (1997) Microgrippers created in microstructurable glass. Microsystem Technol 4:32–34
6. Chronis N, Lee LP (2005) Electrothermally activated SU-8 microgripper for single cell manipulation in solution. J Microelectromech Syst 14(4):857–863
7. Ivanova K, Ivanov T, Badar A, Volland BE, Rangelow IW, Andrijasevic D, Säumecz F, Fischer S, Spitzbart M, Brenner W, Kosti I (2006) Thermally driven microgripper as a tool for micro assembly. Microelectron Eng 83:1393–1395
8. Carrozza MC, Menciassi A, Tiezzi G, Dario P (1997) The development of a LIGA-microfabricated gripper for micromanipulation tasks. In: Proceedings of micro mechanics Europe 1997, Southampton, UK, 31 August–2 September 1997, pp 156–159
9. Menciassi A, Eisinberg A, Carrozza MC, Dario P (2003) Force sensing microinstrument for measuring tissue properties and pulse in microsurgery. IEEE/ASME Trans Mechatron 8 (1):10–17
10. Kohl M, Krevet B, Just E (2002) SMA microgripper system. Sensor Actuator A 97–98:646–652
11. Bellouard Y, Lehnert T, Bidaux JE, Sidler T, Clavel R, Gottardt R, Bellouard Y (1999) Local annealing of complex mechanical devices: a new approach for developing monolithic microdevices. Mater Sci Eng A273–A275:795–798
12. Kim D-H, Kim B, Kang H (2004) Development of a piezoelectric polymer-based sensorized microgripper for microassembly and micromanipulation. Microsyst Technol 10(4):275–280
13. Nah SK, Zhong ZW (2007) A microgripper using piezoelectric actuation for micro-object manipulation. Sensor Actuator A 133:218–224
14. Petrovic D et al (2002) Gripping tools for handling and assembly of microcomponents. In: Proceedings of the 23rd international conference on microelectron, vol 1, pp 247–250
15. Beyeler F, Neild A, Oberti S, Bell DJ, Sun Y, Dual J, Nelson BJ (2007) Monolithically fabricated microgripper with integrated force sensor for manipulating microobjects and biological cells aligned in an ultrasonic field. J Microelectromech Syst 16(1):7–15
16. Butefisch S, Seidemann V, Buttgenbach S (2002) Novel micro-pneumatic actuator for MEMS. Sensor Actuator A Phys 97–98:638–645
17. Molhave K, Hansen O (2005) Electro-thermally actuated microgrippers with integrated force-feedback. J Micromech Microeng 15:1256–1270
18. Arai F, Andou D, Nonoda Y, Fukuda T, Iwata H, Itoigawa K (1998) Integrated microendeffector for micromanipulation. IEEE/ASME Trans Mechatron 3(1):17–23
19. Park J, Moon W (2003) A hybrid-type micro-gripper with an integrated force sensor. Microsyst Technol Micro Nanosyst Inf Storage Process Syst 9(8):511–519
20. Lu MSC, Huang CE, Wu ZH, Chen CF, Huang SY, King YC (2006) A CMOS micromachined gripper array with on-chip optical detection. In: 2006 I.E. sensors, vols 1–3, pp 37–40
21. Kim DH, Lee MG, Kim B, Sun Y (2005) A superelastic alloy microgripper with embedded electromagnetic actuators and piezoelectric force sensors: a numerical and experimental study. Smart Mater Struct 15:1265–1272
22. Zesch W, Brunner M, Weber A (1997) Vacuum tool for handling microobjects with a NanoRobot. In: Proceedings of the IEEE international conference on robotics and automation, Albuquerque, NM, pp 1761–1766
23. Vikramaditya B, Nelson BJ (2001) Modeling microassembly tasks with interactive forces. In: Proceedings of the IEEE international symposium on assembly and task planning, Fukuoka, Japan, pp 482–487

24. Wejinya UC, Shen Y, Xi N, Winder E (2005) Development of pneumatic end effector for micro robotic manipulators. In: Proceedings of the IEEE/ASME international conference on advanced intelligent mechatronics, Monterey, CA, pp 558–563
25. Arai F, Fukuda T (1997) Adhesion-type micro endeffector for micromanipulation. In: Proceedings of IEEE international conference on robotics and automation. Albuquerque, New Mexico, 20–25 April 1997, pp 1472–1477
26. Lambert P, Letier P, Delchambre A (2003) Capillary and surface tension forces in the manipulation of small parts. In: Proceedings of international symposium on assembly and tasks planning (ISATP), Besancon, France, 9–11 July 2003, pp 54–59
27. Lambert P, Delchambre A (2005) A study of capillary forces as a gripping principle. Assem Autom 25(4):275–283
28. Grutzeck H (2005) Investigations of the capillary effect for gripping silicon chips. Microsyst Technol 11:194–203
29. Bark C, Binnenbose T, Vogele G, Weisener T, Widmann M (1998) Gripping with low viscosity fluids. In: Proceedings of the 11th annual international workshop micro electro mechanical system, Heidelberg, Germany, pp 301–305
30. Sinan Haliyo D, Regnier S, Guinot J-C (2003) MAD, the adhesion based dynamic micromanipulator. J Mech A/Solids 22:903–916
31. Saito S, Motokado T, Obata KJ, Takahashi K (2005) Capillary force with a concave probe-tip for micromanipulation. Appl Phys Lett 87(23):234103-1–234103-3
32. Grutzeck H, Kiesewetter L (1998) Downscaling of grippers for micro assembly. In: Proceedings of sixth international conference on micro electro, opto mechanical systems and components, Potsdam, Germany, 1–3 Dec 1998
33. Obata KJ, Saito S, Takahashi K (2003) A scheme of micromanipulation using a liquid bridge. In Proceedings of MRS fall meeting, symposium A, vol 782, Boston, MA, pp A3.6.1–A3.6.6
34. Pagano C et al (2003) Micro-handling of parts in presence of adhesive forces. In: CIRP seminar on micro and nano technology 2003, Copenhagen, Denmark, 13–14 November 2003, pp 81–84
35. Vasudev A, Zhe J (2008) A capillary microgripper based on electrowetting. Appl Phys Lett 93 (10):103503
36. Biganzoli F, Fassi I, Pagano C (2005) Development of a gripping system based on capillary force. In: Proceedings of sixth IEEE international symposium assembly and task planning, Montreal, QC, Canada, pp 36–40
37. Kochan A (1998) European project develops 'ice' gripper for micro-sized components. Assem Autom 17(2):114–115
38. Lang D, Tichem M, Blom S (2006) The investigation of intermediates for phase changing micro-gripping. In: Proceedings of international workshop on microfactories, Besancon, France
39. Changhai R, Xinliang W, Xiufen Y, Shuxiang G (2007) A new ice gripper based on thermoelectric effect for manipulating micro objects. In Proceedings of the 7th IEEE international conference on nanotechnology, Hong Kong, China, pp 438–441
40. Yang Y, Liu J, Zhou Y-X (2008) A convective cooling enabled freeze tweezer for manipulating micro-scale objects. J Micromech Microeng 18(9):095008-1–095008-10
41. López-Walle B, Gauthier M, Chaillet N (2008) Principle of a submerged freeze gripper for microassembly. IEEE Trans Robot 24(4):897–902
42. Fantoni G, Biganzoli F (2004) Design of a novel electrostatic gripper. Int J Manuf Sci Prod 6 (4):163–179
43. Hesselbach J, Wrege J, Raatz A (2007) Micro handling devices supported by electrostatic forces. CIRP Ann Manuf Technol 56:45–48
44. Enikov ET, Lazarov KV (2001) Optically transparent gripper for microassembly. In: Proceedings of SPIE, microrobotics and microassembly III, vol 4568, pp 40–49

45. Lang D, Tichem M (2006) Design and experimental evaluation of an electrostatic microgripping system. In: Proceedings of third international precision assembly seminar, Bad Hofgastein, Austria, 19–21 Feb 2006, pp 33–42
46. White EL, Enikov ET (2007) Self-aligning electrostatic gripper for assembly of millimeter- sized parts. In: IEEE/ASME international conference on advanced intelligent mechatronics, Zurich, Switzerland, 4–7 Sept 2007, pp 1–5
47. Lee SH, Lee KC, Lee SS, Oh HS (2003) Fabrication of an electrothermally actuated electrostatic microgripper. In: The 12th international conference on solid-state sensors, actuators and microsystems, vol 1, Boston, June 2003, pp 552–555

Chapter 9
A Wall-Climbing Robot with Biomimetic Adhesive Pedrail*

Xuan Wu, Dapeng Wang, Aiwu Zhao, Da Li, and Tao Mei

Abstract A prototype of a wall-climbing robot with gecko-mimic adhesive pedrails was developed to demonstrate the adhesive ability of micron adhesive arrays. The robot has two parallel pedrails driven by a DC motor. The outside surfaces of the pedrails were covered by gecko-mimic adhesive hair arrays made by polydimethylsiloxane (PDMS). A two-step template method was used to fabricate gecko-mimic adhesive array in which the hair density, diameter, and length could be adjusted independently. A tail was fixed at the rear of the robot to provide preload for adhesion. Experiment results show that the robot has the ability to climb on vertical wall.

9.1 Introduction

For a long time, men have observed animals like geckos, spiders, flies, etc, climbing or staying on a vertical wall via some kind of mysterious force. In the fourth century BC, Aristotle observed that geckos can "run up and down a tree in any way, even with the head downwards" [1]. With the development of the robot technology, researchers began to fabricate machines that can do the same thing as the mentioned animals.

*This work is supported by National Basic Research Program of China (2011CB302100 and 2006CB300407).

X. Wu
State Key Laboratories of Transducer Technology, Institute of Intelligent Machines, Chinese Academy of Sciences, Hefei, Anhui 230031, China

Department of Precision Machinery and Precision Instrumentation, University of Science and Technology of China, Hefei, Anhui 230027, China
e-mail: wuxuan@mail.ustc.edu.cn

D. Wang • A. Zhao • D. Li • T. Mei (✉)
State Key Laboratories of Transducer Technology, Institute of Intelligent Machines, Chinese Academy of Sciences, Hefei, Anhui 230031, China
e-mail: dpwang@iim.ac.cn; awzhao@iim.ac.cn; lida@iim.ac.cn; tmei@iim.ac.cn

D. Zhang (ed.), *Advanced Mechatronics and MEMS Devices*, Microsystems,
DOI 10.1007/978-1-4419-9985-6_9,

Climbing vertical walls is such a great challenge to the human being from the gravity that robots are needed to take it. Concretely speaking, they can be used in hazardous workspace to work for humans. For example, they can help clean the outside of skyscrapers and do some inspecting in some special terrains, such as storage tanks for petroleum industries and nuclear plants [2]. They may even be used to clean large solar panels and the shells of spacecrafts or space stations in space.

Many different kinds of wall-climbing robots have been fabricated. One kind of wall-climbing robot works via negative pressure, such as a small-scaled robot fabricated by Beijing University of Aeronautics & Astronautics in 2005. The 3.2 kg robot carries a pump, and when it climbs, the pump works at full speed and creates a space of vacuum inside itself. The robot is pressed against the wall by the ambient pressure. It can climb steadily on rough walls, carry a mass of 15 kg, and get over shallow gaps that are less than 2 cm wide [3]. But its noise is unbearable while it cannot get over very wide gaps, because they will make it fail to adhere. Vacuum is not its workspace, either, in which there is no air to provide the pressure.

There are also other traditional ways of adhering, such as the positive pressure of atmosphere and magnetic force. They have different advantages and disadvantages.

Geckos have been observed and admired for centuries. However, men did not get to the point, where geckos can climb so well. So traditional wall-climbing robots could not perform very perfect. In recent years, with the development of material science, biology, chemistry and some other fields, a new method has been put forward to produce materials, whose structures are like that of geckos' toes, to help robots climb. In the year 2000, Dr. Kellar Autumn and his colleagues found out the secret of geckos' climbing. The biological gecko materials have a very complicated structure. Generally speaking, there are about 500,000 setae on each foot. The seta's length is 30–130μm and its diameter is about 5μm. At the top of the seta lie 100–1,000 spatulae, and each spatula terminates in 0.2–0.5μm diameter disk-like pad [4]. The setae and spatulae are made of β-keratin, whose Young's modulus is 4 GPa [5]. When billions of spatulae make contact with the wall, adhesion is created and reaches a remarkable value. Each spatula can provide $194 \pm 25\mu N$ [4] force. Thanks to the micronanostructure, the fibers can conform to different kinds of surfaces smoothly. What is more, they are self-cleaning. The fiber can keep itself clean for a month as long as it is kept moist after it is peeled off a gecko [6].

After many experiments, scientists excluded the atmospheric pressure, friction, electrostatic force, mucilage glue, and capillary force as the adhesion of geckos' climbing [7]. Van der Waal's force turned out to be the answer [1]. Through the study on geckos' movement, scientists also found out that a preload force is necessary for climbing [4].

The MEMS (Micro-Electro- Mechanical System) helps man have geckos' ability to climb. Because geckos' fibers have their special shape, size, texture and are arranged in a certain way, men can make synthetic hairs similar to geckos' fibers and arrange them properly. Given a preload force as high as a certain value, the synthetic material can do the same thing as geckos. This kind of material is the so-called micro/nano structure adhesive array, which can also be called gecko-inspired synthetic dry adhesives.

Metin Sitti's team of Carnegie Mellon University (CMU) fabricated a kind of synthetic dry adhesive in 2003. They used a silicon mold, and got an array of fibers whose diameter is 4μm [8]. Later, several kinds of robots using the dry adhesives were fabricated by Sitti's team, such as the Tankbot [8], Waalbot [8, 9], Geckobot [9, 10], and Wormbot [11]. One of their Geckobots can climb an 85° sloped glass surface. The robot is driven by seven micro servomotors and controlled by a PIC (a kind of Microcomputer) [10].

In 2006, Mark R. Cutkosky and his team of Stanford University fabricated a wall-climbing robot called "Stickybot" [12]. It is equipped with 12 servomotors, a controller, and directional polymer stalks (DPS). DPS is another kind of dry adhesive. The tiny robot can climb vertical smooth surfaces such as glass, plastic, and ceramic at 4 cm/s.

In this work, a kind of micron adhesive array was fabricated. We designed a robot using the adhesives for testing as well as a first step of application of the material. The following sections mainly talk about the process of the design. The adding of a tail helps its climbing. Results of experiments are also presented. The robot can climb steadily on smooth vertical walls, such as glass, iron, wood, and so on. The maximum speed can reach up to 10 cm/s.

9.2 Design of the Climbing-Caterpillar

9.2.1 Aim and Requirements

A climbing-caterpillar is a kind of simple wall-climbing mechanism. Its structure is easy as it is designed to do some testing of the material. We glued several pieces of the synthetic material and a rubber tread together, and then fixed the tread around the wheels so that the material will help the robot move. Due to the simple task of the machine, there is no need adding a controlling module to it for more kinds of movements such as turning or driving back. What is more, the robot should be as light as possible and the center of mass should be low. It has a limited mass estimated to be 200g. A tiny scale limits its functions, and more kinds of movements are impossible to realize.

9.2.2 A Rough Framework

The first step is to choose the motor. For a small robot a micro DC motor is suitable. We looked up the Faulhaber Company's product list, and chose Faulhaber 1516T012SR. Along with the motor, we bought a gearhead which was a module of a certain reduction ratio. The type is 15/5 76:1 and its efficiency is 66%. In addition, we placed a pair of reduction ratio gears at the end of the gearing

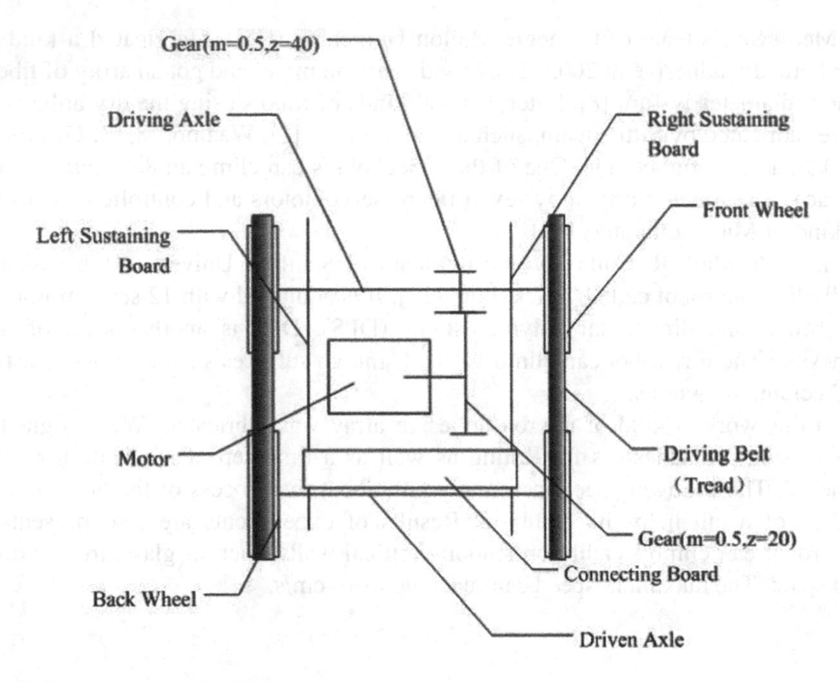

Fig. 9.1 The framework of the robot

chain for the last transmission. Provided that the ratio is 2:1 and its efficiency is 80%, the output torque of 1516T012SR at the driving axle is:

$$T = T_r n\mu l\eta = 0.8 \times 76 \times 0.66 \times 2 \times 0.8 = 64.2 \text{ mN m},$$

T_r is the recommended torque, n & l are the reduction ratio and μ & η are the efficiency. For a robot whose wheels' radius $R = 15$mm and mass $m = 200$g, the required torque on a vertical surface should be at least:

$$T_n = mgR = 0.2 \times 9.8 \times 15 = 29.4 \text{ mN m},$$

T is about 100% beyond the required torque T_n to make an enough margin for detaching of the array. The motor is also very tiny scaled and light. The cylinder-shaped motor's dimension is 34.9mm × Φ16mm and the motor weighs only 34g together with its gearhead. Figure 9.1 gives the framework of the machine.

As Fig. 9.1 shows, the motor provides torque, and the gears send the torque to the driving axle, then the driving axle forces the front wheels to rotate. The driving belt fixed around the wheels makes the back wheels move and also is used as a tread. The adhesive array is fixed on the belt, with its working surface towards the wall the robot climbs.

The dimensions of the robot were also carefully decided, as they were important to the climbing ability. The length of the robot determines the adhesive force and the main gravity of itself. The width also determines the contacting area. It cannot

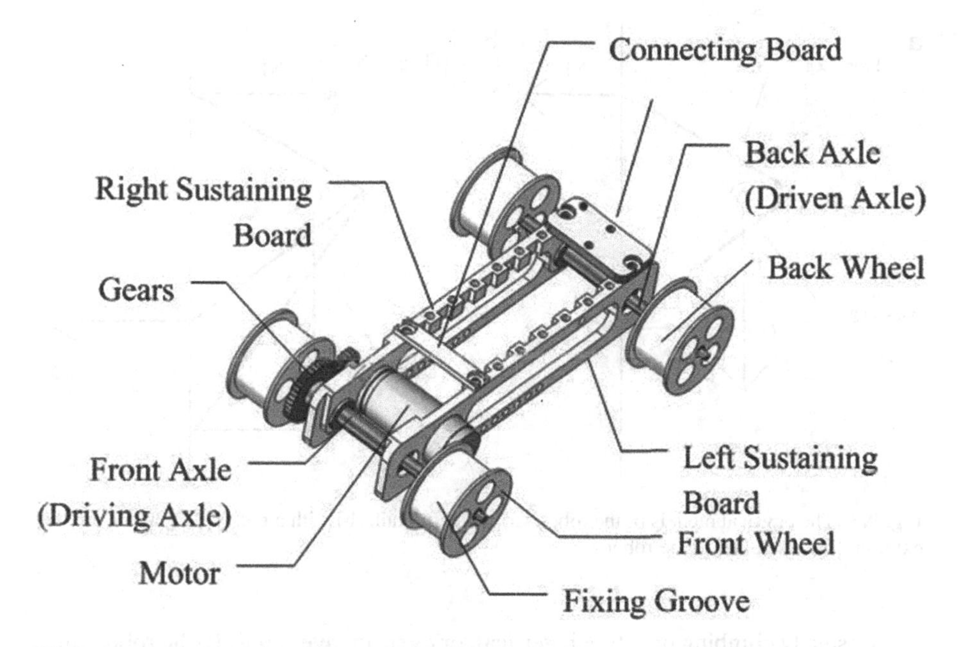

Fig. 9.2 The CAD model given by Solidworks

Table 9.1 Some dimensions of the robot

Dimension	Value (mm)	Dimension	Value (mm)
L_1	100	D_w	16
L_a	120	d_w	14
l_a	92	R	12.5

L_1: the distance between the front and the back axles; L_a: the total length; l_a: the length of the front and back axles; D_w: the width of the wheel; d_w: the width of the fixing groove; R: the radius of the fixing groove

be too small, but if it is too large, the long and thin axle will be hard to manufacture. Then a CAD (Computer Aided Design) model is given in Fig. 9.2, and main dimensions are listed in Table 9.1.

The Solidworks program also gave the property of its mass: The using of light materials such as Aluminum and Nylon provides a rather small mass of 105g. The position of the center of mass was also given. Due to the symmetry of the robot, it is in the middle in the left–right and the up–down direction. As for the front–back direction, it is about 40mm from the front axle.

9.2.3 Adding a Tail

Tails are very important to climbing animals and robots. When climbing, tails help them keep balance, move from one place to another, and fix themselves as a support [9]. Here for gecko-inspired robots, tails are used to offer preload force.

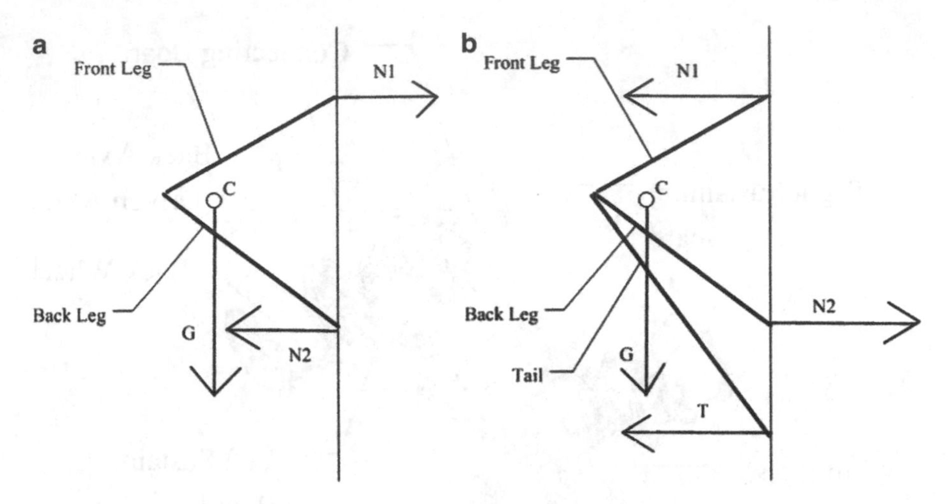

Fig. 9.3 The physical models of the robot: (**a**) without a tail, (**b**) with a tail. N_1, N_2, and T are the external forces applied to the robot

For steady climbing on a titled, vertical, or even an inverted wall, the robot must have a large contacting area with the wall, and when a part of it is about to contact the wall, a preload force must be given. So a tail is needed here and the robot's physical model can be described as a "front leg-back leg-tail" one. Here the front leg means the fore part of the tread and the back leg means the hind part [8].

If there is no tail, when the robot is on the wall, gravity not only tends to pull it down, but also creates a torque to make it tend to overturn. In that case, the front leg will detach from the wall. Nowhere will the preload force come from, without which the array will not adhere to the wall continuously.

With a tail connected at the back, things will be completely different. The tail offers a normal pushing force against the surface. Via the back leg as a support, it makes the front leg press against the surface, which creates a preload force. Meanwhile the back leg, as a support, tends to detach from the surface, and is prevented by the normal adhesive force. Briefly speaking, the tail gives the fore part of the tread a preload force, which is assisted by the array at the hind part. Figure 9.3 shows the models clearly.

Concretely, when the robot is climbing a $\theta°$ sloped surface, the forces applied to it are shown in Fig. 9.4. The tail pushes against the wall and makes T. Theoretically the front wheel contacts the tread at a single line and the force N_1 is created from the line and vertical to the wall. Gravity acts downwards at the center of the mass. An adhesive force as large as $G\sin\,\theta$ is created parallel to the surface, and a distributed force exists all over the tread towards the surface. The equivalent force is N_2. Since the distributing rule is hard to know, considering the trend of the tread's detaching from the surface is stronger at the back, we just suppose that it is distributed following a linear rule from zero to $\sigma_{\max}$. In Fig. 9.5 the robot is put on a horizontal surface and all equivalent forces are shown.

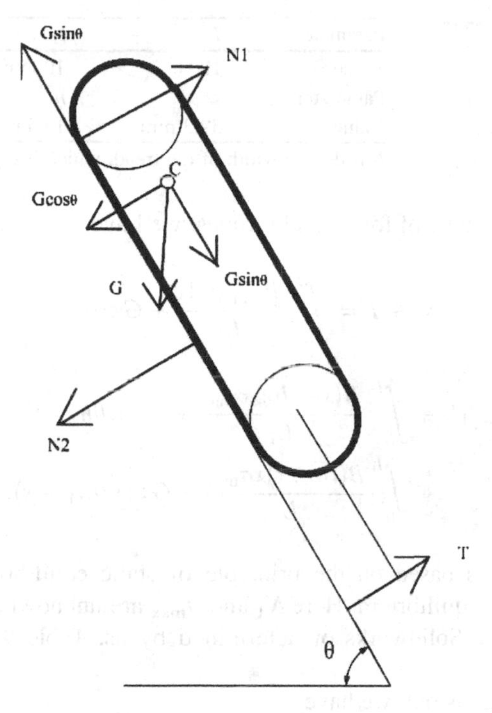

Fig. 9.4 Forces applied to the robot

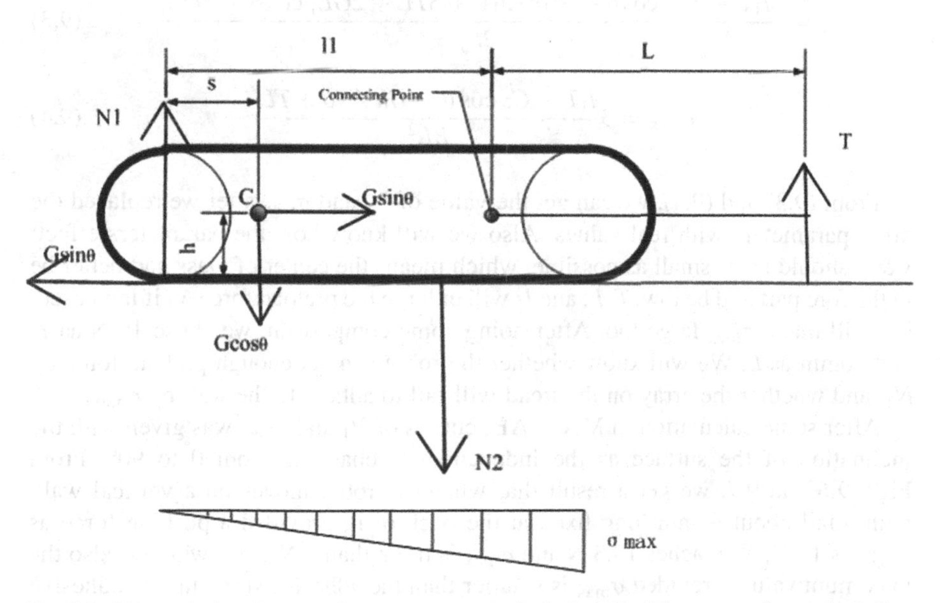

Fig. 9.5 The equivalent forces and some parameters. h: the height of the center of mass, including the thickness of the tread and the adhesive array; s: the horizontal distance between the center of mass and the front axle; l_1: the horizontal distance between the point where the tail and the robot (the connecting point) is connected and the front axle; L: the horizontal length of the tail

Table 9.2 The parameters of the equations

Parameter	T	l_1	L
Value	10 N	100 mm	90 mm
Parameter	s	h	B
Value	39.3 mm	13.5 mm	28 mm

B is double width of the tread, which is equal to double d_w

Based on the balance of forces and torques, we have equations as follows:

$$N_1 + T = \int_0^{L_1} \frac{B\sigma_{max} x dx}{L_1} + G\cos\theta, \tag{9.1}$$

$$\begin{aligned} N_1 l_1 + \int_{l_1}^{L_1} \frac{B(x-l_1)x\sigma_{max}}{L_1} + G\sin\theta h = TL \\ + \int_0^{l_1} \frac{B(l_1-x)x\sigma_{max}}{L_1} + G\cos\theta(l_1-s). \end{aligned} \tag{9.2}$$

Equation (9.1) is based on the principle of static equilibrium and (9.2) the principle of torque equilibrium. Here N_1 and σ_{max} are unknown. Other parameters are either given by Solidworks or determined by us. Table 9.2 gives the other parameters.

Work the equations out, we have

$$N_1 = \frac{3l_1T - 3Gs\cos\theta - 3Gh\sin\theta + 3TL + 2GL_1\cos\theta - 2TL_1}{2L_1}, \tag{9.3}$$

$$\sigma_{max} = 3\frac{l_1T - Gs\cos\theta - Gh\sin\theta + TL}{BL_1^2}. \tag{9.4}$$

From (9.3) and (9.4), we can get the value of N_1 and σ_{max} after we replaced the other parameters with real values. Also we will know how the parameters effect: s & h should be as small as possible, which means the center of mass had better be in the fore part and be low. T, L, and l_1 will enlarge the preload force N_1 if increased, but will make σ_{max} large too. After doing some comparison, we chose 10 N as T, and 90mm as L. We will know whether the robot can get enough preload force by N_1, and whether the array on the tread will fail to adhere to the wall by σ_{max}.

After some calculation in MATLAB, curves of N_1 and σ_{max} was given with the inclination of the surface as the independence changing from 0 to 90°. From Figs. 9.6 and 9.7, we get a result that when the robot moves on a vertical wall, with a tail about 90mm long fixed at the back of it, provided a pushing force as large as 10 N, N_1 reaches 18.3 N and σ_{max} is more than 2 N/cm^2 , which is also the maximum value. Provided σ_{max} is smaller than the adhesive strength, the adhesive part will not decrease. As N_1 is not very small, the preload force will be enough. So we can conclude that the robot can move smoothly and continuously on a vertical surface.

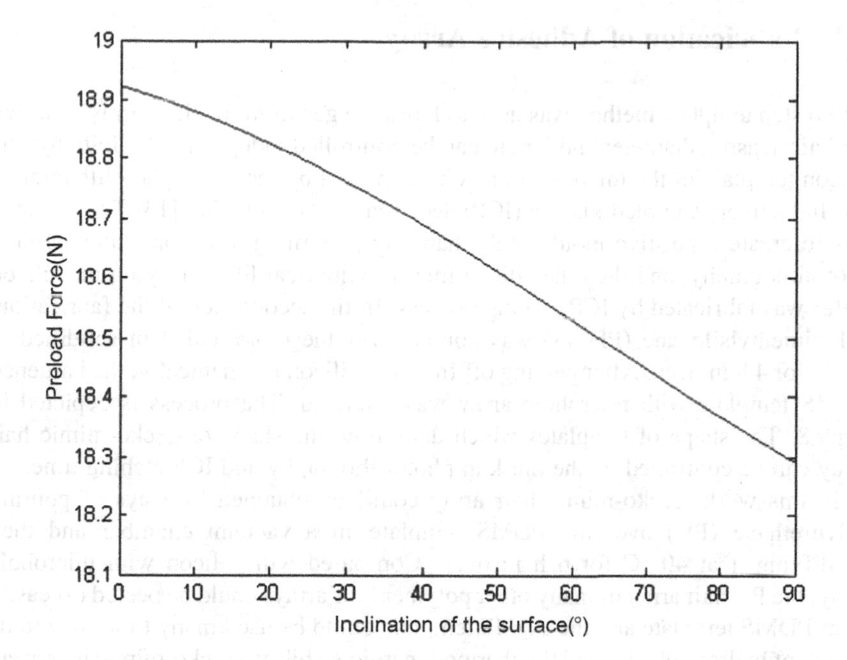

Fig. 9.6 Curve of the preload force N_1

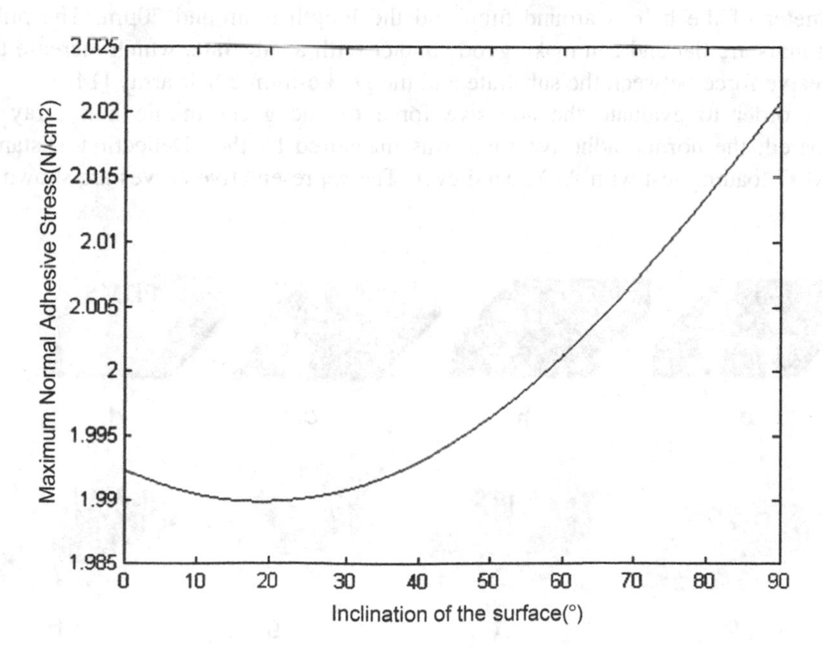

Fig. 9.7 Curve of $\sigma_{\max}$

9.3 Fabrication of Adhesive Array

A two step template method was used to fabricate gecko-mimic hair array in which the hair density, diameter and length can be controlled independently. Initially, the silicon template of the microfiber array was formed by means of photolithography and Inductively Coupled Plasma (ICP) deep reactive ion etching [13].The first step was to create a positive mold for the hairs by patterning a silicon wafer through photolithography, and then the silicon micro cylindrical fiber array on the silicon wafer was fabricated by ICP etching process. In the second step of the fabrication, polydimethylsiloxane (PDMS) was poured over the mold and then solidified at 70 °C for 4 h in oven. After peeling off from the silicon mechanically, the hardened PDMS template with microhole array was obtained. The process is depicted in Fig.9.8. The shape of templates which determine the shape of gecko-mimic hair array can be controlled by the mask in photolithography and ICP etching time.

In this work, gecko-mimic hair array could be obtained by ways of pouring polyurethane (PU) over the PDMS template in a vacuum chamber and then solidifying it at 40 °C for 6 h in oven. Compared with silicon with microhole array, the PU hair array or many other polymer hair arrays could be peeled off easily from PDMS template and the PDMS template could be used many times due to its excellent hydrophobicity and the thermodynamic stability. Gecko-mimic hair array was formed successfully as is shown in Fig. 9.9. It can be clearly seen that the diameter of the hair is around 5μm and the length is around 20μm. The pillar terminals are flat and can make good contact with a substrate, which increase the adhesive force between the substrate and the gecko-mimic hair array [14].

In order to evaluate the adhesive force of the gecko-mimic hair array as prepared, the normal adhesive force was measured by the "Deflection–Distance curves" loading test with AFM cantilever. The representative curves are shown in

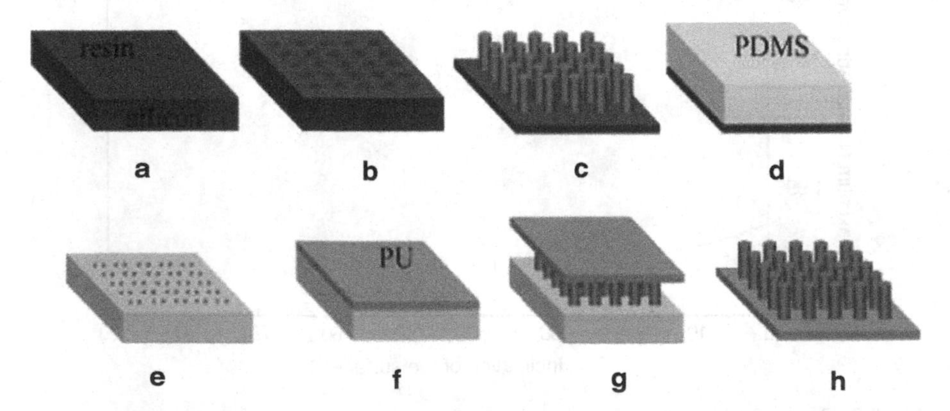

Fig. 9.8 Fabrication process of gecko-mimic hair array: (**a**) Resin on silicon wafer, (**b**) photolithography, (**c**) ICP etching, (**d**) cast and curing PDMS, (**e**) obtain template, (**f**) cast and curing PU, (**g**) remove template, (**h**) obtain hair array

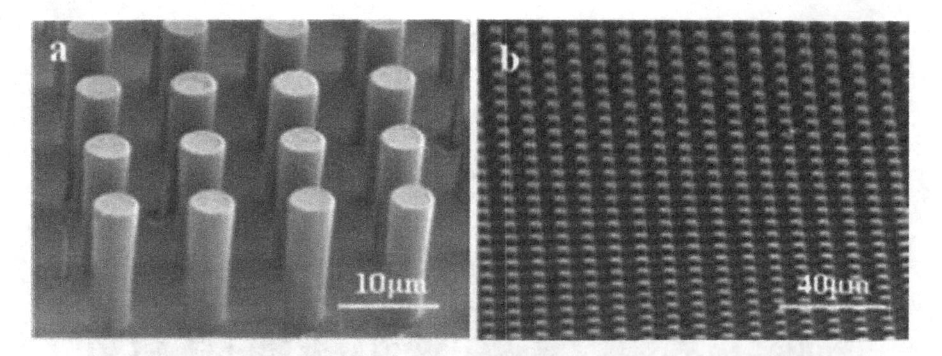

Fig. 9.9 SEM photographs of the as-prepared hair array: (**a**) high magnification; (**b**) low magnification

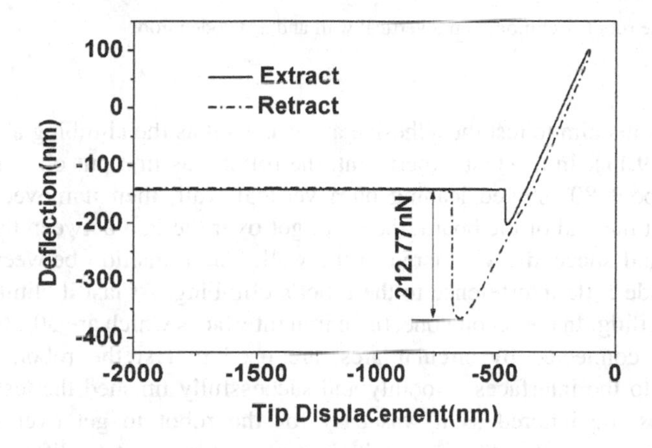

Fig. 9.10 Deflection–distance curves for the gecko-mimic hair array as prepared

Fig. 9.10 and the adhesive force can be obtained from the difference between the measured forces in the extracting and retracting directions. As is shown in the figure, the normal adhesion force is 212.77 nN.

9.4 Prototype and Experiments

A prototype was made without complex manufacture. The axles were made of 45# steel. Wheels were made of Nylon. They were both manufactured on a lathe. The sustaining boards and connecting boards were made of Aluminum and completed by wire-electrode cutting and milling machines. The tail is a part of spring steel. The actual mass is 145g. A rubber belt was used to form the treads, and PDMS is the base of the adhesive array.

Fig. 9.11 The robot is climbing on a vertical wall and a wooden door

Experiments aim to test the adhesive array as well as the climbing ability of the robot (Fig. 9.11). In the first experiment, the robot was first put on a glass board which is about 80° sloped leaning on a vertical wall, then it moved straightly upwards. At the end of the board, the robot got over the gap between the wall and the board and successfully adhered to the wall. The transition between different textures made little interference to the robot's climbing. At last it climbed until it got to the ceiling. In the second one, three iron interfaces which are 30°, 60° and 90° sloped and connected by circular arcs, are used to test the robot. The robot conformed to the interfaces smoothly and successfully finished the test. The arcs could almost be ignored as it was easy for the robot to get over them. The experiments proved that the robot with the array can adapt to different textures and inclinations of the interfaces.

9.5 Conclusion

A wall-climbing robot was presented including its geometrical and mass property and the adhesive arrays. Analysis was given following the supposed distributing rule of the normal adhesive force to predict the stability of the robot's climbing a vertical wall. In the experiments, the adaptation of the robot to both different textures and inclinations of the interfaces was proved. It supported the possibility of the application of the robot in the future. Future works will focus on wall-climbing robots with more functions, such as a robot that can climb as fast as a real gecko, move from the ground to the vertical wall, then to the ceiling, do some inspecting to make its own route, and so on.

References

1. Autumn K, Sitti M, Liang YA et al (2002) Evidence for Van der Waals force in gecko setae. Proc Natl Acad Sci 99:12252–12256
2. Wang Y, Liu SL, Xu DG, Zhao YZ, Shao H, Gao XS (1999) Development and application of wall-climbing robots. In: Proceedings of the IEEE international conference on robotics and automation (ICRA), Detroit, USA, May 1999, pp 1207–1212
3. Wang TM, Meng C, Pei BQ, Dai ZD (2007) Summary on gecko robot research. Robot 29 (3):290–297
4. Autumn K, Liang YA, Hsieh ST et al (2000) Adhesive force of a single gecko foot-hair. Nature 405:681–685
5. Persson BNJ (2003) On the mechanism of adhesion in biological systems. J Chem Phys 118 (16):7614–7621
6. Liang YA, Autumn K, Hsieh ST et al (2000) Adhesion force measurements on single gecko setae. In: Solid-state sensor and actuator workshop, Hilton Head Island, SC, pp 33–38
7. Ren NF, Wang XH, Wang HJ, Shan JH (2006) Progress on the micro/nano-structures adhesive array mimicking gecko foot-hair. In: MEMS device & technology, Shijiazhuang, China, August 2006, pp 386–392
8. Menon C, Murphy M, Sitti M (2004) Gecko inspired surface climbing robots. In: IEEE international conference on robotics and biomimetics (ROBIO), Shenyang, China, August 2004
9. Unver O, Murphy MP, Sitti M (2005) Geckobot and Waalbot: small-scale wall-climbing robots. In: AIAA 5th aviation, technology, integration, and operations conference (ATIO), Arlington, VA, September 2005
10. Unver O, Uneri A, Aydemir A, Sitti M (2006) Geckobot: a gecko inspired climbing robot using elastomer adhesive. In: IEEE international conference on robotics and automation (ICRA), Orlando, FL, May 2006
11. The website of the Nanolab of CMU, US. http://nanolab.me.cmu.edu/
12. Kim S, Spenko M, Trujillo S, Heyneman B, Santos D, Cutkosky MR (2007) Smooth vertical surface climbing with directional adhesion. In: IEEE international conference on robotics and automation (ICRA), Rome, Italy, April 2007
13. Shan JH, Mei T, Ni L, Chen SR, Chu JR (2006) Fabrication and adhesive force analysis of biomimetic gecko foot-hair array. In: Proceedings of the first IEEE international conference on nano/micro engineered and molecular systems, pp 1546–1549
14. Reddy S, Arzt E, Campo A (2007) Bioinspired surfaces with switchable adhesion. Adv Mater 19:3833–3837

Chapter 10
Development of Bioinspired Artificial Sensory Cilia

Weiting Liu, Fei Li, Xin Fu, Cesare Stefanini, and Paolo Dario

Abstract Given inspiration from the natural hair receptors of animals, sensors based on micro/nanofibers are considered as a significant and promising solution for improving the intelligence and automation of microrobots in the future. Thus, we introduce in this chapter the concept and design of some novel artificial hair receptors for the sensing system of microintelligent robots. The natural hair receptor of animals, also called cilium or filiform hair by different research groups, is usually used as a sensitive element for slight disturbance by insects, mammals and fishes, such as a detector for ambient vibration, flow or tactile information. At first, focusing on the development of biomimetic sensory abilities for an undulatory soft-body lamprey-like robot, piezoresistive sensory elements based on highly soft silicone rubber matrix are presented. On the other hand, micro-artificial hair receptor based on suspended PVDF (polyvinylidene fluoride) microfibers is also designed to address useful applications for microrobots working in unstructured environments. Both these cilia shaped sensors show a reliable response with good sensibility to external disturbance, as well as a good prospect in the application on sensing system of mini/microbiorobots.

W. Liu (✉) • F. Li • X. Fu
The State Key Laboratory of Fluid Power Transmission and Control, Zhejiang University, Hangzhou 310027, China
e-mail: liuwt@zju.edu.cn; lifei.zju@gmail.com; xfu@zju.edu.cn

C. Stefanini • P. Dario
CRIM Lab, Polo Sant'Anna Valdera, Pontedera (Pisa) 56025, Italy
e-mail: cesare.stefanini@sssup.it; dario@sssup.it

D. Zhang (ed.), *Advanced Mechatronics and MEMS Devices*, Microsystems,
DOI 10.1007/978-1-4419-9985-6_10,

10.1 Introduction

10.1.1 Sensory Hairs in Natural World

Hair cells possess a characteristic organelle which consists of tens of hair-like stereocilia. So-called hair bundle is able to pivot around their base when a force is applied to the tips [1–6]. Such sensory hairs widely exist in the natural world [8, 9, 11, 24]. As the primary mechanotransducer, natural hair cell sensory receptors usually exist widely from the mammalian sense organs of hearing and balance (cochlea and vestibular organ, respectively) to lateral line organ of fishes and amphibians for water motion detection.

For example, high performance detection systems composed of mechanoreceptive cuticular hairs of some arthropods are evolved to sense the slightest air displacement around them, such as that generated by approaching predators. Like the mechanoreceptive cerci on cricket's abdomen which are sensitive to those slight air currents generated by a wasp's wings or a toad's tongue. Such sensory hairs alert the insects when a predator is sneaking around them, and give them a chance to escape from predation [3, 7]. Adult tropical wandering spider (Cupiennius salei) also has hundreds of trichobothria on its ambulatorial legs and pedipalps ranging from 20 to 1,500μm in diameter. It was also found those sensory hairs in different length are able to mechanically couple with different frequencies and receive the medium vibration generated by flying insects.

Another interesting sensory system exists in all primarily aquatic vertebrates, like cyclostomes (e.g., lampreys, eels), fish, and amphibians. They have in their outer skin (epidermis) special mechanoreceptors called lateral line organs able to detect the motion of surrounding water. The name "lateral line" originates from the line running from head to tail, in which the neuromasts are located. Neuromasts [17] have a core of mechanosensory hair cells, surrounded by support cells, and are innervated by sensory neurons that are localized in a ganglion.

In general, three different applications of such natural hair receptors are summarized in Table 10.1. Due to its simple but high efficient sensing strategy, fiber based artificial sensors can be considered as a significant and promising solution for improving the intelligence and automation of microrobots in the future.

10.1.2 Biological Model of Natural Hair Receptor

The mechanism, morphology, and modeling of the hair cell type mechanoreceptor system such as cricket cercal wind receptors have been unveiled thanks to biological research [2]. The hair is modeled as an inverted pendulum (Fig. 10.1).

It can be described by a second-order mechanical system which is determined by the spring stiffness S, the momentum of inertia I and the torsion resistance R. For the angular momentum,

Table 10.1 Applications of natural hair receptors

Animals	Location of hair receptors	Application
Cricket, spider, etc.	Located ranging from different scales to couple with different vibrating frequency to sense slight air disturbance generated by approaching predators	Vibration sensor
Fish, lamprey, etc.	Located within the lateral line organs reflecting the flow field information	Flow sensor
Leafhopper, etc.	Located between each two sections of its limp to detection the tactile information during leg bending and extending	Tactile sensor

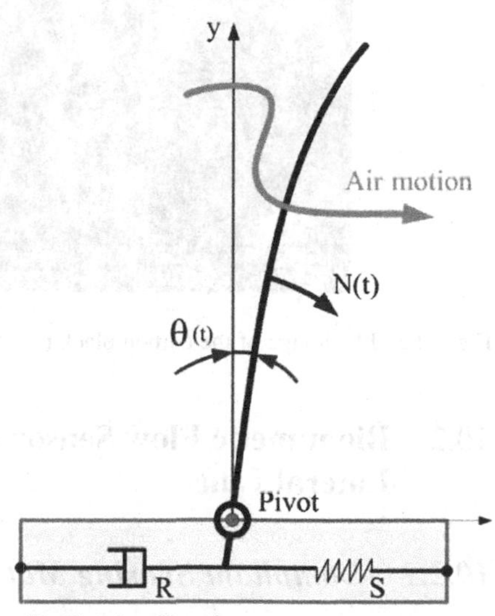

Fig. 10.1 Inverted pendulum model of hair cell [24] (© MDPI 2010), reprinted with permission

$$I\,\frac{\mathrm{d}^2\theta(t)}{\mathrm{d}t^2}+R\,\frac{\mathrm{d}\theta(t)}{\mathrm{d}t}+S\,\theta(t)=N(t). \tag{10.1}$$

The hair is deflected by the drag force on the air shaft due to the airflow surrounding the cercus. The total external torque $N(t)$ can be calculated by the integration of the drag force along the hair shaft.

$$N(t)=\int_0^{L_0}F(y,t)\cdot y\cdot \mathrm{d}y. \tag{10.2}$$

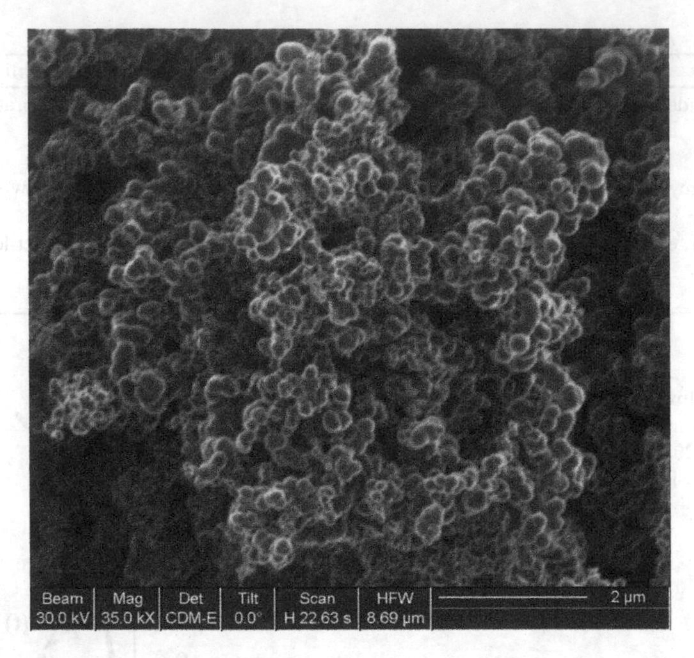

Fig. 10.2 FIB image of the carbon black particles (© Elsevier 2010), reprinted with permission

10.2 Biomimetic Flow Sensor Inspired from Natural Lateral Line

10.2.1 Compliant Sensing Material

Highly flexible sensing elements are necessary in the development of biomimetic sensing systems. Therefore, easy processed compliant polymer and its composites with forces sensory abilities are attractive in robot sensing application. For example, in our research, compliant piezoresistive composite is prepared by mixing conductive micro/nano particles (e.g., carbon black micro/nano powder with diameter about 200 nm as shown in Fig. 10.2) into a silicone matrix together with flexible Kapton and applied to develop a biomimetic cupula receptor through molding process.

The conductive particles are homogeneously distributed into the insulating matrices. Therefore, in the case of low fraction of the conductive particles, there is no contact between the conductive particles. When the volume fraction increases, particles come closer together and small agglomerates begin to grow. When pressure is applied on the composites, the conductive particles come into contact with each other more easily so that the resistance is decreased. The mechanical stimulation changes the volume fraction of conductive particles, thus leading to a change in resistivity. The electrical resistance of the conductive filler-dispersed

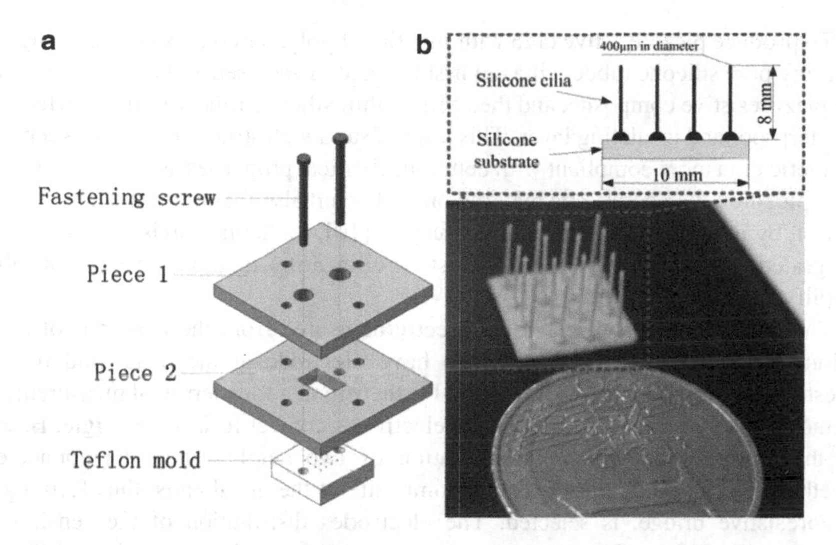

Fig. 10.3 (**a**) Molding device of silicone cilia matrix fabrication. (**b**) A 4 × 4 matrix of silicone cilia, 400 μm in diameter and 8 mm in length (© Elsevier 2010), reprinted with permission

composite depends on the deformation it undergoes [12]. The composites show gradual change in electrical resistivity with applied pressure.

In the meanwhile, the electronic quantum tunneling effect occurs when conductive particles are close enough under mechanical stimulation. In this case, the electrons overcome the potential energy barrier and move from one particle to another which decreases the resistivity of the composite. Classical resistance is linear (proportional to distance), while quantum tunneling effect is exponential with decreasing distance, allowing the resistance to change by a factor 10^{12} between pressured and unpressured states. Therefore, dramatical resistance change occurs even with minor space variation between conductive particles, which is not really applicable in stretch sensing in case of relative large deformation situation. It is necessary to introduce the principle in order to answer an interesting resistive change behavior under mechanical stimulation which is presented in following sections. The estimated conductive particle distance of our working composite is about 185 nm under no mechanical stimulation condition.

10.2.2 Fabrication of Biomimetic Cupula Receptor

It is not trivial to obtain compliant piezoresistive tiny cilia with high aspect ratio. Although cilia matrix made of pure silicone rubber can be successfully fabricated by simple molding technology with Teflon mold (Fig. 10.3), the piezoresistive one is failed due to the change of mechanical properties resulting from adding the conductive particles.

To produce piezoresistive cilia with functional polymer composites, such highly soft tiny pure silicone rubber cilia are first fabricated and used as a core to coat with the piezoresistive composite, and then form a thin silicone rubber film on surface as a waterproof and insulating layer. This kind of sandwich structure helps to keep the synthetic cilia more compliant with better mechanical properties as compared to the fully piezoresistive composite material ones. Then it also facilitates the fabrication. At last, by imitating from the natural structure [10], a silicone cup is fabricated and integrated on the top of the piezoresistive cilia array in order to mechanically amplify the cilia strain.

The change in resistance can be directly measured from the two ends of each cilium. However, it is problematic to have electrode at the distal end of the freestanding cilium in real applications. Furthermore, a four-terminal measurement method has to be used to eliminate the electrodes contact resistance joggle. Based on these considerations, the configuration of two neighboring cilia connected together with the same piezoresistive composite at the distal ends, thus forming a piezoresistive bridge, is selected. The electrodes distribution of the sensilia is shown in Fig. 10.4a. The electrodes are tiny springs which made from 100μm-diameter copper wire and the coil's diameter is about 500μm including the wire size. Each spring has only three coils about 300μm in height when they are not elongated. Two springs are inserted closely to each cilium and locate on the silicone substrate. Another end of these springs' copper wire going through the substrate and is connected to an electrical connection support (Kapton). The reason to use the tiny springs is to increase electrode contact area and have tougher electrodes being able to follow elongation without damage.

The cilia piezoresistive layer coating scheme, silicone cup mold, and prototype of artificial cupula receptor are shown in Fig. 10.4. The coated piezoresistive layer is about 0.2 mm and the length between bottom of the cup and the substrate is 7 mm.

10.2.3 Testing Result of Artificial Cupula Receptor

In order to check the fluidic flowing speed sensibility of the artificial cupula receptor, one test rig has been setup as shown in Fig. 10.5. One continuous working pump (ULTRA zero utility pump from SICCE S.P.A., Italy) drives the water into a 25 mm in diameter tube; the flux in the tube is monitored by a flow meter and controlled by a throttle valve; the artificial cupula receptor is fixed into the tube with the top cup positioned at the center area of the tube (Fig. 10.5b).

Resistance changing under deformation by the flowing medium is transformed into voltage changing when four-terminal measurement is applied. We recorded the voltage outputs of the artificial cupula receptor under different flowing speed. Then calculation is done to obtain voltage difference between the output value and the initiative value, which is the sensor output voltage value that water has been filled in the testing chamber but no substantial medium flowing. Finally, voltage difference is normalized (divided by the initial output value) in order to reduce the variance by

Fig. 10.4 (a) The electrodes distribution of the cupula receptor; (b) scheme of conductive layer coating; (c) the mold for the top cup; (d) prototypes of artificial cupula receptor (© Elsevier 2010), reprinted with permission

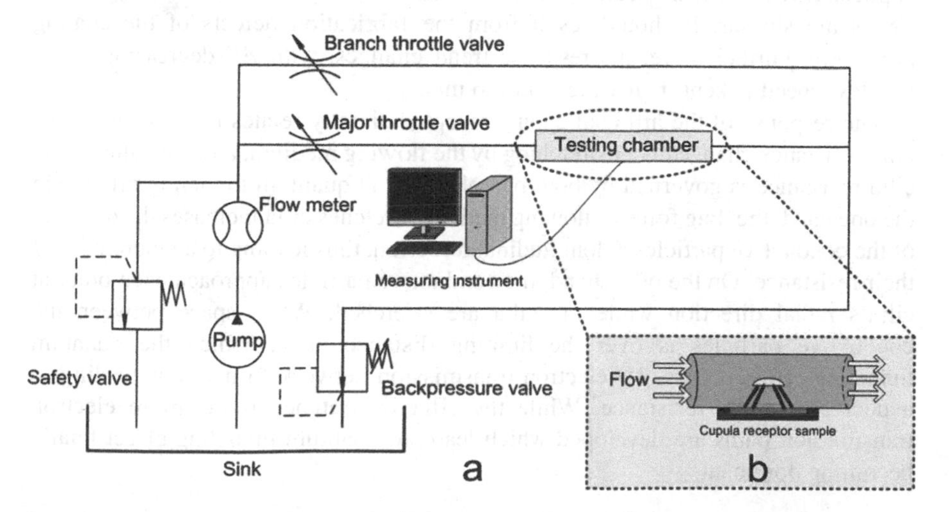

Fig. 10.5 (a) Scheme of the test rig; (b) the cupula receptor in fluid filled tube (© Elsevier 2010), reprinted with permission

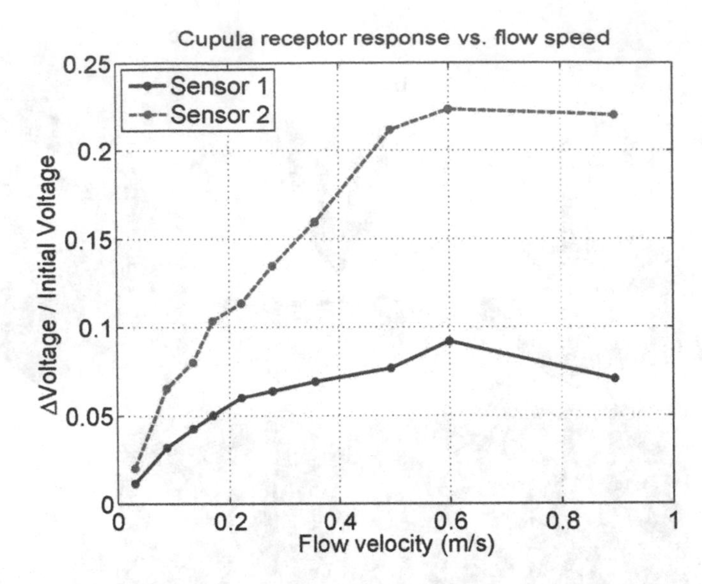

Fig. 10.6 Cupula receptor response vs. flow speed (© Elsevier 2010), reprinted with permission

the fabrication procedure. After the above steps, the obtained value is actually the value of $\Delta R/R$ (where: ΔR is the resistance change of sensor under deformation, R is the initiative value of the sensor resistance). The sensor responses vs. flow speed are shown in Fig. 10.6.

It shows that the artificial cupula receptor is more sensitive at lower speed range because of the linear deformation property at that range. The measurable range is from 0.05 to 0.6 m/s according to our measurement. It is possible that the receptor may also be applicable to lower flow speed if more precisely experiment setup applies. However, it is clearly shown the different response curves though their trends are similar. It should result from the fabrication defects of the coating procedure. Further more, the response trend changes, with ΔR decreasing while the flow speed is kept increasing over 0.6 m/s.

The response of the artificial cupula receptor directly relates to the behavior of cilia resistance change under stretching by the flowing medium, and the value of the cilia resistance is governed by both percolation and quantum tunneling effect. On the one hand, the drag force of flowing medium stretches cilia increases the distance of the conductive particles at longitudinal direction, thus leading to an increment of their resistance. On the other hand, the conductive particles approach each other at cilia's radial direction while the cilia are stretched. When space between the conductive particles is over the limiting distance under which the quantum tunneling effect occurs, 3D electron transmission network forms, thus leading to a decrease of the resistance. While the cilia is stretched more, more electron transmission paths are developed which leads to quantum tunneling effect finally becoming dominant.

Along with the flowing speed increasing (drag force increasing), the resistance of the sensor keeps increasing until the electron transmission activated and finally becoming dominant which reverse the resistance developing trends. The response behavior is exactly showed in Fig. 10.6.

10.3 Drawing Artificial Cilia from Polymer Solution

10.3.1 Setup of Microfiber Drawing

In another research, sensitive cilia can be produced by extending viscous volatile PVDF solution within drawing time window and formed after solidifying [16, 18, 19], which can be called as thermo-direct fiber drawing technique. Not like other existing approaches such as lithography and etching technology based MEMS processes [13], electrospinning [14], and dry spinning [15], this method allows us to fabricate aligned piezoelectric fibers on specific locations with simply processing.

In order to produce such a slim structure in microscale, a fabrication platform shown as Fig. 10.7 is first setup. Such system is composed of an automatic 3D nano positioner for defining the drawing path, a 3D micromanipulator used to preliminarily position the micropipette with respect to the video camera, a Sony CXD-V50 video system for monitoring, and a specific designed work platform fixed with the positioner as well as a local heated micropipette fixed on the pipette holder and vertical to the work platform.

During the polymer stretching, the cohesiveness should be always balanced with the stresses in order to successfully carry on the fiber drawing; thus, proper

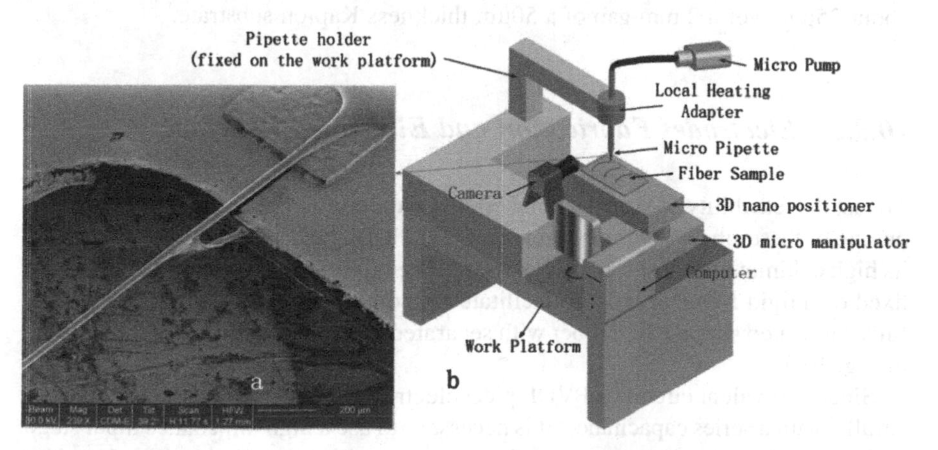

Fig. 10.7 The thermo-direct drawing system [24]; (**a**) suspended PVDF aligned micro/nano fibers (about 25 μm in diameter) on Kapton (thickness 50 μm), (**b**) scheme of PVDF direct drawing setup (© MDPI 2010), reprinted with permission

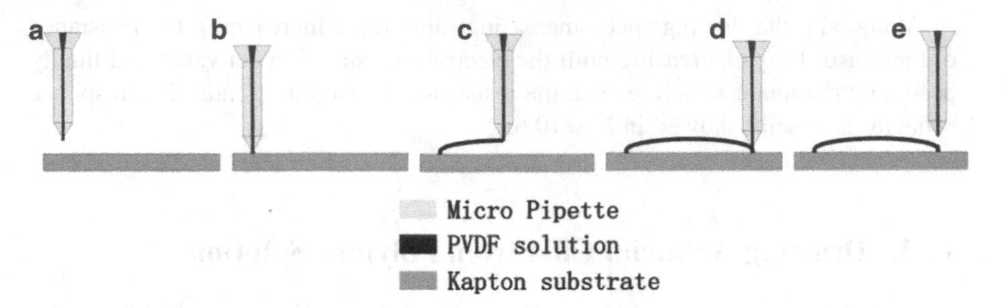

Fig. 10.8 Schematic diagram of the drawing fabrication process of suspended aligned PVDF microfibers: (**a**-**b**) Drawing preparations; (**c**-**d**) suspended microfiber formed by substrate moving; (**e**) a single PVDF fiber formed by pipette moving away [24] (© MDPI 2010), reprinted with permission

viscoelastic behavior of the material is required. Furthermore, drawing processes always accompany with the solidification and transfer the spinning polymer to a solid fiber, which make the situation even more complicated. Therefore, the local heating device is required to adjust both the solvent evaporation and the polymer mobility. The detailed drawing process is shown in Fig. 10.8.

To produce sensitive cilia with piezoelectric property, PVDF/DMF viscous solution system is prepared by dissolving 20% weight of polymer in DMF (viscosity 1,176 cP) because at this viscosity the droplets on the micropipette tip can maintain stable. PVDF/DMF viscous solution is prepared by dissolving 25% weight of PVDF (Polyvinylidene Fluoride) polymer in DMF (Dimethylformamide) solvent and 2-h ultrasonic stirring after that, and the drawing is processing at 70°C. By introducing proper control parameters (temperature, drawing velocity, pump flux, etc.), we successfully produce suspended aligned PVDF microfibers ranging from 2 to 30μm. Figure 10.8a shows one of this piezoelectric fibers with diameter about 25μm over a 2 mm gap of a 50μm thickness Kapton substrate.

10.3.2 Electrodes Fabrication and Electronic Interface

To test the feasibility of such artificial cilia, separated electrodes are then fabricated on one of these fibers (20μm in diameter) with thermo evaporation [20] because of its high collimated deposition path capacity. The fabricated aligned PVDF fibers are fixed on a rigid frame in order to facilitate evaporation on both sides. The successfully fabricated single PVDF fiber with separated electrodes with 1μm gap is shown in Fig. 10.9.

Since equivalent circuit of PVDF piezoelectric material is a charge generator in parallel with a series capacitance, it is necessary to use a high impedance input stage to interface the proposed sensor. Furthermore, parasitic capacitances introduced by connection wires from the sensor to the interface must be eliminated; thus, as the first stage of the electronic interface design, the charge amplifier configuration

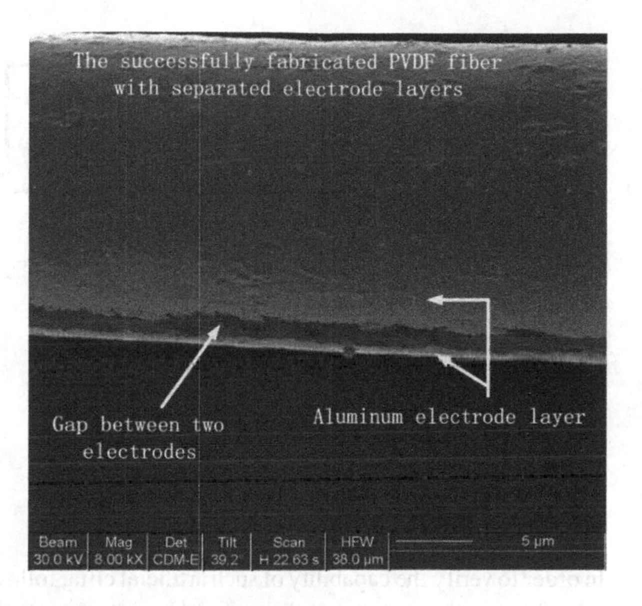

Fig. 10.9 FIB image of the separated electrodes (1 µm gap) fabricated by thermo evaporation technique on a single PVDF micro/nano fiber [24] (© MDPI 2010), reprinted with permission

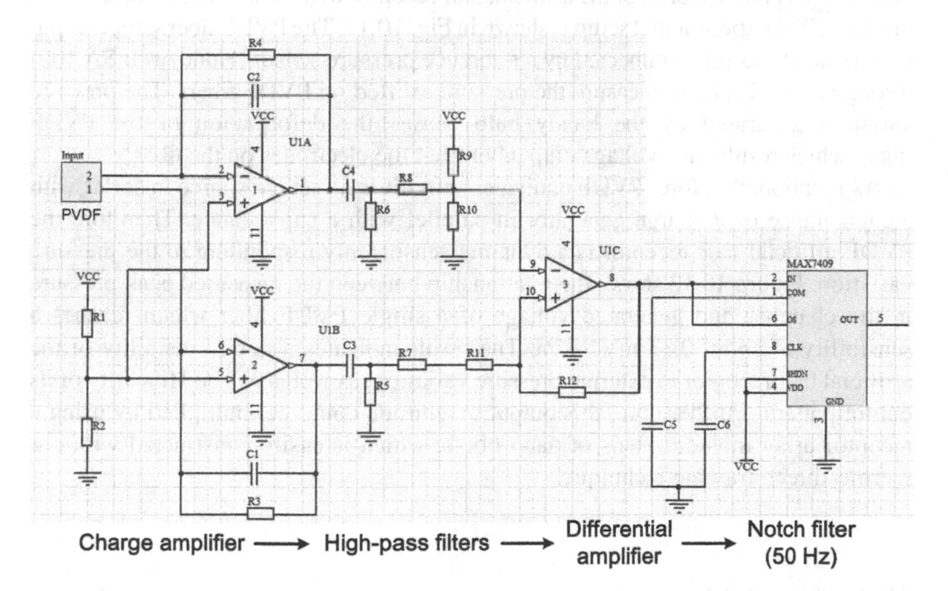

Fig. 10.10 The electronic interface of the PVDF micro/nano fiber sensor [24] (© MDPI 2010), reprinted with permission

described in [21, 22] and illustrated in Fig. 10.10 is selected. This configuration output is only sensitive to the feedback capacitance rather than the input one. The selected feedback capacitance is 1 pF due to the low value of the PVDF micro/nano fiber capacitance in our case, and the notch filter is 50 Hz in order to eliminate the affection of the line power frequency interference.

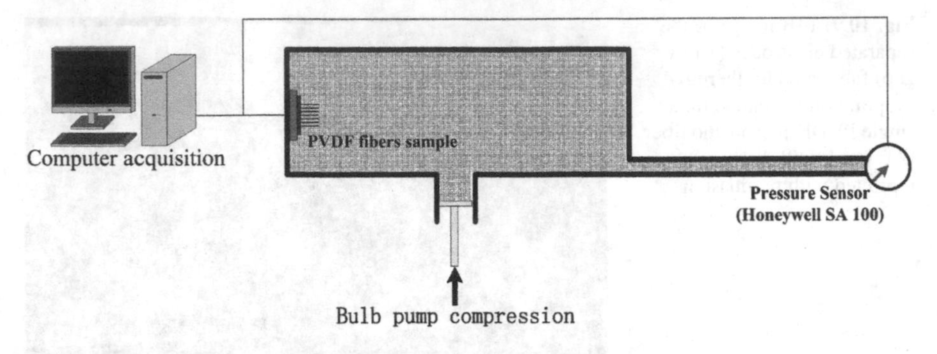

Fig. 10.11 The experimental setup of external pressure testing

10.3.3 Response to Ambient Disturbance

In order to verify the capability of such artificial cilia, following validation experiment is built to test the response of the artificial hair receptor to the changing of the external pressure. The experimental setup is shown in Fig. 10.11. The PVDF fiber sample is put into a sealed chamber connected by a reference pressure sensor (Honeywell SA 100) through a small tube to measure the pressure exerted on PVDF fibers. The pressure variation generated by the handy bulb causes the deformation of the PVDF fibers which results in a voltage output between the electrodes on the fiber.

As mentioned before, PVDF equivalent circuit is a voltage source in series with a capacitance or a charge generator in parallel with a capacitance. Therefore the PVDF artificial hair receptor is a dynamic sensor only responding to the pressure variation. Figure 10.12 shows the relationship between the measured peak pressure in the chamber and generated voltage of a single PVDF fiber which reaches a sensibility of about 0.33 mV / kPa. The result indicates a good sensibility of the artificial hair receptor to external pressure variation, as well as a good linearity of its output pattern. Furthermore, this output magnitude could be multiplied by using a receptor array including tens of nano fibers, which is easily constructed with our thermo-direct drawing technique.

10.4 Conclusion

Learning from the natural hair receptors of animals, the concept of microfiber based sensors is introduced in this chapter. Such design can also be considered as a significant and promising solution for future robotic research to improve their intelligence and automation ability.

Based on this idea, two novel artificial hair receptors inspired from fish and insect respectively are presented in the chapter, including the piezoresistive sensory

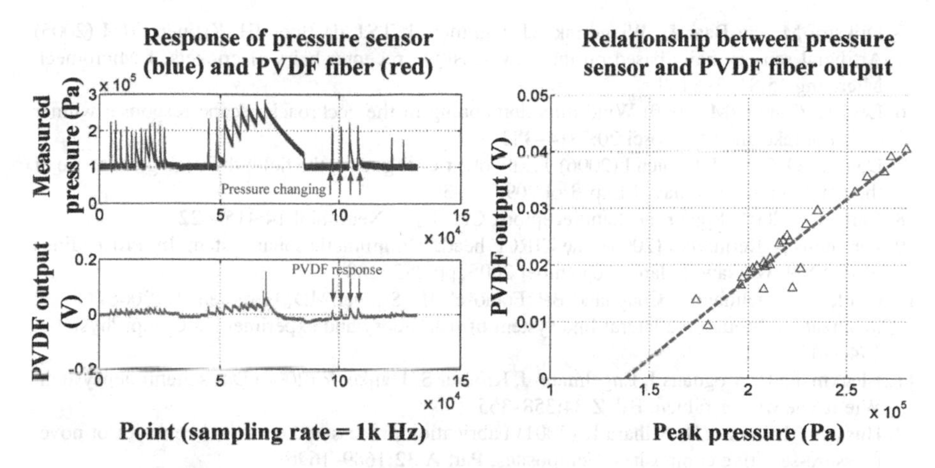

Fig. 10.12 Setup and results of pressure experiment [24]: (*left*) the response of Honeywell pressure sensor (*left-top*) and PVDF fiber sensor (*left-bottom*), (*right*) relationship between the peak pressure measured by the Honeywell pressure sensor and the PVDF fiber output (© MDPI 2010), reprinted with permission

elements fabricated with highly soft silicone rubber matrix, and the piezoelectric sensitive microfiber based on molding technology and direct drawing method. Both these biomimetic cilia sensors show a reliable response with good sensibility to external disturbance, as well as a good prospect in the application on future robotic research.

Contents in this chapter are summarized from our work on the development of bioinspired cilia sensors, and other detailed information of relevant research can be found in our previously published papers [23, 24].

Acknowledgements The activity presented in this chapter is partially supported by LAMPETRA Project (EU Contract No. 216100), the Fundamental Research Funds for the Central Universities of China, Scientific Research Foundation for the Returned Overseas Chinese Scholars and Zhejiang Province Qianjiang Talents Project (2010R10012).

References

1. van Netten SM (1997) Hair cell mechano-transduction: its influence on the gross mechanical characteristics of a hair cell sense organ. Biophys Chem 68:43–52
2. Shimozawa T, Kumagai T, Baba Y (1998) Structural scaling and functional design of the cercal wind receptor. J Comp Physiol A 183:171–186
3. Dangles O, Magal C, Pierr D, Olivier A, Casas J (2004) Variation in morphology and performance of predator-sensing systems in wild cricket populations. J Exp Biol 208:461–468
4. Li J, Chen J, Liu C (2000) Micromachined biomimetic sensor using modular artificial hair cells. In: Presented at nanospace conference, Houston, TX, USA, January 2000.

5. Dijkstra M, van Baar JJ, Wiegerink RJ, Lammerink TSJ, de Boer JH, Krijnen GJM (2005) Artificial sensory hairs based on the flow sensitive receptor hairs of crickets. J Micromech Microeng 15:S132–S138
6. Levi R, Camhi JM (2000) Wind direction coding in the cockroach escape response: winner does not take all. J Neurosci 20:3814–3821
7. Dangles O, Casas J, Coolen I (2006) Textbook cricket goes to the field: the ecological scene of the neuroethological play. J Exp Biol 209:393–398
8. Barth FG (2004) Spider mechanoreceptors. Curr Opin Neurobiol 14:415–422
9. Peremans H, Reijniers J (2005) The CIRCE head: a biomimetic sonar system. In: Proceedings of ICANN, Warsaw, Poland, September 2005, pp 283–288
10. Goulet J, Engelmann J, Chagnaud BP, Franosch JP, Suttner MD, Hemmen JL (2008) Object localization through the lateral line system of fish theory and experiment. J Comp Physiol A 194:1–17
11. Bleckmann H, Mogdans J, Engelmann J, Kröther S, Hanke W (2004) Das seitenliniensystem: Wie fische wasser fühlen. BIUZ 34:358–365
12. Hussain M, Choa YH, Niihara K (2001) Fabrication process and electrical behavior of novel pressure-sensitive composites. Composites: Part A 32:1689–1696
13. Chen N, Tucker C, Engel JM, Yang Y et al (2007) Design and characterization of artificial haircell sensor for flow sensing with ultrahigh velocity and angular sensitivity. J Microelectromech Syst 16:999–1014
14. Seoul C, Kim YT, Back CK (2003) Electrospinning of poly (vinylidene fluoride)/dimethylformamide solutions with carbon nanotubes. J Polym Sci 41:1572–1577
15. Ohzawa Y, Nagano Y, Matsuo T (1969) Studies on dry spinning I. Fundamental equations. J Appl Polym Sci 13:257–283
16. Ondarcuhu T, Joachim C (1998) Drawing a single nanofibre over hundreds of microns. Europhys Lett 42:215–220
17. Harfenist S, Cambron S, Nelson S, Berry S et al (2004) Direct drawing of suspended filamentary micro- and nanostructures from liquid polymers. Nano Lett 4:1931–1937
18. Nain AS, Wong JC, Amon C, Sitti M (2006) Drawing suspended polymer micro-/nanofibers using glass micropipettes. Appl Phys Lett 89:183105
19. Nain A, Amon C, Sitti M (2006) Proximal probes based nanorobotic drawing of polymer micro/nanofibers. IEEE Trans Nanotechnol 5:499–510
20. Hunt TP, Westervelt RM (2006) Dielectrophoresis tweezers for single cell manipulation. Biomed Microdevices 8:227–230
21. Shen Y, Xi N, Li WJ (2003) Contact and force control in microassembly. In: Presented at IEEE international symposium on assembly and task planning, Besanpon, France, July 2003.
22. Liu W, Menciassi A, Scapellato S, Dario P, Chen Y (2006) A biomimetic sensor for a crawling minirobot. Robot Autonom Syst 54:513–528
23. Liu W, Stefanini C, Sumer B, Li F, Chen D, Menciassi A, Dario P, Sitti M (2009) A novel artificial hair receptor based on aligned PVDF micro/nano fibers. In: Proceedings of IEEE international conference on robotics and biomimetics, Bangkok, Thailand, February 2009, pp 49–54
24. Li F, Liu W, Stefanini C, Fu X, Dario P (2010) A novel bioinspired PVDF micro/nano hair receptor for a robot sensing system. Sensors 10:994–1011

Chapter 11
Jumping Like an Insect: From Biomimetic Inspiration to a Jumping Minirobot Design

Weiting Liu, Fei Li, Xin Fu, Cesare Stefanini, Gabriella Bonsignori, Umberto Scarfogliero, and Paolo Dario

Abstract Locomotion is a key issue for autonomous robots, moreover if we consider gait efficiency in exploration and monitoring application. Despite the implicit mechanical and kinematic complication, legged locomotion is often preferred to the simpler wheeled version in unstructured environment, e.g., difficult terrains. Focusing on microrobot, lessons from nature often provide us a good insight of profitable solutions and suggest bioinspired design for small legged robots.

According to the biological observation experiment, it was found that a specific leg configuration maps the nonlinear muscle-like force into a constant force at feet–ground interface so as to minimize the risk of both leg ruptures and tarsus slippage, which represents an optimum design of jumping insects. That gives us the bionic inspiration to optimize the saltatorial legs by reproducing the dynamic characteristics of insect jumping. Based on this idea, jumping robot prototype GRILLO is designed and tested with different ways.

In this chapter, we present the bioinspired design of such a jumping mini robot including the dynamically optimized saltatorial leg which is designed to imitate the characteristics of a real jumping insect, kinematically and dynamically, and proposed to mitigate the peak contact force at tarsus–ground interface during jumping acceleration; the overall design of the jumping robot prototype; and as a part of the biomimetic research, the measuring and comparing of the jumping characteristics between the robot and animal so as to show the dynamic similarity and optimization

W. Liu (✉) • F. Li • X. Fu
The State Key Laboratory of Fluid Power Transmission and Control, Zhejiang University, Hangzhou 310027, China
e-mail: liuwt@zju.edu.cn; lifei.zju@gmail.com; xfu@zju.edu.cn

C. Stefanini • G. Bonsignori • U. Scarfogliero • P. Dario
CRIM Lab, Polo Sant'Anna Valdera, Pontedera (Pisa) 56025, Italy
e-mail: cesare.stefanini@sssup.it; gabriella.bonsignori@gmail.com; umberto.scarfogliero@encrea.com; dario@sssup.it

D. Zhang (ed.), *Advanced Mechatronics and MEMS Devices*, Microsystems,
DOI 10.1007/978-1-4419-9985-6_11,

results between them. The finally energy integrated jumping robot prototype is able to move by continuous jumping, of which a single one reaches 100 mm high and 200 mm long, about twice and four times of its body length respectively.

11.1 Introduction

For animals in the natural world, jumping is more than a method that helps to overcome obstacles higher than themselves; it also improves their moving ability and offers a solution of emergency escaping. As one of the most efficient natural locomotion modes, jumping is applied by animals varying from mammals to amphibians to insects (Fig. 11.1), especially when they are traveling on tough terrains and able to take advantage of the scale effects [1]. Due to the importance of locomotion in animal surviving, gait has been a central subject field for the optimization process actuated by evolution pressure [2]. In past and recent years, jumping has been studied both in vertebrates and invertebrates, insects like fleas [3], locusts [4–6], crickets [7], flea-beetles [8], froghoppers [9–12], and leafhoppers [13] have been revealed as the best jumpers in terms of takeoff velocity, acceleration, and covered distance.

On the other hand, for autonomous robot research, locomotion design is also a key issue especially when we consider the gait efficiency and stability in exploration and monitoring. Therefore, many different legged-robot prototypes have been built in order to find the best efficient and stable gaits during the past years because of its high efficiency, low energy consumption and better stability on unstructured tough terrains. For example, a simple but efficient leg-wheeled locomotion developed based on the traditional wheel structure offers good passing ability for robots [14–16]; walking locomotion inspired from mammals provides robots low energy consumption and better moving stability on tough terrains despite a complex

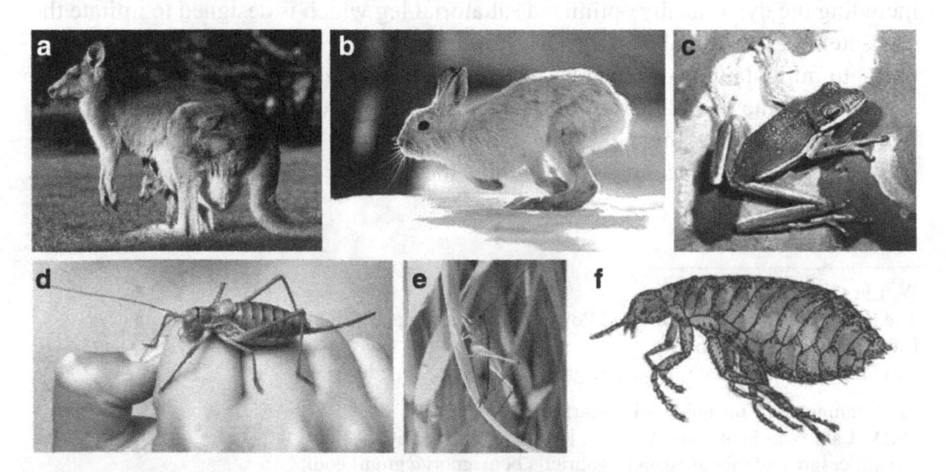

Fig. 11.1 Jumping locomotion in the natural world: (**a**) kangaroo; (**b**) rabbit; (**c**) frog; (**d**) cricket; (**e**) leafhopper; (**f**) flea (© Springer 2010), reprinted with permission

controlling strategy and kinematic complication [17–21]; novel jumping robots inspired from kangaroo [22] and fleas [23, 24] are also developed and proves the feasibility of using jumping as the primary moving mode for robots.

Focusing on the design of such mini autonomous robots, natural lessons often provide us a good insight of profitable solutions and suggest bioinspired design for small legged robots [25]. Based on the observation of natural insects, several things finally became the initial power which drives us to develop a jumping robot prototype named as "GRILLO."

11.1.1 Small Animals Are Easier to Jump

Although people thought that insects are tiny and weak, the truth is that scale effects actually bring those small animals a born advantage of jumping. That can be explained by the following equation:

$$\frac{F_{\max}}{mg} \propto \frac{l^2}{l^3}, \tag{11.1}$$

where $F_{\max}$ is the maximum exerted force, m is the mass of the animal, g is the gravity acceleration, and l is a characteristic body length. The model is built based on the hypothesis that considering a fixed maximum yield stress, the maximum exerted force should be proportional to l^2, while according to the geometrical similarity, the weight is usually proportional to l^3. It suggests that when dimensions are reduced, the relative value between maximum exerted force and body weight increased, which means that it is easier to jump or hop for small animals rather than those animals in large size, and can be proved by the observations in the natural world [25, 26].

On the other hand, scale effect influence on the air friction like the situation described by (11.2),

$$\begin{cases} \ddot{x} + \frac{C_{\mathrm{D}}\rho A}{2m}\dot{x}^2 + g = 0, \\ \dot{x}_0 = \sqrt{2gh}, \end{cases} \tag{11.2}$$

where C_{D} is the drag coefficient and can be treated as a constant when considering a fixed Reynolds number. A is the reference area perpendicular to the motion direction (i.e., cross-sectional area). ρ is the density considered as a fixed coefficient for a given object. h is the maximum height in vacuum of the tossing motion which is determined by the initial vertical velocity. By solving the differential equation, the relation between the maximum height of jumping in vacuum and jumping in air can be written like:

$$\begin{aligned} \frac{h_{\mathrm{air}}}{h} &= C\log(C+1), \\ C &= \frac{m}{C_{\mathrm{D}}\rho hA}, \end{aligned} \tag{11.3}$$

where C is intermediate variable for equation simplification.

Based on previous discussion about geometrical similarity the following relation can be obtained,

$$C \propto \frac{l}{C_D \rho h}. \tag{11.4}$$

It means that both a smaller dimension and a larger density have the effects on reducing the influence of drag friction.

Therefore, we hope, like those small insects, such a jumping-robot, as a millimeter-sized mobile robot with power supply integrated, could also offer a lot of advantages because of its less energy consumption and better practicability.

11.1.2 *Jumping Helps Improve Robot's Moving Ability*

There are many solutions for a robot to move on unstructured terrain and pass different obstacles, even in the natural world, jumping is not the only way for an insect to surmount the obstacles. However, the utilization of a jumping mechanism for biorobots is mainly due to its simplicity and practicality compared with an entirely wheeled design or a flying-robot.

Although many people apply themselves to the improvement of wheeled locomotion so as to make it more adaptive to a rough terrain, a wheeled-robot is only able to pass an obstacle lower than its wheel diameter. Take for example the Mini-Whegs^{TM} 7, a novel wheeled-robot that can climb an obstacle 25 % greater in size than the length of each leg spoke [15], and the robot RHex designed by Uluc Saranli in 2001 who has four rotatable legs correspond to a variation of wheels [16]. Even that, considering the proportion between the body of the robot and its wheels, most wheeled-robot is impossible to pass an obstacle higher than its body, which makes it improper for a centimeter-robot.

On the other hand, hovering robots, which may have no limitations on the maximum obstacle height, is limited by the high energy consumption result in the continuous energy supply. Take for example the four-rotor flying robot described by Daniel Gurdan, which has a diameter of 36.5 mm in size and flying for 30 min without payload [27].

Thus we put more attention on a robot with legged design applying continuous jumping as its primary locomotion mode. The miniaturization and lightweight design make such a centimeter-robot can jump more higher with low energy consumption. Like some insects in the natural world, such a biorobot is able to pass the obstacle some times higher than its body.

11.1.3 Control and Energy Consumption

Flying-robot can be made very light, like the fly-inspired robot introduced by Jean-Christophe Zufferey in 2006 which is only 30 g weight and capable of navigating in a small indoor environment [28], but it also means that this kind of robot is sometimes too light to carry payload, and easily affected by the air flow turbulence (e.g., wind), that makes it more difficult to be used in an outdoor-environment. On the other hand, when a hovering-robot is used in outdoor exploration, the heavy body and large inertia may make this robot more difficult to steer and need a larger turning radius, so it is unfavorable when explore in some narrow spaces [29, 30]. For wheeled-robots, they may move slower than a flying one, but the simple controlling strategy makes pivot steering can be easily executed, if the wheels on both sides are driven respectively [3, 4].

What is more important is the low energy consumption of a walking-robot. Some researches indicated that for a ground covered by 10 in. layer of plastic soil, the propulsive power required by various vehicles are respectively, 15 hp/ton for a wheeled-vehicle, and 10 hp/ton for a tracked-vehicle, but only 7 hp/ton for a walking-vehicle [31]. When a wheel or a track running, it brings more soil to its back and produce a depression for out of which it is continuously trying to climb. On the other hand, when a foot of a walking-robot running, it makes the soil behind it tamped, which increases the propulsive force of the vehicle [32]. Considering the less ground contact of a jumping locomotion, the energy consumption should be more efficient.

11.1.4 Environmental Compatibility

Considering an interplanetary exploration, in which such environmental compatibility becomes more meaningful, the jumping locomotion shows higher energy efficiency than others and could be a better choice for future explorer, because ruts left after the pass of a tracked-robot or a wheeled-robot, but only some discrete footsteps are created when a walking or jumping robot is passing. That makes a great sense for exploring a virgin land. The continuous ruts divide the land into some separate zone, which may bring a huge damage to the environment of this land, while some discrete footsteps are more inconspicuous [20].

11.2 Mechanical Design

11.2.1 Saltatorial Legs

In the natural world, jumping motions can be divided into two styles, countermovement jumping and catapult jumping, which are usually adopted by vertebrates and insects respectively. The former one is carried on by directly

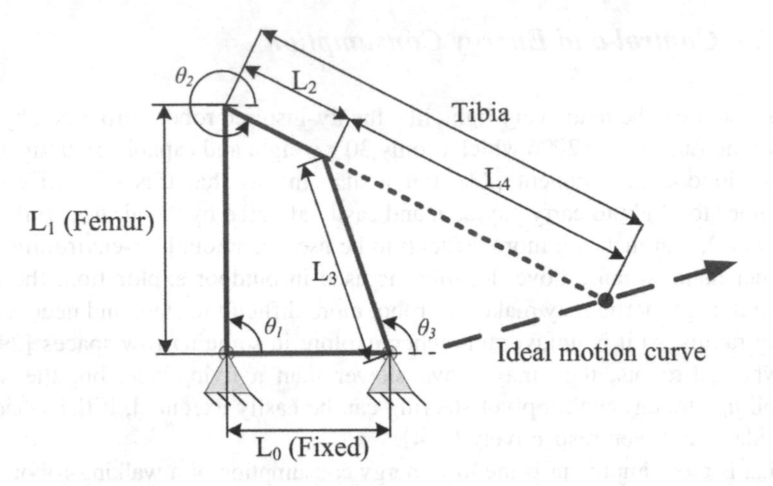

Fig. 11.2 Planar four-bar-linkage mechanism model (© Springer 2010), reprinted with permission

contracting the muscles [7], while the latter by previous energy storing and subsequent fast releasing [33], which is able to generate ground force multiples of body mass and enable small insects to overcome obstacles even hundreds of its characteristic length [10]. For example, leafhopper insects achieve their impressive jumping ability by extending the tarsus along a straight line during leg elongation. According to some research, the extensor of a 19 mg weight and 3.5 mm long Aphrodes can release 77 μJ energy in 2.75 ms which leads to a take off velocity of about 2.9 m/s [33].

Considering the mechanical simplicity and efficiency, such a catapult design is more applicable for the jumping robot prototype around tens of millimeters in size. In order to imitating the jumping kinetics of a natural saltatorial leg, a planar four-bar linkage mechanism as shown in Fig. 11.2 is investigated firstly.

By trying different relative value among L_1, L_2, and L_3, the linear elongation of the tarsus can be achieved. Then the planar mechanism is rebuilt to a three-dimensional structure by rotating L_1 and L_3 along x-axis and elongating L_4 to a proper position in order to let the feet positioned directly beneath the robot body and pressed against each other during jumping acceleration in order to reduce the chance of asymmetric thrust forces between two legs and the destabilizing rotation during the flight phase. The ratio among femur (L_1), tibia (L_2, L_4) and the auxiliary bar (L_3) we used for our jumping robot prototype is about 1:1.72:1.31. In a meanwhile, according to the statistical data from our biological observation of leafhoppers, the ration between femur and tibia is about 1:1.72 ± 0.15, which is very close to the computational optimization result [1]. From a biological viewpoint, this may represent an evolutionary optimization of this specific species.

Figure 11.3 shows the mechanically practicable design of prototype's saltatorial legs. Each leg is composed of nine parts, and has seven rotating joints. Femur is the driving part of this mechanism, whose rotation generates a straight motion of the tip

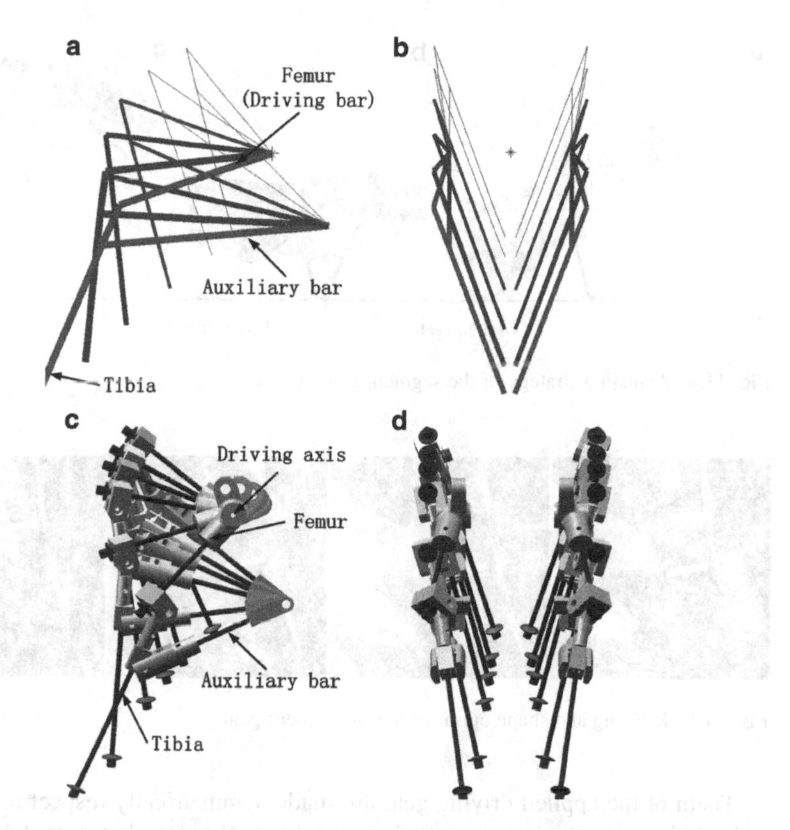

Fig. 11.3 Comparison between the kinematic simulation results of the specific leg design and the fabricated leg mechanism. (**a**, **b**) Front and left view of the kinematic simulation results of the specific leg design; (**c**, **d**) front and left view of the extension of the fabricated leg mechanism. The fabricated leg composes of nine parts and has seven rotating joints. The femur is the driving part of this mechanism whose rotation generates a straight motion of the tarsus. Tarsus motions in (**b**) and (**d**) are different. That results from the difficulties of sphere joint machining (© Springer 2010), reprinted with permission

of tibia and brings the body to the takeoff velocity. The leg elongation is about 27.4 mm, which is obtained through an 80° rotation of the femur, and almost equal to its 27 mm long tibia.

11.2.2 Jumping Actuation

For a mechanical design, a spring is used as the energy storing part, and a segment-gear mechanism is used to imitate the natural muscle-tendon system adopted by insects such as fleas and leafhoppers [3, 33]. By means of the transmitting from motor's continuous rotation to tarsus' reciprocal loading-releasing motion, this mechanism generates continuous jumps for the robot. Figure 11.4 shows the actuating strategy of the system.

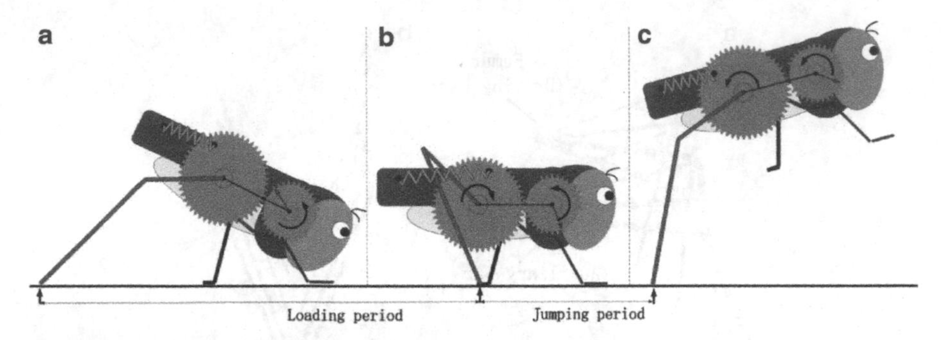

Fig. 11.4 Actuating strategy of the segment gear system

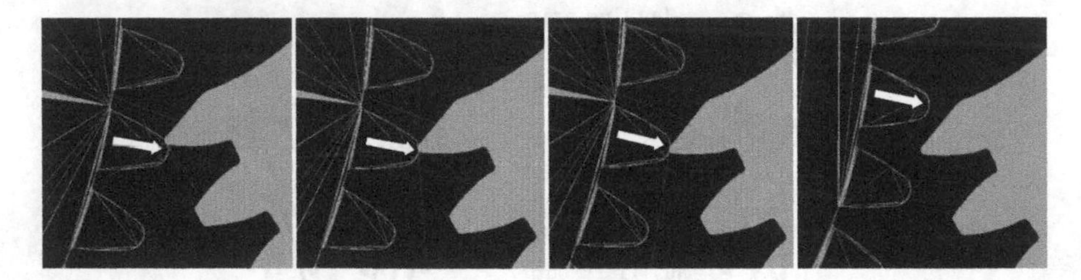

Fig. 11.5 Meshing and shape optimization of segment gear

Teeth of the applied driving gear are made symmetrically respect to a circle in order to improve the jumping efficiency and the other four holes are fabricated for assembling. Based on the results of gear meshing simulation and some tests on previous prototypes, we found that some meshing problems occur at the releasing moment due to the deformation of plastic teeth and may affect the efficiency of leg elongating (Fig. 11.5). To avoid this problem, stainless steel segment gears shown in Fig. 11.6 fabricated by wire electroerosion machining are used to reduce the teeth deformation, as well as some modifications on the last tooth of each teeth group in order to reduce the meshing period. Table 11.1 reports more details of the applied segment gears.

11.2.3 *Jumping Robot Prototyping*

In our GRILLO project, except the first jumping robot based on a cam mechanism [34], there are two other prototypes based on above design are developed (Fig. 11.7). The previous one shown in Fig. 11.7a is a 10 g weight jumping robot with a jump height of about 100 mm (six times of its own height), and the forward speed is over 100 mm/s, which corresponds to about three body length per second. The major problem of this prototype is the instability of the transmission and jumping.

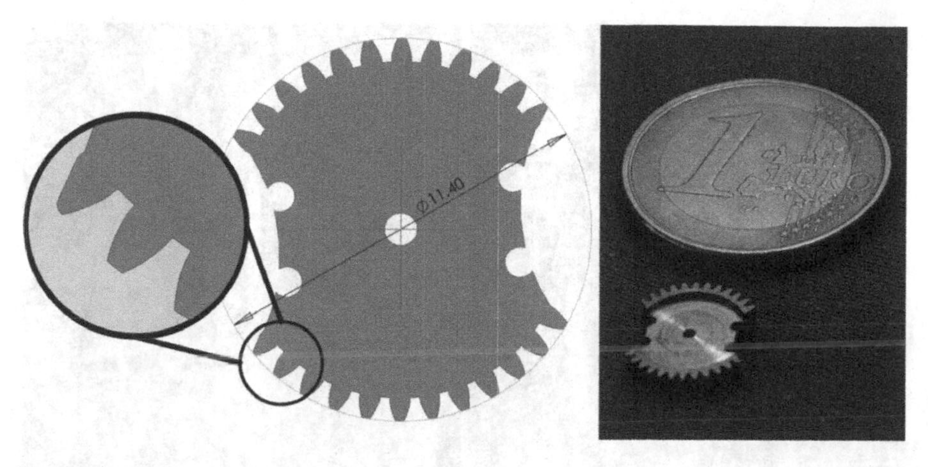

Fig. 11.6 Applied segment gear

Table 11.1 Parameters of segment gear

Gear type	Involute gear
Module number	0.3
Teeth number	12 × 2 (36 for a whole gear)
Addendum diameter	11.40 mm
Reference diameter	10.80 mm
Root diameter	9.99 mm
Teeth thick	2.0 mm
Gear material	Stainless steel
Weight	0.80 g

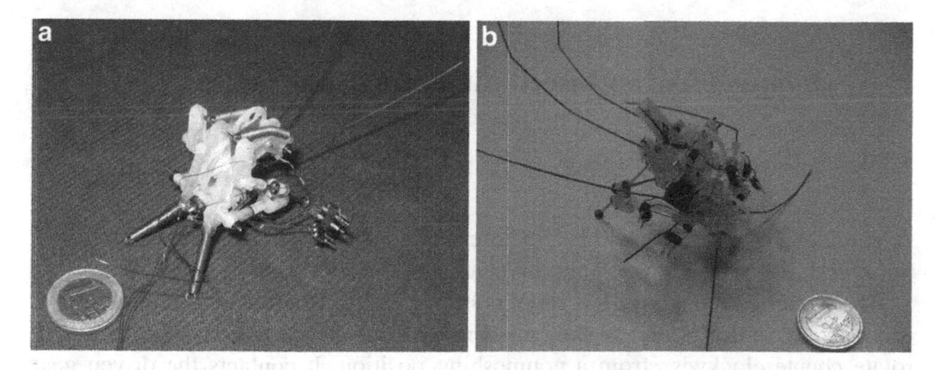

Fig. 11.7 Various jumping robot prototypes: (**a**) The previous jumping robot prototype (10 g weight with a jump height of about 100 mm and a forward speed of 100 mm/s); (**b**) current jumping robot prototypes (22 g weight including the integrated battery with a jump performance of about 100 mm high and 200 mm long)

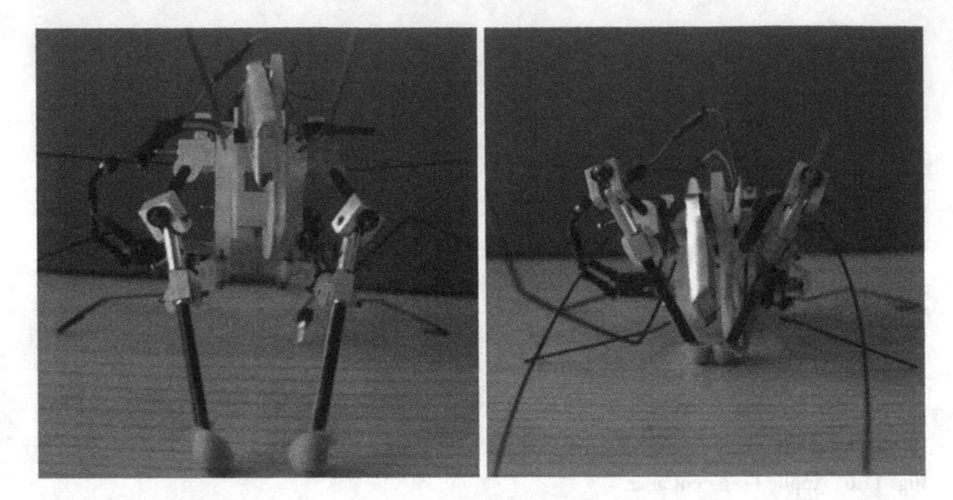

Fig. 11.8 Different status of current jumping robot prototype with bioinspired rear legs (loading and unloading)

By mechanically redesigning the whole body structure, another jumping robot prototype (shown in Fig. 11.7b) was then integrated with wings, energy source, and electrical testing (controlling) board to improve its practicability and reliability. The dimension of this prototype is 50 mm × 20 mm × 25 mm, and the weight is 22 g including a battery and some metal parts (aluminum leg elements and the stainless steel segment gear). Based on some brief tests, for one single jump, the length is about 200 mm (four times of its body length) and the height is about 100 mm (four times of its own height). Figure 11.8 shows its loading and unloading status.

11.3 Jumping like an Insect: Simulation and Testing

11.3.1 Working Sequence

A simulation model of the whole structure (Fig. 11.9) is then built with ADAMS^{TM} in order to learn the detailed jumping characteristics of the prototype.

The working sequence of this robot is shown in Fig. 11.10. At the beginning, the robot stands on the ground as shown in Fig. 11.10a, and the segment-gear starts to rotate counterclockwise from a nonmeshing position. It contacts the driven gear later and starts to load it to an extreme status (Fig. 11.10b, c). The driven gear is fixed with the femurs and a spring which performs like natural tendons of animals. After the last tooth finishing its meshing, driven gear (the big one) rotates backwards and legs elongate rapidly to force the robot jump as a result of the spring shortening (Fig. 11.10d).

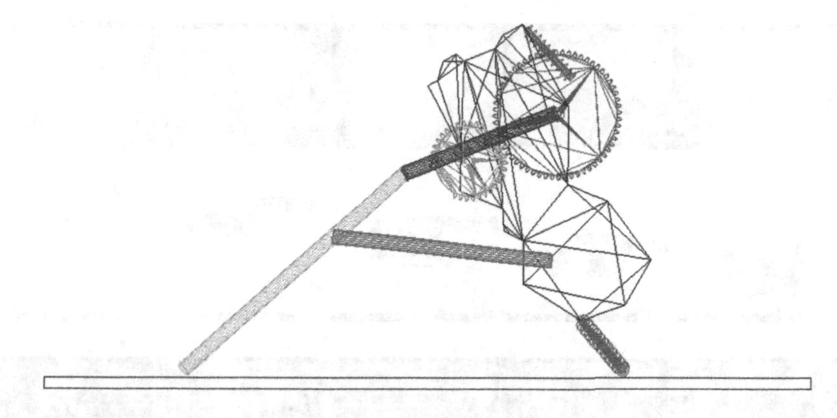

Fig. 11.9 Whole structure ADAMS model of the latest prototype

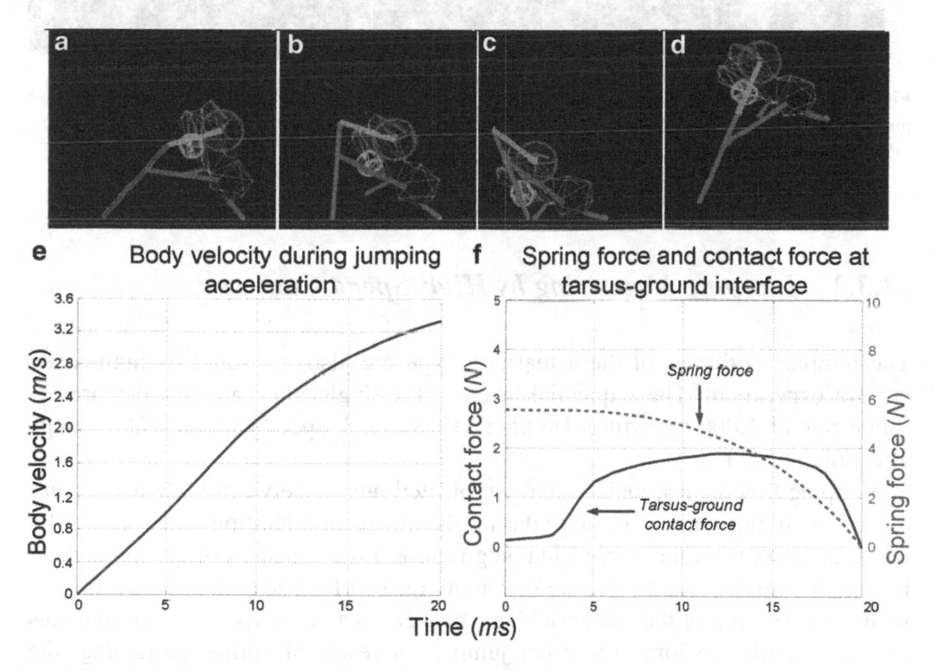

Fig. 11.10 Working sequence and jumping dynamics of current prototype

It also reveals that, with the given leg design, contact force at tarsus–ground interface resulting from leg elongation tends to remain as a nearly constant output during the launching period (Fig. 11.10e, f), which can be considered as an optimized actuating strategy regarding a given takeoff velocity because of the minimization of structure rupture and slippage risk avoidance. From a bionic viewpoint, such optimized jumping characteristics are mainly profited from the successful imitation of insect mechanics and achieved by an increasing effective mechanical advantage (EMA) of the leg structure [1].

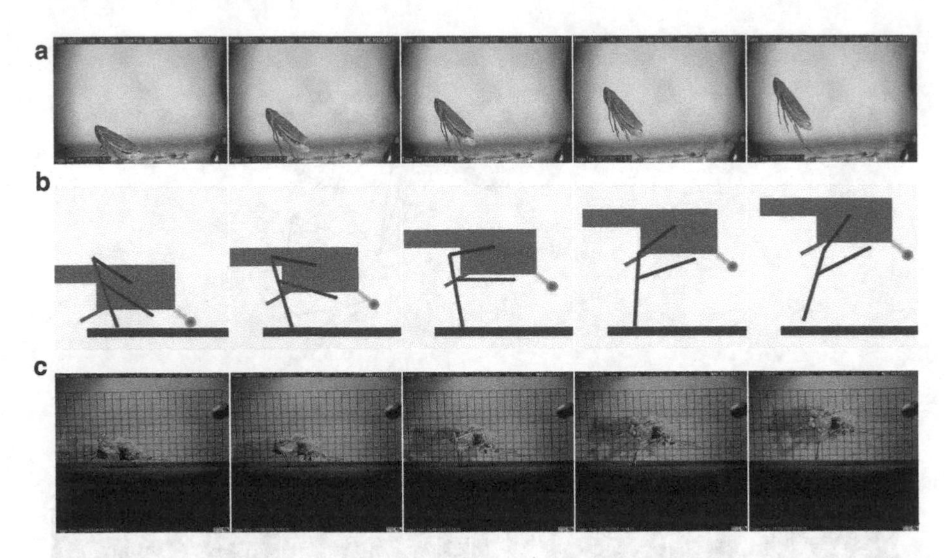

Fig. 11.11 Representative frames of the robot jumping (**a**) model simulation (**b**) and prototype jumping (**c**) captured by high speed camera at 1,000 fps, as well as the computer simulation result (© Springer 2010), reprinted with permission

11.3.2 Jumping Measuring by High Speed Camera

The jumping activities of the actual prototype are also measured by high speed camera experiment. The sequential images of a single jump are first captured at frame rate of 1,000 fps with a HotShot 512 SC high speed camera (NAC Image Technology Inc.).

Jumping kinematics, including *x*–*y* projected angles between body and femur, femur and tibia, jumping velocity during the 40-ms acceleration, are obtained by frame analysis from the start of jumping motion till the instant of ground contact loosing. Representative frames are shown in Fig. 11.11a. The jumping starts at 0 ms with the extending of the saltatorial legs; between 0 and 40 ms, the saltatorial legs elongate rapidly to force the robot jump as a result of spring shortening; the acceleration period ends at 40 ms when the robot is launched into a flight with a measured jumping velocity of about 1.7 m/s.

The limb kinematics of a leafhopper during its jumping measured in our previous biological experiments and the measuring results of the prototype are shown in Fig. 11.12 for comparing. Due to the successful imitation of insect mechanics, the jumping characteristics (body–femur angle and femur–tibia angle vs. time) show a similarity to the one derived from theoretical analysis and dynamic simulation [1, 34]. The little difference between insect and robot kinematical curve may result from the DOF discrepancy at both body-coxa and coxa-trochanter joints.

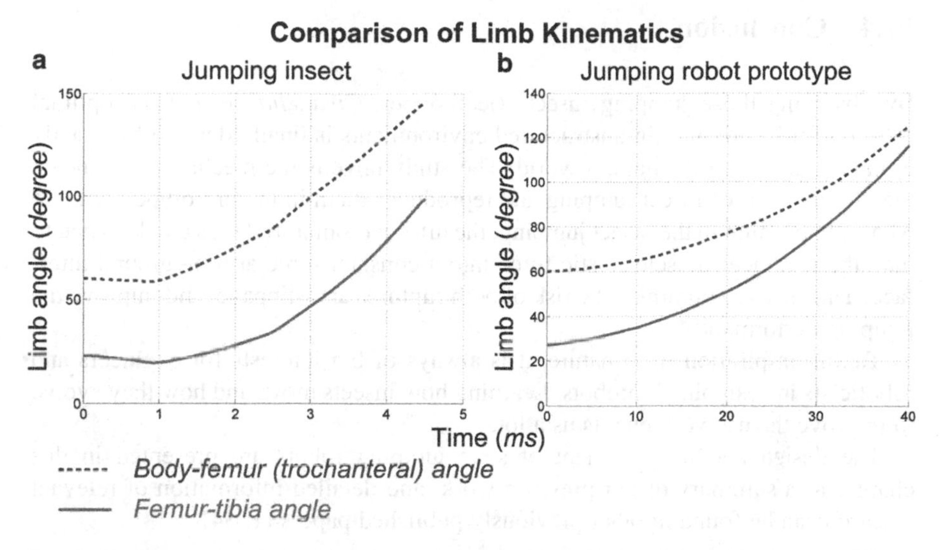

Fig. 11.12 Measuring results of the insect jumping: (**a**) limb kinematics of the jumping insect; (**b**) limb kinematics of jumping robot prototype

Table 11.2 Prototype design and jumping performance (Fig. 11.7c)

Length	50 mm
Width	20 mm
Height	25 mm
Total weight	22 g
Motor power	0.3 W
Battery duration	0.5 h
Takeoff velocity	1.7 m/s
Accelerating period	40 ms
Jumping length	About 200 mm
Jumping height	About 100 mm

Due to the designed structure, two springs with same stiffness of 1 N/mm placed as mechanical tendons are able to restore 56.3 mJ jumping energy at 7.5 mm elongation. Then considering the 31.8 mJ kinetic energy derived from the finally achieved take off velocity and the 8.8 mJ potential changing of the robot body during the jumping acceleration, the total mechanical efficiency of the saltatorial legs should be around 70 % and the energy lost mainly results from the friction at each hinge joint because barely slippage is observed at tarsus–ground interface.

Actually, considering the conservative design required by measuring convenience and robot durability, the jumping performance can be further developed by replacing applied springs with stiffer ones. Details of prototype design and jumping performance are also reported in Table 11.2.

11.4 Conclusion

By observing those jumping insects (leafhopper, *Cicadella viridis*), an optimal solution for locomotion in unstructured environments is finally derived by exactly imitating insects in the natural world. The study unveils the mechanisms beneath the optimization of insect jumping and reproduces them in the prototype. Just like what is happening in the insect jumping, the artificial saltatorial legs are designed to map the nonlinear muscle/elastic force into a constant force at foot–ground interface. This helps to minimize the risk of both ruptures and slippage, and improve the jumping performance.

Taking inspiration from nature, it is always of big interests for engineers and roboticists in designing biorobots. Learning how insects move and how they evolve to improve themselves benefits us a lot.

The design and improvement of such jumping robots are presented in this chapter as a summary of our previous work, and detailed information of relevant research can be found in other previously published papers [1, 34].

Acknowledgments The activity presented in this chapter is partially supported by LAMPETRA Project (EU Contract No. 216100), the Fundamental Research Funds for the Central Universities of China, Scientific Research Foundation for the Returned Overseas Chinese Scholars, and Zhejiang Province Qianjiang Talents Project (2010R10012).

References

1. Scarfogliero U, Bonsignori G, Stefanini C, Sinibaldi E et al (2009) Bioinspired jumping locomotion in small robots: natural observation, design, experiments. In: Springer tracts in advanced robotics, vol 54, pp 329–338
2. McNeill Alexander R (2000) Hovering and jumping: contrasting problems in scaling. Oxford University Press, Oxford, UK, pp 37–50
3. Bennet-Clark HC, Lucey ECA (1967) The jump of the flea: a study of the energetics and a model of the mechanism. J Exp Biol 47:59–76
4. Heitler WJ, Burrows M (1977) The locust jump. I. The motor programme. J Exp Biol 66 (1):203–219
5. Scott J (2005) The locust jump: an integrated laboratory investigation. Adv Physiol Edu 29 (1):21–26
6. Brown RHJ (1967) Mechanism of locust jumping. Nature 214(5091):939
7. Burrows M, Morris O (2003) Jumping and kicking in bush crickets. J Exp Biol 206 (6):1035–1049
8. Brackenbury J, Wang R (1995) Ballistics and visual targeting in flea-beetles (Alticinae). J Exp Biol 198(9):1931–1942
9. Gorb SN (2004) The jumping mechanism of cicada Cercopis vulnerata (Auchenorrhyncha, Cercopidae): skeleton-muscle organisation, frictional surfaces, and inverse-kinematic model of leg movements. Arthropod Struct Dev 33(3):201–220
10. Burrows M (2003) Biomechanics: froghopper insects leap to new heights. Nature 424 (6948):509

11. Burrows M (2006) Jumping performance of froghopper insects. J Exp Biol 209 (23):4607–4621
12. Burrows M (2006) Morphology and action of the hind leg joints controlling jumping in froghopper insects. J Exp Biol 209(23):4622–4637
13. Burrows M (2007) Anatomy of the hind legs and actions of their muscles during jumping in leafhopper insects. J Exp Biol 210(20):3590–3600
14. Birch M, Quinn R, Hahm G, Phillips SM et al (2001) A miniature hybrid robot propelled by legs. In: IEEE/RSJ international conference on intelligent robots and systems, Maui, USA.
15. Lambrecht BGA, Horchler AD, Quinn RD (2005) A small, insect-inspired robot that runs and jumps. In: IEEE international conference on robotics and automation, Barcelona, Spain.
16. Saranli U (2001) RHex: a simple and highly mobile hexapod robot. Int J Robot Res 20:616–631
17. Clark JE, Cham JG, Bailey SA, Froehlich EM et al (2001) Biomimetic design and fabrication of a hexapedal running robot. In: IEEE international conference on robotics and automation, Seoul, Korea
18. Nelson GM, Quinn RD, Bachmann RJ, Flannigan WC et al (1997) Design and simulation of a cockroach-like hexapod robot. In: IEEE international conference on robotics and automation, Albuquerque, USA
19. Berns K, Ilg W, Deck M, Albiez J et al (1999) Mechanical construction and computer architecture of the four-legged walking machine BISAM. IEEE/ASME Trans Mechatron 4:32–38
20. Kimura H (2007) Adaptive dynamic walking of a quadruped robot on natural ground based on biological concepts. Int J Robot Res 26:475–490
21. Geyer H, Seyfarth A, Blickhan R (2006) Compliant leg behaviour explains basic dynamics of walking and running. Proc R Soc B 273:2861–2867
22. Hyon SH, Mita T (2002) Development of a biologically inspired hopping robot–"Kenken". In: IEEE international conference on robotics and automation, IEEE Press, Washington, DC, pp 3984–3991
23. Kovac M, Fuchs M, Guignard A, Zufferey JC et al (2008) A miniature 7 g jumping robot. In: IEEE international conference on robotics and automation, IEEE Press, Pasadena, pp 373–378
24. Kovac M, Schlegel M, Zufferey JC, Floreano D (2010) Steerable miniature jumping robot. Auton Robot 28:295–306
25. Alexander RM (2003) Principles of animal locomotion, vol 371. Princeton University Press, Princeton
26. Scholz MN, Bobbert MF, Knoek van Soest A (2006) Scaling and jumping: gravity loses grip on small jumpers. J Theor Biol 240:554–561
27. Gurdan D, Stumpf J, Achtelik M (2007) Energy-efficient autonomous four-rotor flying robot controlled at 1 kHz. In: Proceeding of the 2007 I.E. international conference on robotics and automation, April 2007, pp 361–366
28. Zufferey J-C, Floreano D (2006) Fly-inspired visual steering of an ultralight indoor aircraft. IEEE Trans Robot 22(1):137–146
29. Arikawa K, Mita T (2002) Design of multi-DOF jumping robot. In: Proceedings of IEEE international conference on robotics and automation, Washington, DC, USA, pp 3992–3997
30. Hyon SH, Mita T (2002) Development of a biologically inspired hopping robot—"Kenken". In: Proceedings of IEEE international conference on robotics and automation, Washington, DC, USA, pp 3984–3991
31. Bekker MG (1960) Off-the-road locomotion. Research and development in terramechanics. The University of Michigan Press, Ann Arbor
32. Song DS-M, Waldron KJ (1989) Machines that walk: the adaptive suspension vehicle. MIT Press, Cambridge, MA
33. Burrows M (2007) Kinematics of jumping in leafhopper insects (Hemiptera, Auchenorrhyncha, Cicadellidae). J Exp Biol 210:3579–3589
34. Li F, Bonsignori G, Scarfogliero U, Chen D et al (2009) Jumping mini-robot with bio-inspired legs. In: Proceedings of IEEE international conference on robotics and biomimetics, Bangkok, Thailand, pp 933–938

Chapter 12
Modeling and H_∞ PID Plus Feedforward Controller Design for an Electrohydraulic Actuator System

Yang Lin, Yang Shi, and Richard Burton

Abstract This work studies the modeling and design of a proportional-integral-derivative (PID) plus feedforward controller for a high precision electrohydraulic actuator (EHA) system. The high precision positioning EHA system is capable of achieving a very high accuracy positioning performance. Many sophisticated control schemes have been developed to address these problems. However, in industrial applications, PID control is still the most popular control strategy used. Therefore, the main objective of this work is to design a PID controller for the EHA system, improving its performance while maintaining and enjoying the simple structure of the PID controller. An extra feedforward term is introduced into the PID controller to compensate for the tracking error especially during the transient period. The PID plus feedforward control design is augmented into a static output feedback (SOF) control design problem and the SOF controller is designed by solving an H_∞ optimization problem with bilinear matrix inequalities (BMIs).

12.1 Introduction

Hydraulic positioning systems play an important role in transportation, earth moving equipment, aircraft, and industry machinery with heavy duty applications [8, 16]. Traditional hydraulic transmission control systems are mainly valve controlled [23]. Pump-controlled hydraulic systems, known as hydrostatic systems, are

Y. Lin • R. Burton
Department of Mechanical Engineering, University of Saskatchewan, 57 Campus Drive, SK, S7N 5A9 Canada
e-mail: yang.lin@usask.ca; richard.burton@usask.ca

Y. Shi (✉)
Department of Mechanical Engineering, University of Victoria, Victoria, BC, V8N 3P6 Canada
e-mail: yshi@uvic.ca

D. Zhang (ed.), *Advanced Mechatronics and MEMS Devices*, Microsystems,
DOI 10.1007/978-1-4419-9985-6_12,

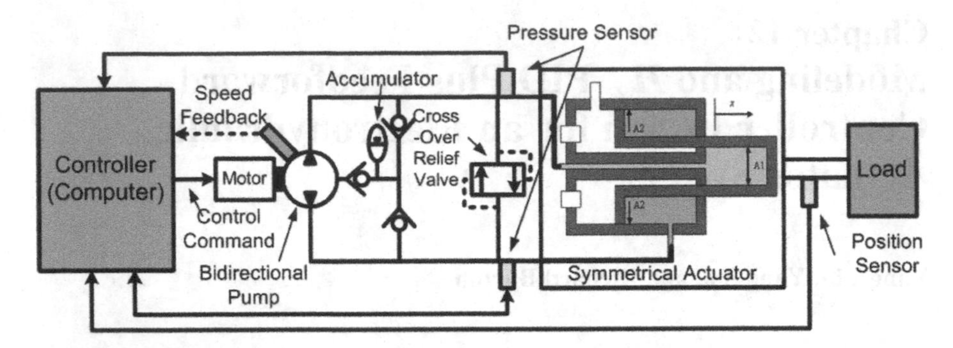

Fig. 12.1 Schematic of the whole hydraulic circuit of the system

used as an alternative for valve controlled systems in applications in which higher energy efficiencies are preferred. One configuration for hydrostatic systems uses a fixed displacement pump which regulates the flow in the hydraulic circuit by changing the rotational speed and direction of the pump [11]. In this study, a particular electrohydraulic actuator (EHA) system, shown in Fig. 12.1, is considered in which high precision position control has been achieved.

Because the EHA is a closed hydraulic system, the modeling for the EHA is more complex than the case of traditional open pump/motor hydrostatic systems. This work introduces such a model which was developed using the power bond graph (PBG). PBG technique has been widely used in modeling and control of mechanical systems because of its ability to handle causality issues for complex modeling problems. In [19], similar approach was applied to analyze the dynamic characteristics of different mechatronic systems.

For this particular EHA system, various control methods such as fuzzy control [17], sliding mode control [22], and robust sliding mode control [15] were proposed and acceptable tracking performance were achieved. But the controller setups for these advanced algorithms were complicated and difficult to implement physically. Once a problem occurs, only the expert in certain areas can fix it. Thus, the first motivation of this work was to design a controller that can be easily implemented for the EHA system.

The proportional-integral-derivative (PID) control has been the most adopted control method for industrial applications. It has been estimated that over 90 % of the controllers in use today are PID controllers, even though other advanced control theories and practical design methods exist [1]. One of the most famous PID tuning methods proposed by Ziegler and Nichols (ZN method) has been extensively used by control engineers over the past several decades [12]. However, when extremely high control performance is required, the traditional PID controllers may not be adequate to deal with disturbance, noises, and model uncertainties. Thus, advanced PID tuning or design method is required for high performance control purpose. For example, in [4], a robust PID controller was designed for a mechanical axis by adding extra flexible pole-placement constraints on the model. In [18], a nonlinear PID controller was designed for a pneumatic artificial muscle (PAM) actuator system by combining neural network PID tuning. Thus, the second motivation for this research was to design a new type of PID controller which can also deal with these problems.

In many studies, PID control design was transformed into a static output feedback (SOF) problem [3, 6, 24]. It should be noted that SOF problem is one of the most important open research topics in control theory and applications [2]. The linear matrix inequality (LMI) technique was applied to solve the SOF problems. Because of the nonconvex property of the SOF control, there have been many research studies during the past decade dedicated to trying to solve this problem. However, to date, there is still no general solution solving this problem for all types of systems. In this work, LMI conditions with nonlinear constraints are developed to find an H_∞ performance SOF controller for the EHA system. An additional problem occurs with the controllers designed in [15, 22], all the states of the system are required to design controllers for the EHA system. State estimators were designed to estimate all the states which makes the design process even more complicated. SOF controller design does not require all the states of the system and hence makes it much easier to be realized in practice.

In the proposed control design, not only the PID feedback control design is augmented into a SOF problem, but a new feedforward term is also introduced into the system to compensate for the tracking error that appeared in [15, 22]. In [7, 14, 20, 21], different types of PID plus feedforward controllers have been presented. In this work, the control system containing both feedback PID controller and the feedforward control is augmented into a new type of SOF control design problem by introducing new state and output vectors. The SOF controller design is transformed into bilinear matrix inequality (BMI) optimizations. As such, then, the main objectives of this work are threefold:

1. To build a model for the EHA system physically using the PBG technique
2. To introduce a feedforward term into the traditional PID feedback controller setup and transform the problem into a SOF problem by defining new state and output vectors
3. To solve the augmented SOF design problem with H_∞ performance by solving a set of LMIs

The chapter is organized as follows. The mathematic model of the EHA system is developed using the PBG technique in Sect. 12.2. In Sect. 12.3, the feedback PID plus feedforward controller design is transformed into a SOF problem, LMI conditions solving the H_∞ optimization problem for the SOF are developed. Simulation studies and experimental results for the EHA system position tracking are illustrated in Sect. 12.4. Finally, conclusion remarks are offered in Sect. 12.5.

The notation used throughout this work is fairly standard. The superscript T stands for the matrix transposition, $\mathbb{R}^n$ denotes the n-dimensional Euclidean space, and $\mathbb{R}^{m \times n}$ is the set of all $m \times n$ real matrices. The notation $P > 0$ means that P is real symmetric and positive definite, and I and 0 represent identity matrix and zero matrix, respectively, with appropriate dimensions. In symmetric block matrices or long matrix expressions, the asterisk $(*)$ represents a term that is induced by symmetry. Matrices, if their dimensions are not explicitly stated, are assumed to be compatible for algebraic operations. The physical meanings of some important variables involved are elaborated in the nomenclature at the end of the chapter.

12.2 Modeling of the EHA System

The layout of the EHA system is shown in Fig. 12.1. The actuator is driven by a bidirectional fixed displacement gear pump. A special symmetrical actuator is connected with the load and the motion of the load is controlled by varying the speed of the electric motor. In this section, the hydraulic circuit is studied from a bond graph point of view. The bond graph of each part of the hydraulic system is presented. To accommodate the linearized model, approximations to several non-linear equations are presented.

12.2.1 Linear Symmetrical Actuator

The bond graph of the hydraulic actuator including the load is given in Fig. 12.2.

The flow in and out of the linear symmetrical actuator are described by the following equations:

$$Q_1 = A\dot{x} + \frac{V_0 + Ax}{\beta}\frac{dP_1}{dt} + L_e P_1 + L_i(P_1 - P_2), \tag{12.1}$$

$$Q_2 = A\dot{x} - \frac{V_0 - Ax}{\beta}\frac{dP_2}{dt} - L_e P_2 - L_i(P_2 - P_1). \tag{12.2}$$

Here, V_{O} is the mean pipe plus actuator chamber volume, Q_1 represents the flow goes into the actuator, Q_2 is the flow goes out from the actuator, P_1 and P_2 are the pressures in the actuator chambers, L_{e} is the actuator external leakage coefficient, and L_{i} is the actuator internal leakage coefficient.

12.2.2 Hydraulic Pump

For the hydraulic pump, the effect of the case drain leakage was ignored, so only the effect of the cross-port leakage was taken into account. The PBG for the hydraulic pump is given in Fig. 12.3.

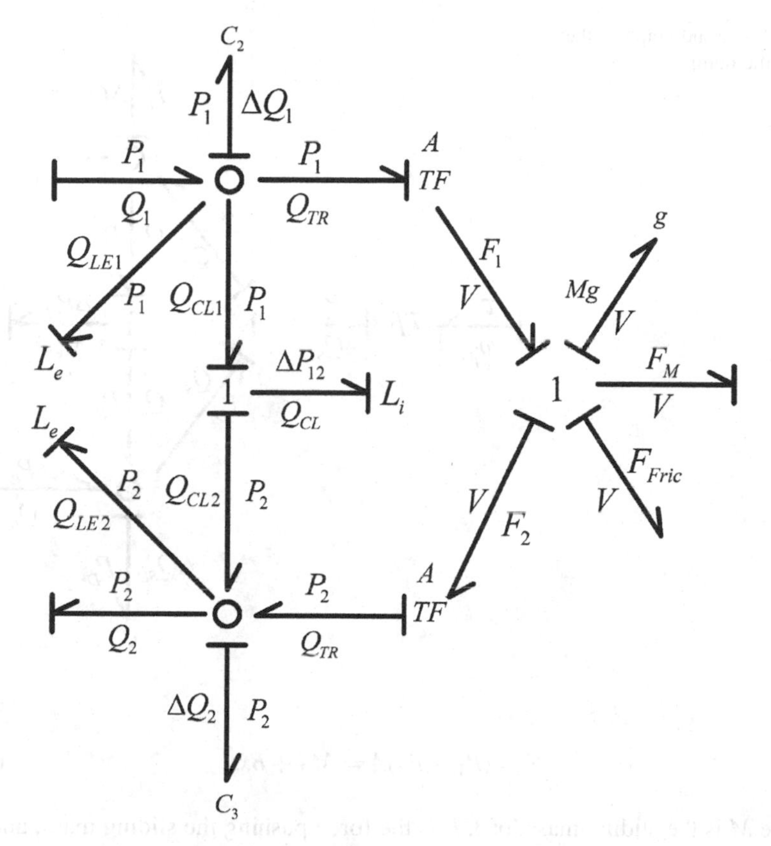

Fig. 12.2 Bond graph of the symmetrical actuator

The pump flow can be modeled as:

$$Q_U = D_p\omega_p - K_{LCP}(P_U - P_D) - \frac{V_U}{\beta}\frac{dP_U}{dt}, \tag{12.3}$$

$$Q_D = D_p\omega_p - K_{LCP}(P_U - P_D) + \frac{V_D}{\beta}\frac{dP_D}{dt}, \tag{12.4}$$

where D_p is the fixed pump displacement, ω_p is the pump angular velocity, K_{LCP} is the pump cross-port leakage coefficient, Q_U and Q_D are the flow coming out and going into the bidirectional pump, P_U and P_D are the pressures at the outlet and inlet port of the pump.

12.2.3 Pump/Actuator Connection and Overall Hydraulic Model

In this EHA system, the hydraulic actuator is connected to a horizontal movement sliding mass, so the motion equation of the hydraulic actuator is:

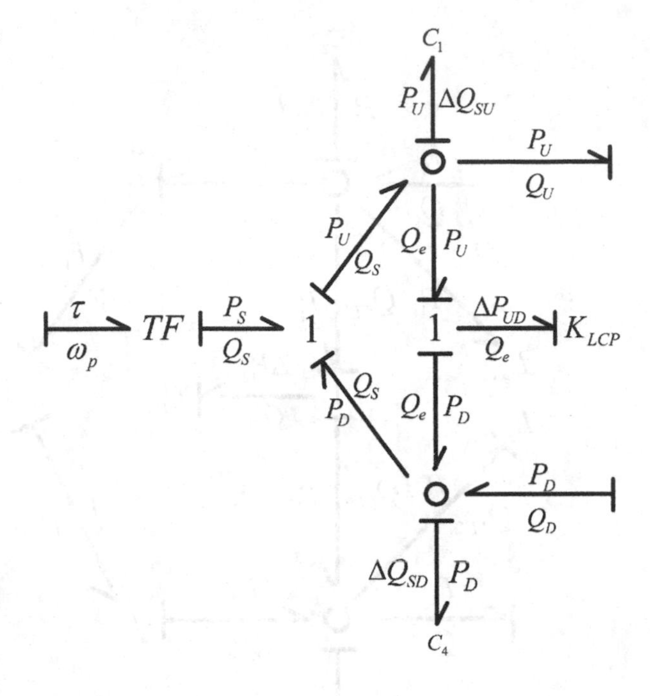

Fig. 12.3 Bond graph of the hydraulic pump

$$F = (P_1 - P_2)A = M\ddot{x} + B\dot{x}, \tag{12.5}$$

where M is the sliding mass load, F is the force pushing the sliding mass, and B is the actuator viscous friction coefficient.

The pump and actuator pipe connection can be modeled as a pressure drop [10] which is represented by $P_{\text{pipe}} \approx 2K_{\text{pipe}}D_{\text{p}}{}^2(\omega_{\text{P}}{}^2) + P_{\text{ele}}$ (using Darcy's pipe flow equation). This pressure drop can be approximated by linearizing this equation at the operating point ω_{Pop}:

$$\Delta P_{pipe} \approx 2K_{pipe}D_p^2\omega_{Pop}(\Delta\omega_p). \tag{12.6}$$

The relationship between the pump and actuator port pressure can be approximated as $P_1 = P_U - P_{pipe}$ and $P_2 = P_D + P_{pipe}$. Assuming that the flow from the accumulator to be zero then $Q_1 = Q_U$ and $Q_2 = Q_D$. Because of the symmetry of the pump and actuator, under steady-state conditions, the exiting flow from the pump equals the inlet flow to the actuator, and the flow exiting the actuator equals the flow back into the pump. Then the load flow Q_L can be approximated to be:

$$Q_L = \frac{Q_1 + Q_2}{2} = \frac{Q_U + Q_D}{2}.$$

Substituting the pump and actuator flow equations (12.1–12.4):

$$D_p\omega_p - K_{LCP}(P_U - P_D) - \frac{V_0}{2\beta}\left(\frac{dP_U}{dt} - \frac{dP_D}{dt}\right) - \frac{Ax}{2\beta}\left(\frac{dP_U}{dt} + \frac{dP_D}{dt}\right)$$
$$= A\dot{x} + \frac{V_0}{2\beta}\left(\frac{dP_1}{dt} - \frac{dP_2}{dt}\right) + \frac{Ax}{2\beta}\left(\frac{dP_1}{dt} + \frac{dP_2}{dt}\right) + \left(L_i + \frac{L_e}{2}\right)(P_1 - P_2)$$

Because the pressure drop across the pipeline is very small when compared to the pump and actuator port pressures, $(dP_U/dt) \approx (dP_1/dt)$ and $(dP_D/dt) \approx (dP_2/dt)$. Also due to the symmetry of the actuator, $(dP_1/dt) \approx -(dP_2/dt)$. Defining a new parameter $L_{ie} = 2L_i + L_e$, the approximated model of the pump and actuator connection is obtained as:

$$D_p\omega_p = A\dot{x} + \frac{V_0}{\beta}\left(\frac{dP_1}{dt} - \frac{dP_2}{dt}\right) + \left(K_{LCP} + \frac{L_{ie}}{2}\right)(P_1 - P_2)$$
$$+ 2K_{LCP}\Delta P_{pipe}. \tag{12.7}$$

Substituting (12.5) into (12.7) and take Laplace Transform on the equation:

$$x(s) = \frac{D_p}{s^3\left(\frac{MV_0}{\beta A}\right) + s^2\left(\frac{L_{ie}M/2 + K_{LCP}M + BV_0/\beta}{A}\right) + s\left(\frac{A^2 + BL_{ie}/2 + BK_{LCP}}{A}\right)}$$
$$- \frac{2K_{LCP}P_{pipe}(s)}{s^3\left(\frac{MV_0}{\beta A}\right) + s^2\left(\frac{L_{ie}M/2 + K_{LCP}M + BV_0/\beta}{A}\right) + s\left(\frac{A^2 + BL_{ie}/2 + BK_{LCP}}{A}\right)}.$$

Recall (12.6), the transfer function of the hydraulic part of the system is:

$$G_H(s) = \frac{\Delta x(s)}{\Delta\omega_p(s)}$$
$$= \frac{D_p(1 - 2K_{pipe}K_{LCP}D_p\omega_{Pop})}{s^3\left(\frac{MV_0}{\beta A}\right) + s^2\left(\frac{L_{ie}M/2 + K_{LCP}M + BV_0/\beta}{A}\right) + s\left(\frac{A^2 + BL_{ie}/2 + BK_{LCP}}{A}\right)}.$$

Because $2K_{\mathrm{pipe}}K_{\mathrm{LCP}}D_{\mathrm{p}}\omega_{\mathrm{Pop}} \ll 1$ is negligible, the approximated hydraulic system transfer function can be presented as:

$$G_H(s) = \frac{\Delta x(s)}{\Delta\omega_p(s)}$$
$$= \frac{D_p}{s^3\left(\frac{MV_0}{\beta A}\right) + s^2\left(\frac{L_{ie}M/2 + K_{LCP}M + BV_0/\beta}{A}\right) + s\left(\frac{A^2 + BL_{ie}/2 + BK_{LCP}}{A}\right)}.$$

A discrete-time state space model is required for this PID plus feedforward controller design. Define states for the state space model: $x = [x_1 \;\; x_2 \;\; x_3] = [x \;\; \dot{x} \;\; \ddot{x}]$ and generate the discrete-time state space model by using the approximation given by [9]:

$$\dot{x}_i(t) = \frac{x_i(k+1) - x_i(k)}{T_s}, \tag{12.8}$$

where T_s is the sampling period. Then the discrete-time state space model can be represented by:

$$\begin{aligned} x_{k+1} &= A_d x_k + B_d u_k, \\ y_k &= C_d x_k, \end{aligned}$$

where

$$A_d = \begin{bmatrix} 1 & T_s & 0 \\ 0 & 1 & T_s \\ 0 & -\frac{T_s\beta[A^2 + B(K_{LCP} + L_{ie}/2)]}{MV_0} & 1 - \frac{T_s[\beta M(K_{LCP} + L_{ie}/2) + BV_0]}{MV_0} \end{bmatrix},$$

$$B_d = \begin{bmatrix} 0 \\ 0 \\ \frac{\beta AD_p T_s}{MV_0} \end{bmatrix}, \quad C_d = [1 \quad 0 \quad 0]$$

12.3 Discrete-Time Robust PI Plus Feedforward Controller with H_∞ Performance

In this section, the design of a discrete-time PI plus feedforward controller for the EHA system is presented. The derivative term of the PID controller is neglected in order to avoid the derivative kicks and reduce the computational load of the optimization calculation. The PI plus feedforward control platform is shown in Fig. 12.4.

The discrete-time model of the EHA system is represented by the following linear system:

$$\begin{aligned} x_{k+1} &= A_d x_k + B_d(u_k + v_{1k}), \\ y_k &= C_d x_k + v_{2k}, \end{aligned} \tag{12.9}$$

where k is the sampling instant, $x_k \in \mathbb{R}^n$ is the state vector of the EHA system, $y_k \in \mathbb{R}^p$ is the measurement output, $u_k \in \mathbb{R}^m$ is the input control signal, v_{1k} is the disturbance or noise at the input, and v_{2k} represents the measurement noise. The PI plus feedforward control law is given by

$$u_k = K_p e_k + K_i \sum_{i=0}^{k-1} e_i + K_{ff}(r_{k-1} - r_k). \tag{12.10}$$

Here, K_{p} and K_{i} are the feedback proportional and integral gains and K_{ff} is the feedforward gain to be designed.

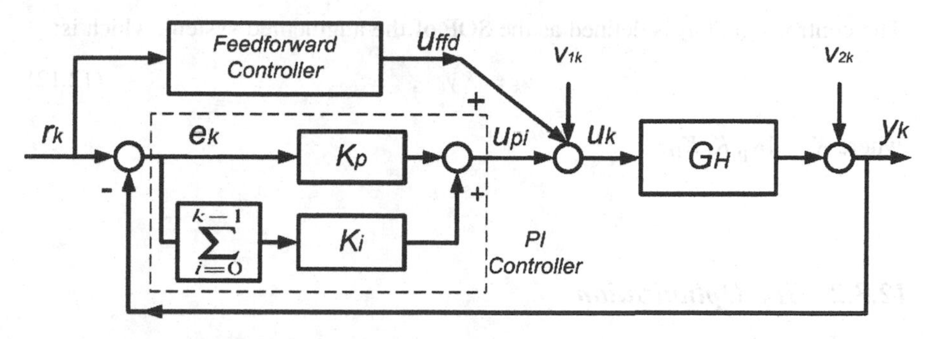

Fig. 12.4 PI plus feedforward controller platform

12.3.1 *Transforming the PI Plus Feedforward Controller into a SOFController*

In order to formulate the problem of PI plus feedforward controller design into a problem of SOF control, a new state vector and a new output sequence $\bar{x}_k = [x_k^{\mathrm{T}},\ \sum_{i=0}^{k-1} e_i^{\mathrm{T}},\ r_{k-1}^{\mathrm{T}}]^T$, $\bar{y}_k = [e_k^{\mathrm{T}},\ \sum_{i=0}^{k-1} e_i^{\mathrm{T}},\ G(r_{k-1} - r_k)^{\mathrm{T}}]^{\mathrm{T}}$ are defined. Here, $e_k = r_k - y_k$ is the controlled tracking error of the EHA system and G is a weighting factor. Then the augmented system is written by the following form:

$$\begin{aligned} \bar{x}_{k+1} &= \bar{A}\bar{x}_k + \bar{B}u_k + \bar{B}_1\omega_k, \\ \bar{y}_k &= \bar{C}\bar{x}_k + \bar{D}_k, \end{aligned} \tag{12.11}$$

where

$$\bar{A} = \begin{bmatrix} A_d & 0 & 0 \\ -C_d & I & 0 \\ 0 & 0 & 0 \end{bmatrix},\quad \bar{B} = \begin{bmatrix} B_d \\ 0 \\ 0 \end{bmatrix},$$

$$\bar{B}_1 = \begin{bmatrix} 0 & B_d & 0 \\ I & 0 & -I \\ I & 0 & 0 \end{bmatrix},\quad \bar{C} = \begin{bmatrix} -C_d & 0 & 0 \\ 0 & I & 0 \\ 0 & 0 & GI \end{bmatrix},$$

$$\bar{D} = \begin{bmatrix} I & 0 & -I \\ 0 & 0 & 0 \\ -GI & 0 & 0 \end{bmatrix},\quad \omega_k = \begin{bmatrix} r_k \\ v_{1k} \\ v_{2k} \end{bmatrix}$$

The control signal u_k is defined as the SOF of the augmented system, which is:

$$u_k = K\bar{y}_k \quad , \tag{12.12}$$

where $K = [K_p \, K_i \, K_{ff}]$.

12.3.2 H_∞ Optimization

Now the design problem of the PI plus feedforward controller with H_∞ performance is investigated. The SOF H_∞ control problem is to find a controller of the form as (12.12) such that the closed-loop control system is stable and to minimize the ∞-norm ($\| \cdot \|_\infty$) of the system from the external input ω_k to the controlled output $\bar{z}_k$ ($T_{\omega_k \bar{z}_k}$).

The controlled output sequence is chosen as

$$\bar{z}_k = Ex_k + Fu_k, \tag{12.13}$$

where $E = R[0 \; I \; 0]$, R and F are another two weighting matrices. Substituting (12.12) into (12.11) and combining (12.13), the closed-loop control system for H_∞ optimization can be represented by the following form

$$\begin{aligned} \bar{x}_{k+1} &= [\bar{A} + \bar{B}K\bar{C}]\bar{x}_k + [\bar{B}K\bar{D} + \bar{B}_1]\omega_k, \\ \bar{z}_k &= [E + FK\bar{C}]\bar{x}_k + FK\bar{D}\omega_k \end{aligned} \tag{12.14}$$

Using the well-known Bounded Real Lemma for discrete-time system [5] and the congruence transformation, the H_∞ optimization problem for the SOF controller design leads to the following theorem.

Theorem 12.1 *Given the plant (12.9), the PI plus feedforward controller (12.10) guarantees that the closed-loop system (12.14) is stable and $\|T_{\omega\bar{z}}\|_\infty < \gamma$ if there exist symmetrical matrices $P > 0$, $Q > 0$ and matrix K, with appropriate dimensions, which satisfy the following LMIs*

$$\begin{bmatrix} -Q & [\bar{A} + \bar{B}K\bar{C}] & [\bar{B}K\bar{D} + \bar{B}_1] & 0 \\ * & -P & 0 & [E + FK\bar{C}]^T \\ * & * & -\gamma I & [FK\bar{D}]^T \\ * & * & * & -\gamma I \end{bmatrix} < 0, \tag{12.15}$$

$$\begin{bmatrix} P & I \\ I & Q \end{bmatrix} > 0, \tag{12.16}$$

and a constraint

$$PQ = I. \tag{12.17}$$

Proof From the closed-loop system represented by (12.14), apply the discrete-time Bounded Real Lemma [5], the H_∞ controller exist if and only if there exist positive symmetric matrix P and matrix K satisfying

$$\begin{bmatrix} -P & P[\bar{A}+\bar{B}K\bar{C}] & P[\bar{B}K\bar{D}+\bar{B}_1] & 0 \\ * & -P & 0 & [E+FK\bar{C}]^T \\ * & * & -\gamma I & [FK\bar{D}]^T \\ * & * & * & -\gamma I \end{bmatrix} < 0. \tag{12.18}$$

This is a BMI problem which is very complicated to be solved. Taking congruence transform on both sides of (12.18) by diag$[P^{-1}\ I\ I\ I]^{\mathrm{T}}$ and diag$[P^{-1}\ I\ I\ I]$, and define a new LMI matrix $Q = P^{-1}$. Then (12.18) is transformed into

$$\begin{bmatrix} -Q & [\bar{A}+\bar{B}K\bar{C}] & [\bar{B}K\bar{D}+\bar{B}_1] & 0 \\ * & -P & 0 & [E+FK\bar{C}]^T \\ * & * & -\gamma I & [FK\bar{D}]^T \\ * & * & * & -\gamma I \end{bmatrix} < 0. \tag{12.19}$$

This completes the proof.

Remark 12.1 The conditions of (12.15–12.17) are in fact an LMI problem with nonconvex constraints. It can be conveniently solved by using the cone complementarity linearization (CCL) algorithm [13]. The results of solving the LMI problem give rise to the matrix K, and then PI plus feedforward controller gains can be determined.

12.4 Simulation Studies and Experimental Test

According to the *Theorem 12.1* presented in the previous section, the CCL algorithm is employed to solve the LMI problem in (12.15–12.17), and as a result, an optimized solution of K is obtained by minimizing the γ as:

$$K = [2428.6 \quad 20.2 \quad 2.8 \times 10^4].$$

Then the SOF control signal equals to:

$$u_k = K\left[e_k^T,\ \sum_{i=0}^{k-1} e_i^T,\ G(r_{k-1} - r_k)^T\right]^T.$$

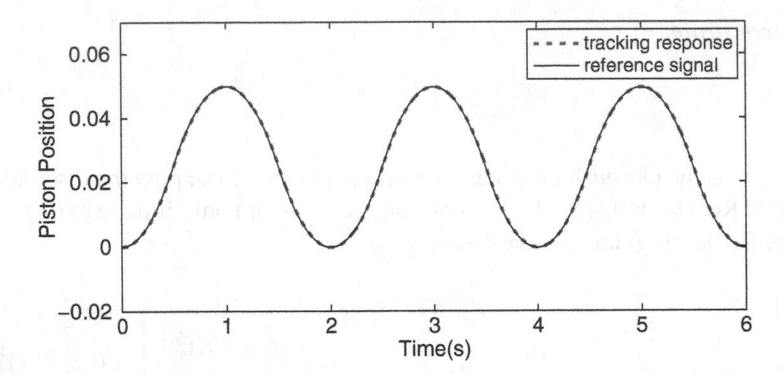

Fig. 12.5 Simulation tracking response with the feedforward

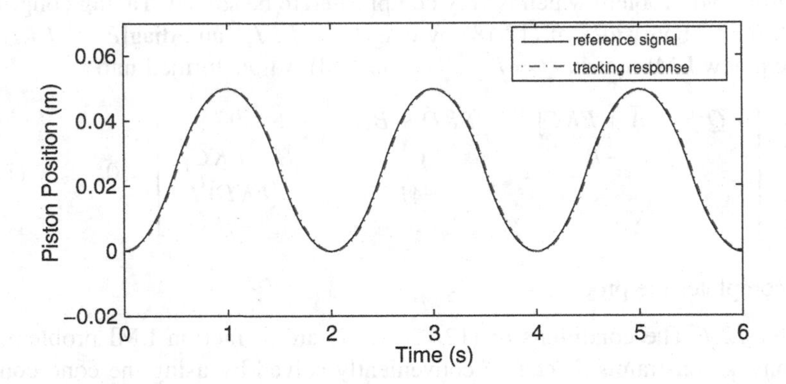

Fig. 12.6 Simulation tracking response without the feedforward

12.4.1 Simulation Study

The simulation tracking response of the EHA system using the proposed PI plus feedforward controller is shown in Fig. 12.5.

For the purpose of comparison, the simulation tracking response of the EHA system using the proposed H_∞ PI controller without including the feedforward pass is also shown in Fig. 12.6.

It is hard to tell a big difference from the two simulation tracking responses. The tracking error of these two simulation results are shown in Fig. 12.7.

Remark 12.2 The H_∞ PI controller without the feedforward term was developed using the same optimization method in *Theorem 1*. Using the proposed PI plus feedforward controller, the tracking error gets significantly smaller than the one without a feedforward term.

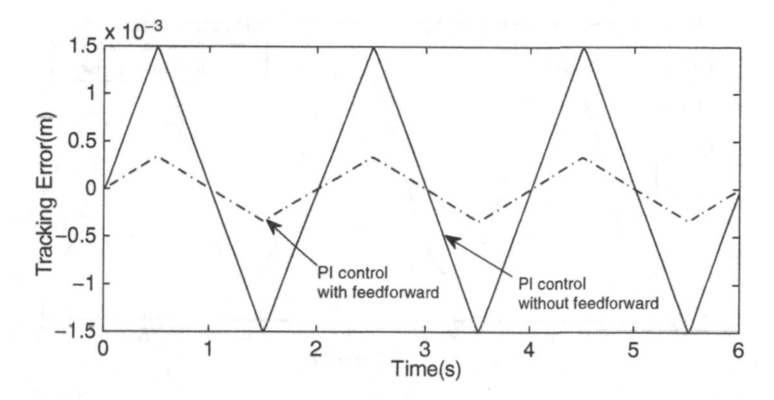

Fig. 12.7 Simulation tracking errors

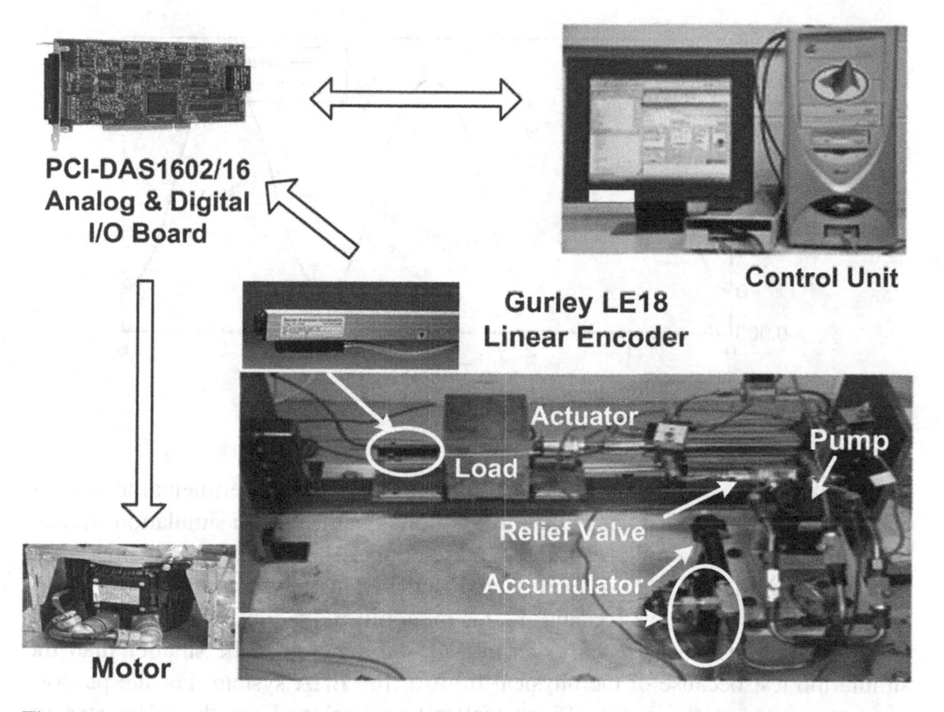

Fig. 12.8 Experimental setup of the EHA system

12.4.2 Experimental Test

The experimental EHA system under study is shown in Fig. 12.8. The hardware-in-the-loop experimental test system includes the following components: Pentium IV computer, PCI-DAS1602/16 Analog and Digital I/O Board, Gurley LE18 Linear

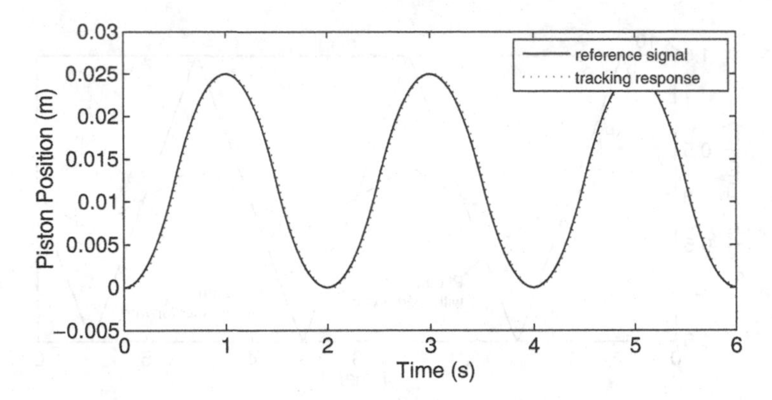

Fig. 12.9 Experimental tracking response with the feedforward

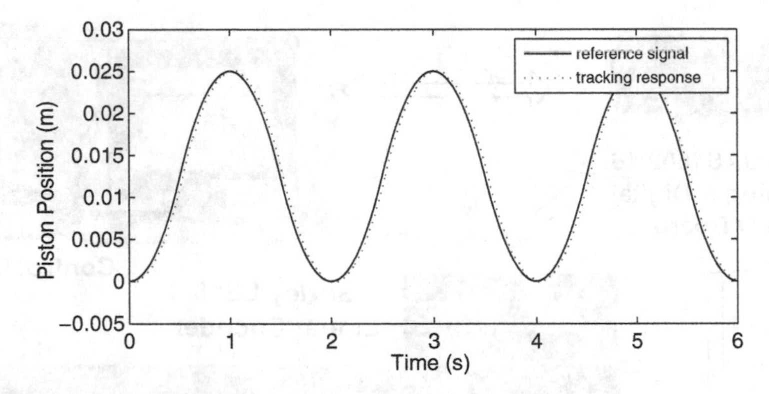

Fig. 12.10 Experimental tracking response without the feedforward

Encoder, and the designed bidirectional hydraulic circuit. Experimental tests were performed to confirm and verify the observations obtained from simulation studies. Same tracking tests are done on the experimental EHA system as the simulation results. Experimental studies for this proposed PI plus feedforward controller are shown in Figs. 12.9, 12.10, and 12.11 to confirm the observations made by simulations. The stroke of the experiment test is chosen to be smaller than the simulation test because of the physical limit of the EHA system. For the purpose of comparison, a Z-N tuning PI controller was designed for the EHA system. The experimental tracking error is also shown in Fig. 12.11.

From the experimental results, it can be summarized that:

1. The experimental tracking error looks very close to the simulation tracking errors using the same controllers.
2. From Fig. 12.11, the experimental tracking error using the proposed PI plus feedforward controller did decrease comparing with the one without a feedforward loop. However, the tracking error did not decrease that significantly as was shown in the simulation in Fig. 12.7. It is suspected that the main reasons

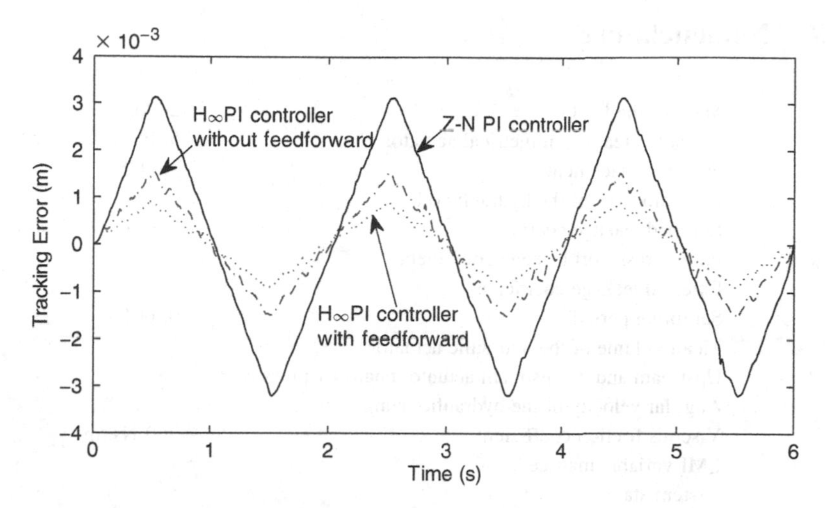

Fig. 12.11 Experimental tracking error without the feedforward

caused this phenomena are: the practical EHA system is naturally a nonlinear system and the linear EHA model used for the controller design must have missed some dynamic characteristics of the system.

3. Despite its simple formulation, the proposed PI plus feedforward controller can still achieve a very accurate tracking performance (tracking error is less than 3.6 % of the full tracking displacement).
4. The H_∞ PI controller outperforms the traditional Z-N PI tuning method which can be easily seen in Fig. 12.11.

12.5 Conclusions

This work has proposed a discrete-time PID plus feedforward controller design for an EHA system. The controller formulation was transformed into a SOF problem with H_∞ performance and LMI optimization technique was applied to solve the controller design problem. To the author's best knowledge, this work is the first of its kind that solving a PID plus feedforward controller using the LMI optimization technique.

Simulation studies and experimental tests on the EHA system verify the effectiveness of the proposed method for position tracking from the application perspective. The extra feedforward term did improve the tracking performance significantly. Despite its simple setup, the proposed PI plus feedforward controller achieved very good tracking performance. However, there is no guarantee that there exists a general form of solution for all types of control systems. If one can find such a general solution, maybe we do not have to turn to other complex control design algorithms when high control performances are preferred.

12.6 Nomenclature

M	Mass of the load	20 kg
A_p	Pressure area in symmetrical actuator	5.05×10^{-4} m^2
D_p	Pump displacement	$1.6925 \times 10^{-7}\ m^3/rad$
β	Bulk modulus of the hydraulic oil	2.1×10^8 Pa
C_T	Lumped leakage coefficient	$5 \times 10^{-13}\ m^3/sPa$
K_{LCP}	Pump cross-port leakage coefficient	
L_e	External leakage coefficient	
T_s	Sampling period	0. 001 s
V_0	Mean volume of the hydraulic actuator	
P_1, P_2	Upstream and downstream actuator chamber pressure	
ω_p	Angular velocity of the hydraulic pump	
B	Viscous friction coefficient	760 Ns./m
P, Q, K	LMI variable matrices	
x	System states	
x_1, x_2, x_3	Position, velocity, and acceleration of the load	

References

1. Aström K, Hagglund, T (2001) The future of PID control. Contr Eng Pract 9(11):1163–1175
2. Bara GI, Boutayeb M (2005) Static output feedback stabilization with H_∞ performance for linear discrete-tiem systems. IEEE Trans Automat Contr 50(2):250–254
3. Bianchi FD, Mantz RJ, Christiansen CF (2008) Multivariable PID control with set-point weighting via BMI optimization. Automatica 44(2):472–478
4. Dieulot JY, Colas F (2009) Robust PID control of a linear mechanical axis: a case study. Mechatronics 19:269–273
5. Gahinet P, Apkarian P (1994) A linear matrix inequality approach to H_∞ control. Int J Robust Nonlinear Contr 4:421–448
6. Ge M, Chiu MS, Wang QG (2002) Robust PID controller design via LMI approach. J Process Contr 12(1):3–13
7. Ge P, Jouaneh M (1996) Tracking control of a piezoceramic actuator. IEEE Trans Contr Syst Technol 4(3):209–216
8. Gomis-Bellmunt O, Campanile F, Galceran-Arellano S, Montesinos-Miracle D, Rull-Duran J (2008) Hydraulic actuator modeling for optimization of mechatronic and adaptronic systems. Mechatronics 18:634–640
9. Grewal MS, Amdrews AP (2001) Kalman filtering theory and practice using MATLAB. Wiley, New York
10. Habibi SR, Goldenberg A (2000) Design of a new high-performance electrohydraulic actuator. IEEE/ASME Trans Mechatron 5(2):158–164
11. Habibi SR, Goldenberg AA (2000) Design of a new high-performance electrohydraulic actuator. IEEE/ASME Trans Mechatron 5(2):158–165
12. Hang CC, Aström K, Ho WK (1991) Refinements of the Ziegler-Nichols tuning formula. IEE Proc Contr Theor Appl 138(2):111–118
13. Laurent EG, Francois O, Mustapha A (1997) A cone complementarity linearization algorithm for static output feedback and related problems. IEEE Trans Automat Contr 42(8):1171–1176

14. Leva A, Bascetta L (2006) On the design of the feedforward compensator in two-degree-of-freedom controllers. Mechatronics 16:533–546
15. Lin Y, Shi Y, Burton R (2009) Modeling and robust discrete-time sliding mode control design for a fluid power electrohydraulic actuator (EHA) system. In: Dynamic system control conference, Hollywood, USA
16. Rovira-Mas F, Zhang Q, Hansen AC (2007) Dynamic behavior of an electrohydraulic valve: typology of characteristic curves. Mechatronics 17:551–561
17. Sampson E (2005) Fuzzy control of the electrohydraulic actuator. Master's thesis, University of Saskatchewan
18. Thanh TDC, Ahn KK (2006) Nonlinear PID control to improve the control performance of 2-axes pneumatic artificial muscle manipulator using neural network. Mechatronics 16:577–587
19. Toufighi MH, Sadati SH, Najaif F (2007) Modeling and analysis of a mechatronic actuator system by using bond graph methodology. In: Proceedings of IEEE aerospace conference, Big Sky, MT, pp 1–8
20. Visioli A (2004) A new design for a PID plus feedforward controller. J Process Contr 14 (5):457–463
21. Wang J, Wu J, Wang L, You Z (2009) Dynamic feed-forward control of a parallel kinematic machine. Mechatronics 19:313–324
22. Wang S, Habibi S, Burton R, Sampson E (2008) Sliding mode control for an electrohydraulic actuator system with discontinuous non-linear friction. Proc IME J Syst Contr Eng 222 (8):799–815
23. Yanada H, Furuta K (2007) Adaptive control of an electrohydraulic servo system utilizing online estimate of its natural frequency. Mechatronics 17:337–343
24. Zheng F, Wang QG, Lee TH (2002) On the design of multivariable PID controller via LMI approach. Automatica 38(3):517–526

Index

D. Zhang (ed.), *Advanced Mechatronics and MEMS Devices*, Microsystems,
DOI 10.1007/978-1-4419-9985-6, © Springer Science+Business Media New York 2013

T